Smart Grid Sensors

Discover the ever-growing field of smart grid sensors, covering traditional and state-of-the-art sensor technologies, as well as data-driven and intelligent methods for using sensor measurements in support of innovative smart grid applications. This book covers recent and emerging topics, such as smart meters, synchronized phasor measurements, and synchronized waveform measurements. Additional advanced topics and future trends are also discussed, such as situational awareness, probing, and working with off-domain measurements. Including real-world examples, exercise questions, and sample data sets, this is an essential text for students, researchers, and scientists, as well as field engineers and practitioners in the areas of smart grids and power systems.

Hamed Mohsenian-Rad is Professor of Electrical and Computer Engineering and a Bourns Family faculty fellow at the University of California, Riverside. He has served as Editor of *IEEE Transactions on Smart Grid* and *IEEE Power Engineering Letters*. He has received several awards in research and teaching, and is a fellow of the IEEE.

Smart Grid Sensors

Principles and Applications

HAMED MOHSENIAN-RAD
University of California, Riverside

CAMBRIDGE
UNIVERSITY PRESS

University Printing House, Cambridge CB2 8BS, United Kingdom

One Liberty Plaza, 20th Floor, New York, NY 10006, USA

477 Williamstown Road, Port Melbourne, VIC 3207, Australia

314–321, 3rd Floor, Plot 3, Splendor Forum, Jasola District Centre, New Delhi – 110025, India

103 Penang Road, #05–06/07, Visioncrest Commercial, Singapore 238467

Cambridge University Press is part of the University of Cambridge.

It furthers the University's mission by disseminating knowledge in the pursuit of education, learning, and research at the highest international levels of excellence.

www.cambridge.org
Information on this title: www.cambridge.org/9781108839433
DOI: 10.1017/9781108891448

First published 2022

A catalogue record for this publication is available from the British Library.

Library of Congress Cataloging-in-Publication Data
Names: Mohsenian-Rad, Hamed, 1979– author.
Title: Smart grid sensors : principles and applications / Hamed Mohsenian-Rad, University of California, Riverside.
Description: Cambridge : Cambridge University Press, 2022. | Includes bibliographical references and index.
Identifiers: LCCN 2021029866 (print) | LCCN 2021029867 (ebook) | ISBN 9781108839433 (hardback) | ISBN 9781108891448 (epub)
Subjects: LCSH: Smart power grids–Equipment and supplies. | BISAC: SCIENCE / Energy
Classification: LCC TK3105 .M655 2022 (print) | LCC TK3105 (ebook) | DDC 621.31028/4–dc23
LC record available at https://lccn.loc.gov/2021029866
LC ebook record available at https://lccn.loc.gov/2021029867

ISBN 978-1-108-83943-3 Hardback

Additional resources for this title at www.cambridge.org/mohsenian-rad

To Hoda, Kian and Liam

Contents

Preface

This book is about *smart grid sensors*, a key pillar in the field of smart grids. As the interest in modernizing the electric power infrastructure continues to grow worldwide, there are also rapid changes and important advancements in the type and extent of sensor measurements that are becoming available in this field. Several new trends, such as the so-called big data applications, have emerged due to these recent advancements. Nonetheless, there is still a gap in the literature to provide us with a clear understanding about the choices of smart grid sensors that are available and the wide variety of existing and emerging smart grid applications that can be supported by different types of sensors. The purpose of this book is to fill this gap and to present a concerted view of the fast-growing field of smart grid sensors, their working principles, and their applications.

Goal

Understanding smart grid sensors is the gateway to understanding the broader subject of smart grid development. Therefore, my goal in this book is to help the reader build a *working knowledge* of the field of smart grid sensors. Achieving this goal requires discussion of not only the sensors and their measurements, but also the methodologies that can be used to transform the raw measurements to actionable information in order to solve important smart grid problems.

Smart grid sensors are defined in this book rather broadly. In fact, here I consider smart grid sensing as a *whole process*. This process starts from sensor technologies and instrumentation to obtain different types of measurements from different types of sensors; but it also goes on to include making use of the obtained measurements in various smart grid monitoring applications by applying adequate data-analytic techniques, such as from signal processing, estimation, and machine learning. All of these techniques are discussed to some extent in this book.

While this book is *not* a text on instrumentation, it does occasionally talk about the physics of certain smart grid sensors. However, such discussions are minimal and introductory. This book is also *not* meant to be a comprehensive survey of methods that can be used to analyze smart grid sensor measurements. Nevertheless, this book covers several useful data-driven methods and mathematical techniques, in order to show how to use the type of tools and techniques that can be applied to various sensor measurements in the field of smart grids.

My hope is that, by reading this book, the reader will appreciate the extensive and diverse (and still growing) choices that we currently have when it comes to smart grid sensors and measurements, and will think about their innovative applications, based not only on the current state of practice but also in the future, and based on the reader's own imagination and interest in the field of smart grids.

Audience

This book is meant for not only the students, researchers, and scientists in the areas of smart grids and power systems, but also the field engineers and practitioners who work in the industry in these fields. This book can be used to learn about the recent advancements in the field of smart grid sensors. It can also be used to set the stage to conduct innovative research and to develop new ideas in the field of smart grids.

The reader should have basic knowledge in power systems and smart grids and also some background in calculus and algebra. The necessary background in power systems and smart grids is provided in Chapter 1. Therefore, this book can be used by a reader whose background is not in power engineering. The necessary background in data-driven and mathematical methodologies is also discussed throughout Chapters 2–7, at the specific locations where they are used and in the context of their applications.

Using This Book in Courses

This book can be used as the primary or supplementary textbook for different types of courses. It can be used as the primary textbook for a course on Smart Grid Sensors and Measurement-based Applications (or Power System Monitoring). This type of course can be taught at the graduate level or the senior undergraduate level. If the students already have the background in power systems, then the instructor may skip Chapter 1 and instead only give homework assignments based on the exercises in this chapter, to help the students quickly review the related subjects to be ready for the course. Also, if the course is taught at the senior undergraduate level, then some of the more advanced subjects can be excluded, such as Sections 3.5 and 3.7 in Chapter 3, Sections 5.6 and 5.7 in Chapter 5, or the entirety of Chapter 6.

This book can also be used as a supplementary textbook for a number of courses. It can be used as a supplementary textbook for a broader Introduction to Smart Grids course, where selected chapters of this book can be used to cover the pillar on smart grid sensors. It can also be used as a supplementary textbook for a course on Power System Data Analysis or Big Data in Power Systems. In that case, selected chapters of this book, such as Chapters 2–5, can be used for about half of the course, while the other half of the course can focus on techniques for advanced data analytics and machine learning. Excerpts of this book, including a subset of the exercises, can also be used as supplemental materials for a more traditional course on Power Systems.

Examples and Data Files as Learning Materials

Throughout this book, examples are used as key learning materials. Many of the examples are based on real-world engineering cases and related sensor measurements. Most exercises have also been designed based on real-world scenarios.

Many exercises in this book are accompanied by data files. The data files are available as part of the Online Resources. These data files provide the reader with an opportunity for hands-on learning. Most cases require some level of computer programming, which can be done in MATLAB or any other programming platform with the capability to work with data files.

Online Resources

The online resources for this title are located at www.cambridge.org/mohsenian-rad. There you will find the data files that accompany the exercises, as well as the solution manual for the exercise questions that is available to instructors.

Acknowledgments

The material for this book was gathered and developed over a period of about one decade, since I first offered a course on smart grids. Over the years, I benefited from feedback provided by my students at the University of California, Riverside, and Texas Tech University. I also benefited from discussions with my colleagues and industry collaborators. I wish to thank everyone who helped me in this journey. I would also like to thank Alireza Shahsavari for our helpful discussions; Alireza Akrami for his help with preparing the solution manual; and Mohammad Farajollahi for his feedback regarding the design of Example 3.11 and Exercise 3.16. Finally, I would like to thank the staff at Cambridge University Press for their help and support.

Feedback about the Book

For any comment or question, please feel free to contact me at hamed@ece.ucr.edu.

1 Background

The purpose of this chapter is to provide some background knowledge on smart grids and power systems that will be useful throughout this book. First, we will look at an overview of the field of smart grids. We will also discuss why we need to study smart grid sensors. More details about the field of smart grids are available in books such as [1–7] and in papers such as [8–11]. Next, we will provide an overview of the basics of electric power systems and the power grid. More details about these subjects are available in books such as [12–15]. At the end of this chapter, we will also briefly go through the topics that are covered in Chapters 2–7.

It should be noted that some of the techniques that we will discuss in this book require some knowledge in signal processing, estimation, and machine learning. However, the background information related to those techniques is *not* discussed in this chapter. Instead, it is discussed throughout Chapters 2–7, at the specific locations in which they are used and in the context of their applications.

1.1 Smart Grids

1.1.1 Definitions

In short, a smart grid (or a smart power grid) is a *modernized* electric grid. Electric grid modernization can be achieved by adopting different technologies, such as with respect to the broad areas of sensing, control, and communication. In other words, what makes the grid "smart" or "smarter" is the use of various advanced technologies, including various advanced sensor technologies.

There are also other, more detailed, definitions for a smart grid. Each definition may highlight certain features or certain applications. For example, the U.S. Department of Energy defines smart grids as the technologies, equipment, and controls that allow for two-way communication between the electric utility and its customers, and the *sensing* along the transmission and distribution lines [16]. Further to the above definition, the U.S. Department of Energy has emphasized how smart grids can help with seamless integration of renewable energy resources, enhanced intelligence in power transmission and distribution, making large-scale energy storage a reality, engaging customers and enhancing consumer choices, and enabling nationwide use of plug-in hybrid electric vehicles [17].

As another example, the European Commission Directorate-General for Energy defines a smart grid as an electricity network that can cost-efficiently integrate the behavior and actions of all users connected to it, including generators and consumers, to ensure an economically efficient, sustainable power system with low losses and high levels of quality, security, and safety [18]. Further to this definition, it is emphasized that smart grids have the ability to automatically *monitor* energy flows and adjust to changes in energy supply and demand accordingly [19].

Several other national and international institutions have also provided definitions for the smart grid; for example, see the definitions in [20–24].

1.1.2 Benefits

A smart grid is expected to provide a wide range of benefits, such as: (1) deployment of intelligent technologies that are real-time and automated to provide enhanced monitoring of grid operation conditions and status; (2) increased use of digital information and controls technology to improve reliability, security, and efficiency of the power grid; (3) dynamic optimization of grid operations and resources for more efficient generation, transmission, and distribution of electricity; (4) increased integration of distributed energy resources, in particular renewable resources, to reduce the environmental impact of the whole electricity supply system; (5) increased use of electric vehicles and other means of transportation electrification; (6) faster diagnosis of failures and faster restoration of electricity after power disturbances; (7) reduced cost of operations and management and ultimately lower power costs for consumers; (8) encouraging consumers to play a part in enhancing the operation of the power system by developing and incorporating technologies for demand response, demand-side flexible resources, and smart appliances; (9) deployment and integration of advanced electricity storage to support various applications, such as peak-load shaving; and (10) provision to consumers of timely information and control options [8–11, 16, 17, 25].

1.1.3 Smart Grid Sensors

Sensors play a crucial role in developing a smart grid. Measurements from various sensor technologies can help us reach a proper level of *understanding* and *awareness* about the conditions of the electric grid and its components, such that we can make *informed decisions* for control and operation of the electric grid.

Smart grid sensors are defined in this book rather broadly. In fact, here we consider smart grid sensing as a *whole process*. The process starts from sensor technologies and instrumentation to obtain different types of raw measurements from different types of sensors and goes on to include *making use* of the obtained measurements in various smart grid monitoring applications by applying adequate data-analytic techniques, such as from signal processing, estimation, and machine learning, all of which are discussed to some extent in this book.

There are three key questions to ask when it comes to smart grid sensors:

(1) What is out there to measure?
(2) What is the right choice of sensor to measure it?
(3) What can we do with the obtained measurements?

We seek to answer these questions throughout this book, in Chapters 2–7. For example, if we need to measure *voltage*, there are multiple options to consider. Depending on the smart grid application that we seek to support, we may use the kinds of sensors and the kinds of analytics that are discussed in Chapters 2, 3, or 4. Conversely, different types of sensors may support the same applications, but with different performance or at different cost. For example, different methods can solve the *phase identification* problem. Each method uses different types of measurements, depending on what is available. Therefore, we may use the methods in Chapters 2, 3, 5, or 6 to solve the same problem but by using different measurements.

Before we dive deep into the above discussions, let us first briefly review some of the relevant basic concepts in power systems and the electric grid.

1.2 Basic Concepts in Power Systems

The electric power grid is the largest man-made machine ever built, and one of the greatest achievements of the twentieth century [17]. In essence, the electric power grid is a large network of interconnected power systems.

Traditionally, a typical power grid is divided into four sectors: *generation*, *transmission*, *distribution*, and *consumption*. The generation sector comprises a wide range of power plants, such as fossil-fueled, hydroelectric, nuclear, and, more recently, wind farms and solar farms. The transmission system is a meshed network of transmission lines that carry bulk power over long distances. Distribution systems are several smaller regional networks of distribution lines that deliver power to customers. The consumption sector comprises all kinds of customers, such as residential, commercial, industrial, municipal, and agricultural load. A potential fifth sector is *storage*, which is an emerging major component in power systems.

Today's power grids are primarily alternating current (AC) power systems. Thus, in this section, we briefly discuss the basics of AC power systems. We will discuss the five aforementioned sectors of the electric power grid in Section 1.3.

1.2.1 AC Voltage and Current

Voltage and current in AC circuits have sinusoidal waveforms; see Figures 1.1(a) and (b), respectively. The equations for these sinusoidal waveforms are as follows:

$$\begin{aligned} v(t) &= \sqrt{2}\, V \cos(\omega t + \theta), \\ i(t) &= \sqrt{2}\, I \cos(\omega t + \phi), \end{aligned} \tag{1.1}$$

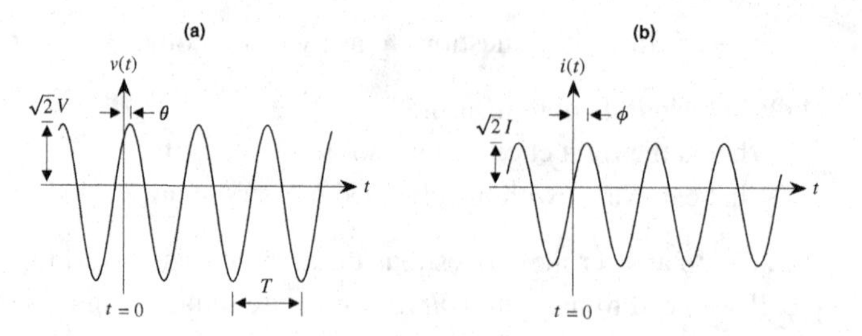

Figure 1.1 Illustration of a sinusoidal voltage wave and a sinusoidal current wave.

where t denotes time. The angular frequency ω is expressed in radian per second. It is also common to express frequency in Hz, i.e., cycle per second:

$$f = \omega/2\pi. \tag{1.2}$$

The typical frequency of the electric grid is either 60 Hz, such as in the United States, or 50 Hz, such as in Europe. The period of the sinusoidal wave in (1.1) is

$$T = 1/f. \tag{1.3}$$

The period indicates the length of each sinusoidal cycle. At 60 Hz, each cycle takes $T = 16.667$ msec (milliseconds). At 50 Hz, each cycle takes $T = 20$ msec.

The Root-Mean-Square (RMS) values for voltage and current in (1.1) are

$$\begin{aligned} V_{\text{rms}} &= \sqrt{\frac{1}{T}\int_0^T v(t)^2\,dt} = V, \\ I_{\text{rms}} &= \sqrt{\frac{1}{T}\int_0^T i(t)^2\,dt} = I. \end{aligned} \tag{1.4}$$

The unit of voltage is volt (V) and the unit of current is amp (A).

1.2.2 Phasor Representation

A *phasor* is a complex number that contains the *magnitude* (in RMS) and *phase angle* information of a sinusoidal function. In the conventional circuit theory literature, the sinusoidal voltage and current in (1.1) are represented by $\sqrt{2}\,V\angle\theta$ and $\sqrt{2}\,I\angle\phi$, respectively. However, in the literature in power systems, the sinusoidal voltage and current in (1.1) are often represented by

$$\begin{aligned} &V\angle\theta, \\ &I\angle\phi. \end{aligned} \tag{1.5}$$

In other words, they are represented by their *effective phasors* [12].

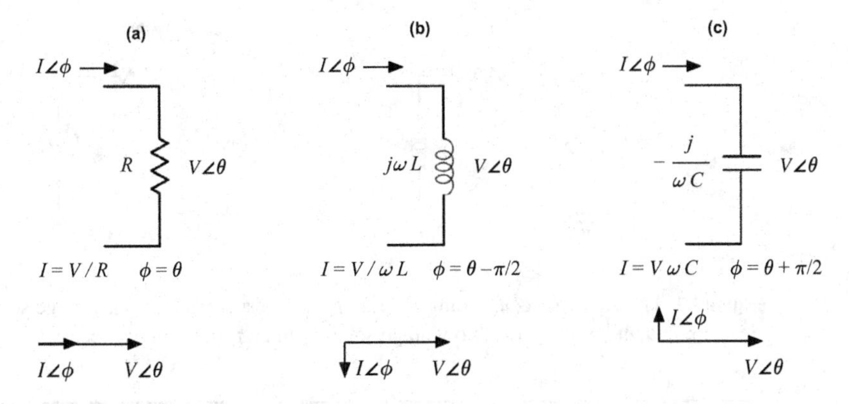

Figure 1.2 Summary of the relationships between voltage phasors and current phasors for RLC circuit elements [13]: (a) resistor; (b) inductor; and (c) capacitor.

Example 1.1 Consider the following sinusoidal voltage wave:

$$v(t) = 169.7\cos(\omega t + \pi/6). \tag{1.6}$$

Its phasor representation is $120\angle 30^\circ$. Note that $169.7 = \sqrt{2} \times 120$.

1.2.3 Impedance and Admittance

The relationship between voltage phasor and current phasor at resistors, inductors, and capacitors can be represented by their *impedance*; see Figure 1.2. Impedance is a complex number. It is often represented in Cartesian form:

$$Z = R + jX, \tag{1.7}$$

where $j = \sqrt{-1}$. The real part, i.e., R, is resistance; and the imaginary part, i.e., X, is reactance. Reactance represents the characteristics of inductors and capacitors. For an inductor with inductance L, we have:

$$X = \omega L. \tag{1.8}$$

For a capacitor with capacitance C, we have:

$$X = -\frac{1}{\omega C}. \tag{1.9}$$

The following relationship holds between voltage and current phasors:

$$Z = \frac{V\angle\theta}{I\angle\phi} = \frac{V}{I}\angle(\theta - \phi). \tag{1.10}$$

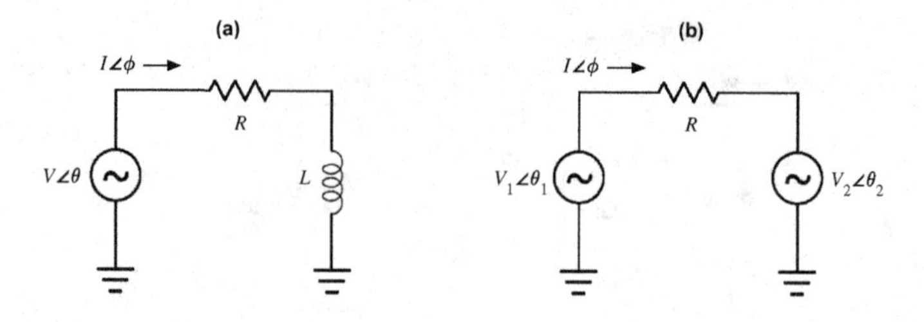

Figure 1.3 The AC circuits in Examples 1.2 and 1.3, respectively: (a) a voltage source, a resistor, and an inductor; (b) two voltage sources and a resistor in between.

Example 1.2 Figure 1.3(a) shows an AC circuit with a voltage source, where $v(t)$ is as in (1.6). The load in this example comprises a resistor $R = 12\ \Omega$ and an inductor $L = 10$ mH [3]. The impedance of the load is

$$Z = 12 + j\,2\pi \times 60 \times 0.01 = 12 + j3.77. \tag{1.11}$$

From (1.10), we can obtain the current phasor as

$$I\angle\phi = \frac{120\angle 30^\circ}{12 + j3.77} = 9.54\angle 12.56^\circ. \tag{1.12}$$

Accordingly, the current that is drawn by the load is obtained as

$$i(t) = 13.49\cos(\omega t + 0.22). \tag{1.13}$$

Here, we used the fact that $\sqrt{2} \times 9.54 = 13.49$ and $12.56^\circ = 0.22$ radian.

The inverse of impedance is *admittance*, which is a complex number:

$$Y = \frac{1}{Z} = G + jB. \tag{1.14}$$

The real part, i.e., G, is conductance; and the imaginary part, i.e., B, is susceptance. From (1.7) and (1.14), we can show that

$$G = \frac{R}{R^2 + X^2}, \qquad B = -\frac{X}{R^2 + X^2}. \tag{1.15}$$

Relative Difference in Voltage Phase Angles

The flow of current between two points in an AC circuit depends on both the magnitude and the phase angle of voltage at those two points. As we will see in the next example, even if the voltage magnitude is the *same* at the two points, current can still flow if there is difference between the two voltage phase angles.

Example 1.3 Figure 1.3(b) shows two AC voltage sources and one resistor [3]. The magnitude of the voltage phasors is the *same* in both voltage sources:

$$V_1 = V_2 = 120V. \tag{1.16}$$

However, the phase angles at the two voltage sources are *different*:

$$\theta_1 = 0^\circ, \quad \theta_2 = 30^\circ. \tag{1.17}$$

The current phasor $I\angle\phi$ that goes through the resistor $R = 12\ \Omega$ is obtained as

$$\frac{120\angle 0^\circ - 120\angle 30^\circ}{12} = 5.1764\angle - 75^\circ. \tag{1.18}$$

Therefore, despite the fact that the RMS values of the two voltage sources are the same, their phase angle differences result in flowing current at the resistor.

1.2.4 Instantaneous Power and Complex Power

The *instantaneous power* that is generated or consumed by a power device is

$$p(t) = v(t)\, i(t). \tag{1.19}$$

By substituting voltage and current from (1.1) into (1.19), we can derive:

$$\begin{aligned} p(t) &= 2VI\cos(\omega t + \theta)\cos(\omega t + \phi) \\ &= VI\cos(\theta - \phi) + VI\cos(2\omega t + \phi + \theta). \end{aligned} \tag{1.20}$$

We can see that the above instantaneous power is a summation of two terms. The first term is a *constant* because it does *not* depend on time t. The second term is a sinusoidal function with angular frequency 2ω.

Average Power

If we take the *average* of the instantaneous power over one cycle, we obtain:

$$P = \frac{1}{T}\int_0^T p(t)\, dt. \tag{1.21}$$

The average power for the instantaneous power in (1.20) is obtained as

$$P = VI\cos(\theta - \phi). \tag{1.22}$$

The unit of instantaneous power and average power is watt (W).

Example 1.4 Suppose $v(t)$ is as in (1.6) and $i(t)$ is as in (1.13). The instantaneous power corresponding to $v(t)$ and $i(t)$ is obtained as

$$p(t) = 1092.2 + 1144.8\cos(2\omega t + 0.7436). \tag{1.23}$$

The average power is $P = 1092.2$ W ≈ 1.1 kW (kilowatts).

The integral of instantaneous power over any period of time also provides the energy consumption or energy generation during that time period. The unit of energy is Wh (watt-hour). For example, if the instantaneous power that is dissipated at a load is as in (1.23) for a period of 30 minutes, then the energy usage of the load is $1092.2/2 = 0.546$ kWh (kilowatt-hour).

Apparent Power and Reactive Power

From (1.22), and based on the definition of voltage and current phasors, we can also express the average power in the following form:

$$P = \mathrm{Re}\left\{ V\angle\theta (I\angle\phi)^* \right\}, \tag{1.24}$$

where Re{·} denotes the real part; and {·}* denotes the *conjugate transpose*. This raises the question of the significance of the imaginary part of the above complex product as well as the significance of the above complex product itself [12]. In fact, both quantities are important. Specifically, reactive power is defined as

$$Q = \mathrm{Im}\left\{ V\angle\theta (I\angle\phi)^* \right\} = VI\sin(\theta - \phi), \tag{1.25}$$

where Im{·} denotes the imaginary part. Complex power is defined as

$$S = V\angle\theta (I\angle\phi)^* = P + jQ. \tag{1.26}$$

The magnitude of complex power is obtained as

$$\sqrt{P^2 + Q^2} = VI = V_{\mathrm{rms}}\, I_{\mathrm{rms}}, \tag{1.27}$$

where the second equality in (1.27) is due to the definition of the RMS value in (1.4).

The unit of reactive power is volt-amp-reactive (VAR). The unit of complex power is volt-amp (VA). Complex power is also referred to as *apparent* power. Furthermore, average power is also referred to as *active* power or *real* power.

Example 1.5 Again consider the voltage and current in Example 1.4. The voltage phasor is $120\angle 30^\circ$ and the current phasor is $9.54\angle 12.56^\circ$. We already know that the active power is $P = 1092.2$ W. From (1.25), the reactive power is obtained as $Q = 342.2$ VAR. From (1.26), the apparent power is obtained as $S = 1092.2 + j342.2$. The magnitude of the apparent power is 1144.6 VA.

Power Factor

The cosine term that occurs in (1.22) is referred to as the *power factor*:

$$\mathrm{PF} = \cos(\theta - \phi). \tag{1.28}$$

It is a key term to explain the average power in (1.22). In direct current (DC) systems, the product of voltage and current gives power. However, in AC systems, we need to multiply the product of voltage and current by the power factor.

We usually indicate whether the power factor is *lagging* or *leading*. The power factor is lagging if $\theta - \phi$ is positive, i.e., current lags voltage. The power factor is leading if $\theta - \phi$ is negative, i.e., current leads voltage. Whether $\theta - \phi$ is positive or negative, the power factor is the same. From (1.28), the power factor in Example 1.5 is obtained as 0.954. The power factor in this example is lagging.

From (1.22), (1.27), and (1.28), we can also express the power factor as

$$\mathrm{PF} = \frac{P}{\sqrt{P^2 + Q^2}}. \tag{1.29}$$

Power Factor Correction

While active power is the power that is *dissipated* by the load, reactive power is the power that is *stored* and *released* in the magnetic field of inductors or electric field of capacitors during the sinusoidal cycles of instantaneous power. Therefore, supplying reactive power takes away part of the power delivery capacity of the transmission and distribution lines without adding to the actual power usage of the customers. It can also increase power loss [12]. Therefore, utilities are concerned about customers who draw excessive amounts of reactive power, i.e., the customers that have *low power factors*. Notice that, from the expression in (1.29), increasing reactive power results in decreasing power factor.

A low power factor is almost always due to large inductive devices, such as large electric motors for commercial and industrial customers. In other words, a low power factor is almost always a lagging power factor. Therefore, power factor can be increased by installing capacitors in locations close to major inductive loads.

Capacitors that are used for power factor correction can be installed by the customer or, more often, by the utility. Power factor correction is also sometimes referred to as power factor compensation or reactive power compensation.

1.2.5 Three-Phase Systems

A typical power outlet at home provides *single-phase* power at a nominal voltage. The nominal voltage in the United States is 120 V. A single-phase power supply is used for residential customers to run most home appliances such as a refrigerator, an iron, etc. However, commercial and industrial customers are supplied with *three-phase* power at a higher nominal voltage. The nominal three-phase voltage in the United States is 277 V on each phase, or 480 V phase-to-phase (line-to-line).

The above options are available to customers because the electric grid is *three-phase*. There are advantages to having a three-phase electric grid over a single-phase electric grid. For example, three-phase equipment is more efficient and makes more effective use of conductors and other materials, and thus costs less, than single-phase equipment of the same total power-handling capability [12].

Three-Phase Connection Configurations

Three-phase power systems may have different configurations. Here, we discuss only one configuration, the four-wire star-connection, as shown in Figure 1.4. This configuration is also called the four-wire wye-connection. Other three-phase connection configurations include the three-wire star-connection, three-wire and four-wire delta connections, and combinations of star and delta connections, such as a star connection on the supply side and a delta connection on the demand side.

The four-wire star-connection in Figure 1.4 includes one wire for each phase, denoted by A, B, and C. There is also one wire for *neutral current*. The *phase-to-neutral* voltages on the three phases are denoted by $v_A(t)$, $v_B(t)$, and $v_C(t)$, respectively. The *phase-to-phase* voltages, also known as *line-to-line* voltages, are denoted by $v_{AB}(t)$, $v_{BC}(t)$, and $v_{CA}(t)$. The phase currents are denoted by $i_A(t)$, $i_B(t)$, and $i_C(t)$, respectively. The neutral current is denoted by $i_N(t)$.

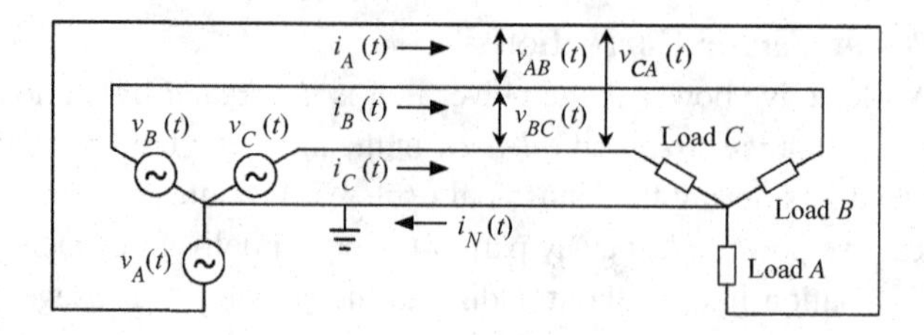

Figure 1.4 A four-wire, star-connection, three-phase circuit, showing sources and loads.

Three-Phase Waveform and Phasor Representations

In a three-phase power system, the voltages have the following waveforms:

$$\begin{aligned} v_A(t) &= \sqrt{2}\, V_A \cos(\omega t + \theta_A), \\ v_B(t) &= \sqrt{2}\, V_B \cos(\omega t + \theta_B), \\ v_C(t) &= \sqrt{2}\, V_C \cos(\omega t + \theta_C). \end{aligned} \tag{1.30}$$

We can represent the above voltages by using the following phasors:

$$\begin{aligned} &V_A \angle \theta_A, \\ &V_B \angle \theta_B, \\ &V_C \angle \theta_C. \end{aligned} \tag{1.31}$$

If the three phases are *balanced*, then we have:

$$V_A = V_B = V_C = V \tag{1.32}$$

and

$$\begin{aligned} \theta_A &= \theta, \\ \theta_B &= \theta - 120^\circ, \\ \theta_C &= \theta + 120^\circ. \end{aligned} \tag{1.33}$$

Thus, a balanced three-phase voltage can be represented by a single phasor $V\angle\theta$.

Example 1.6 Figures 1.5(a) and (b) show examples of balanced and unbalanced three-phase voltages, respectively. The unbalance is due to the three phases having different magnitudes and/or not having 120° differences.

Example 1.7 Consider a balanced three-phase voltage with $V = 277$ V and $\theta = 0^\circ$. The line-to-line voltage between Phase A and Phase B is obtained as

$$\begin{aligned} v_{AB}(t) &= v_A(t) - v_B(t) \\ &= \sqrt{2}\, 277 \cos(\omega t) - \sqrt{2}\, 277 \cos(\omega t - 2\pi/3) \\ &= \sqrt{2}\, 277\, (\cos(\omega t) - \cos(\omega t - 2\pi/3)) \\ &= \sqrt{2}\, (\sqrt{3}\, 277) \cos(\omega t + \pi/6). \end{aligned} \tag{1.34}$$

Therefore, the magnitude of the line-to-line voltage is $\sqrt{3} \times 277 = 480$ V.

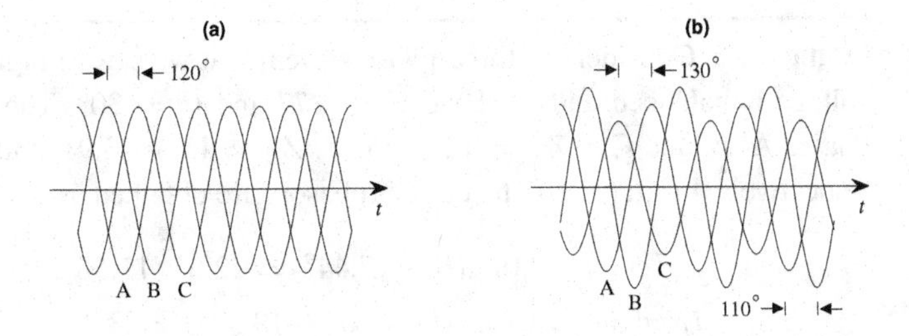

Figure 1.5 Three-phase voltage wave: (a) balanced; (b) unbalanced.

Per-phase currents in three-phase systems can be expressed similarly:

$$\begin{aligned} i_A(t) &= \sqrt{2}\, I_A \cos(\omega t + \phi_A), \\ i_B(t) &= \sqrt{2}\, I_B \cos(\omega t + \phi_B), \\ i_C(t) &= \sqrt{2}\, I_C \cos(\omega t + \phi_C). \end{aligned} \tag{1.35}$$

We can define current phasors similar to (1.31) as

$$\begin{aligned} &I_A \angle \phi_A, \\ &I_B \angle \phi_B, \\ &I_C \angle \phi_C. \end{aligned} \tag{1.36}$$

If the currents on the three phases are balanced, then we have:

$$I_A = I_B = I_C = I \tag{1.37}$$

and

$$\begin{aligned} \phi_A &= \phi, \\ \phi_B &= \phi - 120°, \\ \phi_C &= \phi + 120°. \end{aligned} \tag{1.38}$$

A balanced three-phase current can be represented by a single phasor $I\angle\phi$.

Neutral Current

The neutral current is defined as

$$i_N(t) = i_A(t) + i_B(t) + i_C(t). \tag{1.39}$$

If the three-phase currents are balanced, then $i_N(t) = 0$; see Exercise 1.7.

Power in Three-Phase Systems

Instantaneous power, active power, reactive power, apparent power, and power factor can all also be defined for *each phase* of a three-phase power system. The formulations are exactly the same as those that we saw in Section 1.2.4.

Example 1.8 Consider the three-phase system that we saw in Figure 1.4. Suppose the voltage is balanced and we have $V = 277$ and $\theta = 30°$. The load impedance on Phases A, B, and C is $Z_A = 12 + j3.77$, $Z_B = 11 + j3.91$, and $Z_C = 13 + j3.54$, respectively. From (1.10), the current phasors are obtained as

$$\begin{aligned} I_A \angle \phi_A &= 22.02\angle(\theta - 17.44°) = 22.02\angle 12.56°, \\ I_B \angle \phi_B &= 23.73\angle(\theta - 120° - 19.57°) = 23.73\angle -109.57°, \\ I_C \angle \phi_C &= 20.56\angle(\theta + 120° - 15.23°) = 20.56\angle 134.77°. \end{aligned} \tag{1.40}$$

The apparent power on the three phases is $S_A = 5819.7 + j1828.3$, $S_B = 6192.9 + j2201.3$, and $S_C = 5494.8 + j1496.3$. Even though the voltages are balanced, current, active power, and reactive power are unbalanced due to the unbalanced load.

1.2.6 Transformers

Transformers allow transforming AC voltage and current to levels that are appropriate for generation, transmission, distribution, and consumption.

Figure 1.6(a) shows the schematic representation of a single-phase two-winding transformer. The left-hand side is the *primary* side of the transformer, and the right-hand side is the *secondary* side of the transformer. For an ideal transformer, which has no internal power loss and no leakage, the incoming instantaneous power on the primary side of the transformer is equal to the outgoing instantaneous power on the secondary side of the transformer. Therefore, we have:

$$\frac{v_2(t)}{v_1(t)} = \frac{i_1(t)}{i_2(t)} = \text{Turn Ratio}. \tag{1.41}$$

As an example, if the turn ratio is two, then the voltage on the secondary side is twice the voltage on the primary side, i.e., $v_2(t) = 2\, v_1(t)$. It is also common to show transformers by using the symbol that is shown in Figure 1.6(b).

Step Up Transformation

The generation of power in large power plants is typically in the range of 11 kV–30 kV [12]. However, for efficient transmission of power over long distances, most transmission lines operate at much higher voltages. For example, the major transmis-

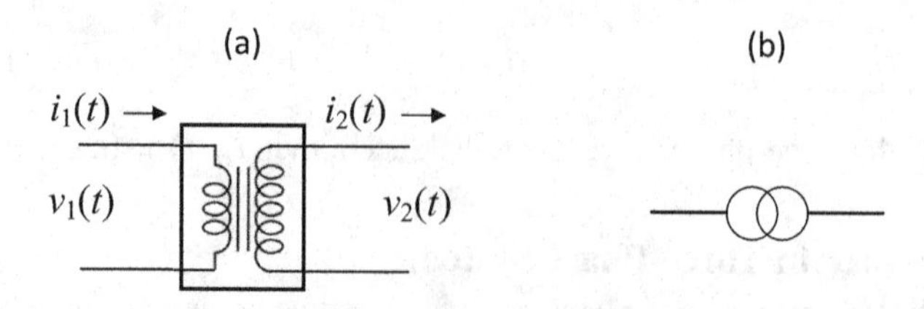

Figure 1.6 A transformer: (a) schematic representation; (b) simplified symbol.

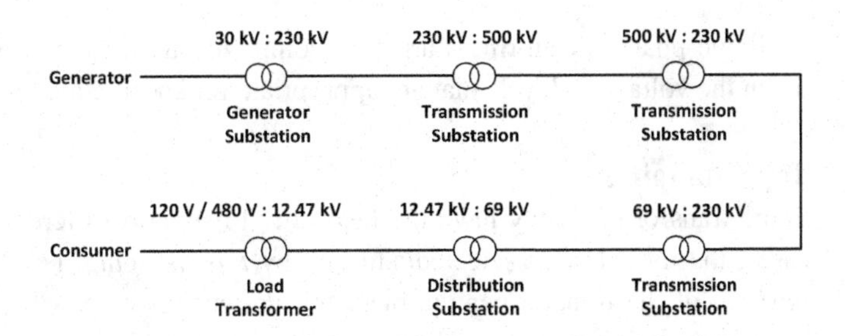

Figure 1.7 Several transformers are used between a generator and a consumer.

sion lines in the United States operate at the 115 kV–230 kV range (high voltage) and the 345 kV–765 kV range (extra high voltage) [26]. Therefore, *step up* transformers are needed to increase voltage both at power plants and at the interconnection of transmission lines with different voltage levels.

Two step up transformers are shown in Figure 1.7. The first transformer increases the voltage at a power plant from 30 kV to 230 kV. The second transformer further increases the voltage from 230 kV to 500 kV.

Step Down Transformation

Recall from Section 1.2.5 that the nominal voltage at the customer side of the electric grid is 480 V (three-phase) and 120 V (single-phase), which is much lower than the voltage at transmission lines. Furthermore, power distribution systems often operate at the voltage range of 4 kV–13 kV. Thus, *step down* transformers are needed to lower voltage both at distribution systems and at customer locations.

Four step down transformers are shown in Figure 1.7. The first step down transformer decreases the voltage from 500 kV to 230 kV. The second step down transformer decreases the voltage from 230 kV to 69 kV. The third step down transformer decreases the voltage from 69 kV to 12.47 kV. The last transformer, which is a *load transformer*, decreases the voltage from 12.47 kV to 480 V or 120 V.

It is worth noting that the power system at the 26 kV–69 kV voltage level is sometimes referred to as the *sub-transmission system*. It is also sometimes referred to as the *medium voltage* transmission system [26], as opposed to the high voltage and the extra high voltage transmission systems that we mentioned earlier.

All the transformers that we discussed above are three-phase; and all the voltage levels are line-to-line. All the transformers in Figure 1.7, except for the 12.47 kV to 480 V load transformer, are located in *substations*. An electric grid may have hundreds of substations. Besides transformers, substations usually have other equipment, such as for switching, protection, and monitoring. Distribution substations may also include capacitor banks to supply reactive power.

Load transformers, such as the 12.47 kV to 480 V transformer in Figure 1.7, are usually installed outside the distribution substations and along the distribution feeders. The load transformers are either *pole-mounted* or *pad-mounted*, i.e., they are installed on power poles or on the ground on concrete pads, respectively.

Single-phase transformers are also commonly used by customers to further step down the voltage to levels that are appropriate for specific appliances.

Tap Changing

Some transformers may have the flexibility to support different turn ratios. In such cases, the turn ratio is selected/adjusted by *tap changing*, i.e., by changing the *tap position* of the transformer via mechanical switches or power electronics switches. Each tap position provides a different turn ratio; cf. [13].

Example 1.9 A load transformer is rated at 12.47 kV on its primary side. The tap changer can set the turn ratio to 26. In that case, the rated voltage at the secondary side is 480 V, i.e., $12.47 \times 1000 / 26$. As we discussed before, this is the nominal voltage level for a three-phase load. However, the tap changer may also set the turn ratio to 25. In that case, the rated voltage on the secondary side increases to 499 V, i.e., $12.47 \times 1000/25$. In this example, tap changing can help with *regulating voltage*. For instance, if the voltage of the distribution feeder drops from 12.47 kV to 12 kV, for reasons such as the increased load, then changing the turn ratio from 26 to 25 can help maintain the customer voltage at the nominal 480 V, because $12 \times 1000 / 25 = 480$ V.

1.2.7 Per Unit Normalization

Quantities such as voltage, current, impedance, power, and so on are usually normalized in power systems calculations. The normalization is called *per unit normalization*, which makes it easier to analyze power systems that have several transformers and different voltage levels. A quantity in *per unit* is defined as

$$\text{Quantity in Per Unit} = \frac{\text{Actual Quantity}}{\text{Base Value of Quantity}}. \tag{1.42}$$

Example 1.10 Consider the following quantities:

$$P = 200 \text{ MW}, \quad Q = 75 \text{ MVAR}, \quad V = 228.8 \text{ kV}, \quad Y = -j0.02067 \text{ ℧}. \tag{1.43}$$

Suppose the base values are as follows:

$$S_{\text{base}} = 100 \text{ MVA}, \quad V_{\text{base}} = 220 \text{ kV}, \quad Z_{\text{base}} = \frac{V_{\text{base}}^2}{S_{\text{base}}} = 484 \ \Omega. \tag{1.44}$$

Accordingly, the quantities in *per unit* (p.u.) are obtained as

$$P = 2, \quad Q = 0.75, \quad V = 1.04, \quad Y = -j10. \tag{1.45}$$

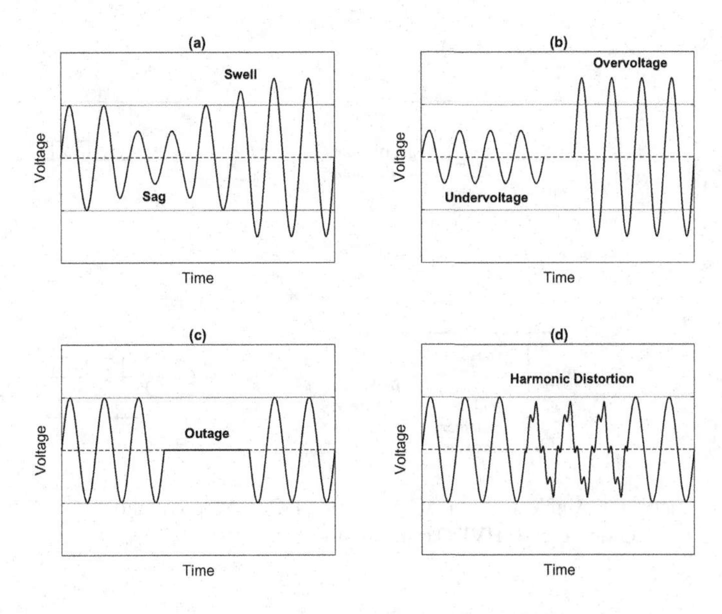

Figure 1.8 Power quality issues: (a) voltage sag and voltage swell; (b) undervoltage and overvoltage; (c) momentary outage; (d) harmonic distortions.

1.2.8 Power Quality

Irregularities in voltage and current are referred to as *power quality* issues [27]. Figure 1.8 illustrates some of these irregularities. If voltage momentarily falls below its nominal level, then it is called a *voltage sag*. If voltage momentarily rises above its nominal level, then it is called a *voltage swell*. If the voltage falls and rises are not momentary and last for several seconds or longer, then they are referred to as *undervoltages* and *overvoltages*, respectively. Downed power lines or other faults in the system can blow fuses or trip circuit breakers, resulting in power interruptions and *outages*. Distortions in the sinusoidal shape of the voltage or current waves are referred to as *harmonic distortions*.

Voltage sags, voltage swells, undervoltages, and overvoltages can all affect the performance, power usage, and aging of customer appliances. Also, some loads are very sensitive to power interruptions, even for a short duration. As for harmonic distortions, they too can create a variety of problems, ranging from blown circuit breakers to transformer failures and overloaded neutral lines in three-phase circuits [3]. Harmonic distortions are usually caused by nonlinear loads and power electronics devices, such as inverters; see Section 1.2.9.

It is necessary for electric utilities to monitor power quality issues and take corrective actions accordingly so as to maintain power quality at acceptable conditions. For example, it is required in the United States to maintain voltage at customer locations within $\pm 5\%$ of its nominal value [28].

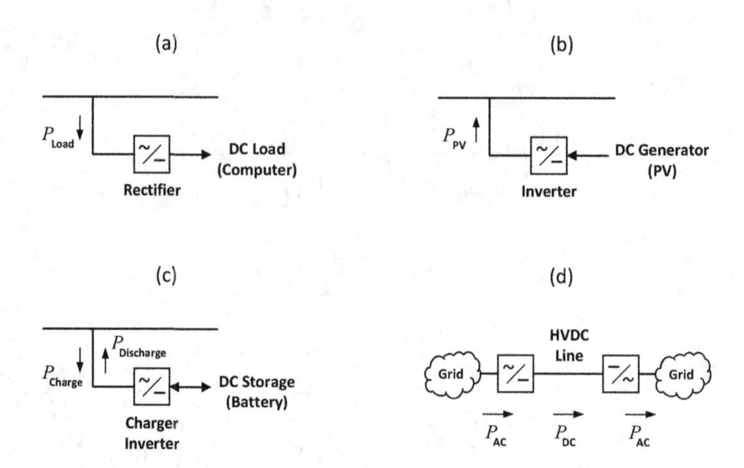

Figure 1.9 Applications of AC to DC and DC to AC conversion: (a) DC load; (b) DC generator; (c) DC storage; (d) HVDC transmission line.

1.2.9 AC to DC Conversion and DC to AC Conversion

Although the electric grid is for the most part an AC power system; it also includes several DC components. For example, some generators, such as solar photo-voltaic (PV) panels, generate DC power. Some loads, such as computers, require a DC power supply. AC to DC conversion and DC to AC conversion are done by *power electronics* devices.

The power electronics device that converts AC to DC is called a *rectifier*. A rectifier can serve as a DC power supply for a DC load. The power electronics device that converts DC to AC is called an *inverter*. Together, a DC generator and an inverter can serve as an AC power supply to inject power to the electric grid.

Figure 1.9 shows examples of how AC to DC converters and DC to AC converters are used in different applications. In Figure 1.9(a), a rectifier is used to supply DC power to a DC load. In Figure 1.9(b), an inverter is used to connect a DC generator to the power grid. In Figure 1.9(c), a charger inverter, i.e., a combination of a rectifier and an inverter, is used to charge and discharge a DC storage unit, such as a battery. In Figure 1.9(d), a rectifier and an inverter are used on two end terminals of a high-voltage DC (HVDC) transmission line.

HVDC lines are economically suitable for power transmission over very long distances, such as 500 miles or longer, due to their lower power loss compared to AC transmission lines [13]. For example, the Pacific HVDC line, also known as the Pacific DC Intertie, which is part of the Western Interconnection in the United States, is 850 miles long and operates at 500 kV. Currently, there are much fewer HVDC lines than AC transmission lines in most electric grids worldwide.

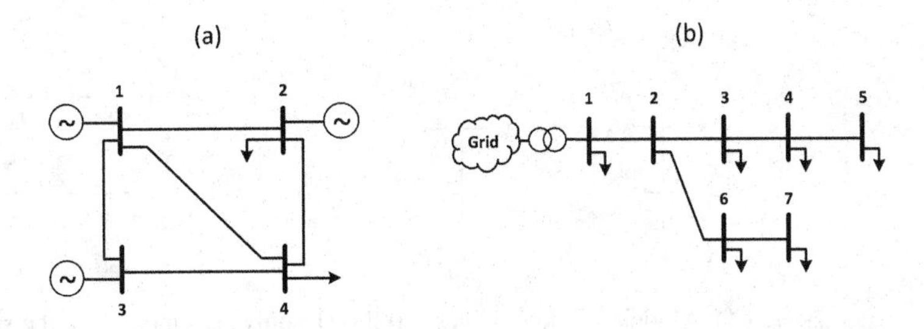

Figure 1.10 Examples of one-line diagrams of transmission and distribution systems: (a) a 4-bus transmission network; (b) a 7-bus distribution network.

1.3 Different Sectors in a Power Grid

Recall from the early discussions in Section 1.2 that the electric power grid includes the following sectors: transmission, distribution, generation, consumption, and storage. In this section, we will overview each of these sectors.

1.3.1 Transmission and Distribution

Power is delivered from generators to customers through a *network* of transmission lines and distribution lines. Transmission lines carry bulk power over long distances. Distribution lines deliver power to customers.

Figure 1.10(a) shows the *one-line diagram* of a 4-bus power transmission system. Each bus is a transmission substation. Each transmission line is represented by a single line. Generators are represented by circles, and loads are represented by arrows. Power transmission systems usually have a *mesh topology*. This increases reliability because power delivery can continue without interruption even if we lose a transmission line; e.g., see the analysis of power system contingencies in [15].

Figure 1.10(b) shows the one-line diagram of a 7-bus power distribution system. Each bus is a load transformer. Each distribution line is represented by a single line. The main source of power is the distribution substation. Loads are represented by arrows. Power distribution systems usually have a *radial topology*, i.e., a tree topology, where the root of the tree is the distribution substation. A distribution feeder usually has a *main line* and a few *laterals*. In Figure 1.10(b), the main line includes buses 1, 2, 3, 4, and 5; and there is one lateral that includes buses 6 and 7. The radial topology by itself does not provide reliability; however, it is common for a distribution feeder to have one or more *tie switches* to other distribution feeders in order to reconfigure the network topology during contingencies.

Modeling Transmission and Distribution Lines

Transmission and distribution lines are typically modeled by their impedance. Figure 1.11(a) shows the Π-equivalent model for a transmission line. It includes

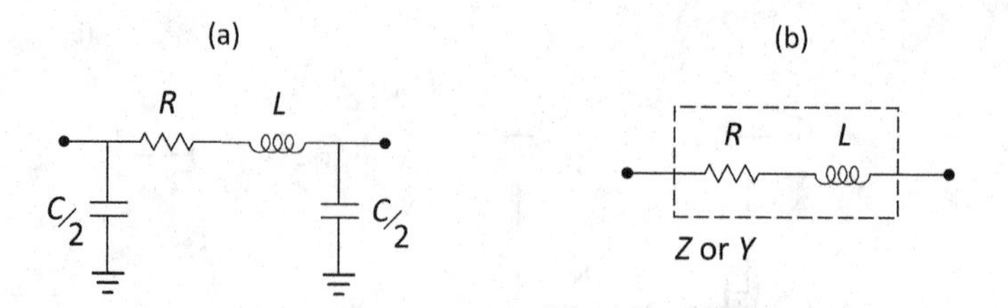

Figure 1.11 Models for power lines: (a) the Π-equivalent model; (b) the simplified model.

a resistor and an inductor in series and two capacitors in shunt. A simplified version of this model is shown in Figure 1.11(b), where the shunt capacitors are removed. This simplified model can also be used for distribution lines.

The X/R ratio is high for high-voltage transmission lines. For example, the X/R ratio for a 500 kV transmission line can be as high as 20 [3]. Accordingly, the model in Figure 1.11(b) can be further simplified to include only the series inductor; i.e., we may remove the series resistor from the model in Figure 1.11(b). In contrast, in distribution lines, the X/R ratio is relatively small. For example, the typical X/R ratio for a 4 kV distribution line is 1.2. Therefore, distribution lines are usually modeled with *both* the series resistor and the series inductor [14].

Network Equations

We can use the transmission and distribution line models, along with Kirchhoff's voltage law (KVL), Kirchhoff's current law (KCL), and Ohm's law, to obtain a system of *network equations* for the electric grid.

Consider the 4-bus transmission system in Figure 1.10(a). Let us denote the voltage phasors and the *injected* current phasors at buses 1 to 4 by

$$V_1\angle\theta_1,\ V_2\angle\theta_2,\ V_3\angle\theta_3,\ V_4\angle\theta_4 \tag{1.46}$$

and

$$I_1\angle\phi_1,\ I_2\angle\phi_2,\ I_3\angle\phi_3,\ I_4\angle\phi_4, \tag{1.47}$$

respectively. Current is injected at each bus by the generators and/or loads at that bus. Current injection is zero at a bus with no generator or load.

Suppose the transmission lines are represented by their admittance Y, as we saw in Figure 1.11(b). The admittance of the transmission line between bus 1 and bus 2 is denoted by Y_{12}, or equivalently, Y_{21}. Similarly, the admittance between any two buses i and k is denoted by Y_{ik}, or equivalently, Y_{ki}. Based on KCL, we can express the injected current phasor at bus 1 as follows:

$$\begin{aligned} I_1\angle\phi_1 = {} & Y_{12}(V_1\angle\theta_1 - V_2\angle\theta_2) \\ & + Y_{13}(V_1\angle\theta_1 - V_3\angle\theta_3) \\ & + Y_{14}(V_1\angle\theta_1 - V_4\angle\theta_4). \end{aligned} \tag{1.48}$$

After reordering the terms, we can rewrite (1.48) as

$$I_1\angle\phi_1 = \begin{bmatrix} Y_{12}+Y_{13}+Y_{14} & -Y_{12} & -Y_{13} & -Y_{14} \end{bmatrix} \begin{bmatrix} V_1\angle\theta_1 \\ V_2\angle\theta_2 \\ V_3\angle\theta_3 \\ V_4\angle\theta_4 \end{bmatrix}. \tag{1.49}$$

We can similarly obtain the expressions for $I_2\angle\phi_2$, $I_3\angle\phi_3$, and $I_4\angle\phi_4$. Accordingly, we can obtain the following system of linear equations to relate the bus voltage phasors to the bus current injection phasors:

$$\mathbf{I} = \mathbf{Y}^{\text{bus}}\mathbf{V}, \tag{1.50}$$

where

$$\mathbf{I} = \begin{bmatrix} I_1\angle\phi_1 \\ I_2\angle\phi_2 \\ I_3\angle\phi_3 \\ I_4\angle\phi_4 \end{bmatrix}, \quad \mathbf{V} = \begin{bmatrix} V_1\angle\theta_1 \\ V_2\angle\theta_2 \\ V_3\angle\theta_3 \\ V_4\angle\theta_4 \end{bmatrix} \tag{1.51}$$

and

$$\mathbf{Y}^{\text{bus}} = \begin{bmatrix} Y_{12}+Y_{13}+Y_{14} & -Y_{12} & -Y_{13} & -Y_{14} \\ -Y_{21} & Y_{21}+Y_{24} & 0 & -Y_{24} \\ -Y_{31} & 0 & Y_{31}+Y_{34} & -Y_{34} \\ -Y_{41} & -Y_{42} & -Y_{43} & Y_{41}+Y_{42}+Y_{43} \end{bmatrix}. \tag{1.52}$$

The above matrix is called the Y-bus matrix. The Y-bus matrix can be obtained similarly for any arbitrary transmission or distribution network. In short, the *diagonal elements* in the Y-bus matrix are the sum of the adimittances that are connected to each bus; and the *off-diagonal elements* in the Y-bus matrix are the *minus* of the admittance of the line between any two buses [13]. We can separate the real part and the imaginary part of the Y-bus matrix as

$$\mathbf{Y}^{\text{bus}} = \mathbf{G}^{\text{bus}} + j\mathbf{B}^{\text{bus}}. \tag{1.53}$$

Let Y_{ik}^{bus} denote the element at row i and column k of the Y-bus matrix. This notation should be distinguished from Y_{ik}, which is the admittance of the line between bus i and bus k. From (1.26), (1.50), and (1.51), the apparent power that is injected at each bus i in Figure 1.11(b) is obtained as

$$\begin{aligned} S_i = P_i + jQ_i &= V_i\angle\theta_i\,(I_i\angle\phi_i)^* \\ &= V_i\angle\theta_i \left(\sum_{k=1}^{4} Y_{ik}^{\text{bus}}\, V_k\angle\theta_k \right)^*. \end{aligned} \tag{1.54}$$

If bus i is a net supplier of active power, then $P_i > 0$, and if it is a net consumer of active power, then $P_i < 0$. Similarly, if bus i is a net supplier of reactive power, then $Q_i > 0$, and if it is a net consumer of reactive power, then $Q_i < 0$.

Given the voltage phasors at all buses, we can similarly obtain the active and reactive power flow from bus i to bus k on transmission line (i,k) as

$$\begin{aligned} S_{ik} = P_{ik} + jQ_{ik} &= V_i \angle\theta_i \, (I_{ik} \angle\phi_{ik})^* \\ &= V_i \angle\theta_i \left(-Y_{ik}^{\text{bus}} \, (V_i \angle\theta_i - V_k \angle\theta_k)\right)^* . \end{aligned} \tag{1.55}$$

The equations in (1.54) and (1.55) are in the domain of complex numbers. However, they can be expressed also in the domain of real numbers [12]:

$$\begin{aligned} P_i &= \sum_{k=1}^{n} V_i V_k \left(G_{ik}^{\text{bus}} \cos(\theta_i - \theta_k) + B_{ik}^{\text{bus}} \sin(\theta_i - \theta_k)\right) \\ Q_i &= \sum_{k=1}^{n} V_i V_k \left(G_{ik}^{\text{bus}} \sin(\theta_i - \theta_k) - B_{ik}^{\text{bus}} \cos(\theta_i - \theta_k)\right) \end{aligned} \tag{1.56}$$

and

$$\begin{aligned} P_{ik} &= -V_i^2 G_{ik}^{\text{bus}} + V_i V_k \left(G_{ik}^{\text{bus}} \cos(\theta_i - \theta_k) + B_{ik}^{\text{bus}} \sin(\theta_i - \theta_k)\right) \\ Q_{ik} &= V_i^2 B_{ik}^{\text{bus}} + V_i V_k \left(G_{ik}^{\text{bus}} \sin(\theta_i - \theta_k) - B_{ik}^{\text{bus}} \cos(\theta_i - \theta_k)\right). \end{aligned} \tag{1.57}$$

Example 1.11 Consider the 4-bus transmission network in Figure 1.10(a). The admittance for each transmission line is $0.5 - j10$ p.u. The Y-bus matrix is

$$\mathbf{Y}^{\text{bus}} = \begin{bmatrix} 1.5 - j30 & -0.5 + j10 & -0.5 + j10 & -0.5 + j10 \\ -0.5 + j10 & 1.0 - j20 & 0 & -0.5 + j10 \\ -0.5 + j10 & 0 & 1.0 - j20 & -0.5 + j10 \\ -0.5 + j10 & -0.5 + j10 & -0.5 + j10 & 1.5 - j30 \end{bmatrix}. \tag{1.58}$$

Suppose the voltage phasors are known and expressed in per unit as

$$\begin{aligned} V_1 \angle\theta_1 &= 1.0332 \angle 38.8884^\circ \\ V_2 \angle\theta_2 &= 0.9974 \angle 31.5332^\circ \\ V_3 \angle\theta_3 &= 1.0499 \angle 37.0645^\circ \\ V_4 \angle\theta_4 &= 0.9755 \angle 28.3416^\circ . \end{aligned} \tag{1.59}$$

From (1.56), we can obtain the power injections in per unit at all buses as follows:

$$\begin{aligned} P_1 &= 3.56, & Q_1 &= 0.88, \\ P_2 &= -0.78, & Q_2 &= 0, \\ P_3 &= 1.26, & Q_3 &= 1.02, \\ P_4 &= -4.02, & Q_4 &= -1.00. \end{aligned} \tag{1.60}$$

From (1.57), the power flow on all transmission lines are obtained in per unit as

$$\begin{aligned}
P_{12} + j\,Q_{12} &= 1.342 + j0.389, & P_{21} + j\,Q_{21} &= -1.333 - j0.206,\\
P_{13} + j\,Q_{13} &= 0.334 - j0.184, & P_{31} + j\,Q_{31} &= -0.336 + j0.198,\\
P_{14} + j\,Q_{14} &= 1.883 + j0.674, & P_{41} + j\,Q_{41} &= -1.865 - j0.300,\\
P_{24} + j\,Q_{24} &= 0.553 + j0.206, & P_{42} + j\,Q_{42} &= -0.552 - j0.172,\\
P_{34} + j\,Q_{34} &= 1.598 + j0.822, & P_{43} + j\,Q_{43} &= -1.584 - j0.530.
\end{aligned} \tag{1.61}$$

The difference between $P_{12} + j\,Q_{12}$ and $P_{21} + j\,Q_{21}$ is due to *power loss* on the transmission line between bus 1 and bus 2. Note that $P_{12} + j\,Q_{12}$ is the apparent power that leaves bus 1 toward bus 2; and $P_{21} + j\,Q_{21}$ is the apparent power that leaves bus 2 toward bus 1. The active power loss is 0.009 p.u., and the reactive power loss is 0.183 p.u. Power loss can be calculated similarly for all other transmission lines. Since X/R is high for transmission lines, reactive power losses are higher on the inductive elements of the transmission lines compared to active power losses on the resistive elements of the transmission lines.

The above analysis can also be done in reverse. If the amount of power injection is known at each bus, then we can use either (1.54) or (1.56) at all buses to obtain the voltage phasors at all buses. In order to obtain a unique solution, we need a *reference bus*, and we need to know the voltage phasor at the reference bus. If the voltage at the reference bus is not explicitly known, then we may assume that the magnitude of the voltage at the reference bus is one per unit and the phase angle of the voltage at the reference bus is zero.

Either the network equations in (1.54) and (1.55) or the network equations in (1.56) and (1.57) can be used in power distribution systems as well; see Exercise 1.18.

Simplified Network Equations

There are different ways to approximate and simplify the network equations. For example, recall the fact that the resistive element of a transmission line is much less than its inductive element. Therefore, one option to simplify the network equations is to remove the resistive element from the transmission line models. From (1.15) and (1.53), this results in setting the real part of the Y-bus matrix to zero, which in turn significantly simplifies the models in (1.56) and (1.57). For example, the active power flow between bus i and bus k becomes:

$$\begin{aligned}
P_{ik} &\approx V_i\, V_k\, B_{ik}^{\text{bus}}\, \sin(\theta_i - \theta_k)\\
&= -V_i\, V_k\, B_{ik}\, \sin(\theta_i - \theta_k)\\
&= \frac{V_i\, V_k}{X_{ik}}\, \sin(\theta_i - \theta_k).
\end{aligned} \tag{1.62}$$

The last equality in (1.62) is due to (1.15) as well as the assumption that there is no resistor component in the simplified transmission line model.

In practice, the difference between the phase angles of the voltage phasors at any two buses in the same power network is usually small, typically less than 10° to 15°. Therefore, another common approximation is to assume the following:

$$\sin(\theta_i - \theta_k) \approx \theta_i - \theta_k. \tag{1.63}$$

For this approximation to hold, the phase angles must be expressed in *radian*.

Another common approximation is based on the fact that, during normal grid operating conditions, the magnitude of voltage phasors is close to one per unit at all buses. Thus, for any two neighboring buses i and k, we can assume that

$$V_i \, V_k \approx 1. \tag{1.64}$$

For this approximation to hold, the voltages must be expressed in *per unit*.

By applying the assumptions in (1.62)–(1.64) to the equations in (1.56) and (1.57), we can obtain the following simplified network equations:

$$P_i = \sum_{k=1}^{n} B_{ik}^{\text{bus}} \, (\theta_i - \theta_k) \tag{1.65}$$

and

$$P_{ik} = B_{ik}^{\text{bus}} \, (\theta_i - \theta_k) \, . \tag{1.66}$$

The above equations are called DC power flow equations [12]. Unlike the original equations in (1.56) and (1.57) that are *nonlinear*, the simplified equations in (1.65) and (1.66) are *linear*. Furthermore, while the magnitude of voltage as well as the reactive power are included in the original equations in (1.56) and (1.57), they are *not* included in the simplified equations in (1.65) and (1.66).

Example 1.12 In Example 1.11, if we consider only the phase angles in (1.59), then we can use (1.65) to obtain the active power injection at all buses:

$$\begin{aligned} P_1 &= 3.44, \\ P_2 &= -0.72, \\ P_3 &= 1.20, \\ P_4 &= -3.92. \end{aligned} \tag{1.67}$$

Furthermore, we can use (1.66) to obtain

$$\begin{aligned} P_{12} &= 1.284, & P_{21} &= -1.284, \\ P_{13} &= 0.318, & P_{31} &= -0.318, \\ P_{14} &= 1.841, & P_{41} &= -1.841, \\ P_{24} &= 0.557, & P_{42} &= -0.557, \\ P_{34} &= 1.522, & P_{43} &= -1.522. \end{aligned} \tag{1.68}$$

The results in (1.67) are an approximation of the results in (1.60), and the results in (1.68) are an approximation of the results in (1.61). Notice that, based on (1.68), there is no power loss on transmission lines. This is because we *assumed* that there is no resistive component in the transmission line models.

The simplified network equations in (1.65) and (1.66) are commonly used in various smart grid applications; e.g., see [29–31].

Note that the above linearized model is typically used only in power transmission systems. It is typically *not* used in power distribution systems because some of the above assumptions are not applicable to power distribution lines. For example, resistance cannot be ignored in power distribution lines. Therefore, we should use either the original models in (1.54)–(1.57) or other models that are designed for power distribution systems, such as in [32, 33].

Power Flow Analysis

The network equations can be used also to solve the *power flow* problem in power transmission or power distribution networks. In this problem, we solve the system of power flow equations in (1.54) and (1.55), or alternatively the system of power flow equations in (1.56) and (1.57), to calculate the amount of power flow on each transmission or distribution line. In this regard, the amount of active power injection and reactive power injection at each bus is assumed to be known, while the amount of active power flow and reactive power flow on each line is assumed to be unknown. We can solve the power flow problem by using a solver for nonlinear equations; such as by using the command `fsolve` in MATLAB [34]; see Exercises 1.12 and 1.16.

The simplified network equations in (1.65) and (1.66) can also be used to obtain an approximate solution for the power flow problem.

1.3.2 Generation

Traditionally, electric power has been generated in bulk in large power plants. In this section, we will briefly discuss different technologies for bulk power generation. We will also discuss different technologies for renewable and distributed power generation. The latter technologies have been among the main driving forces for the development of smart grids in recent years.

Fossil Fuels, Nuclear Power, and Hydro Power

About 90% of the world's electric power is generated by fossil-fuel power plants, nuclear power plants, and hydroelectric power plants [35]. The most common fossil-fuel power plants are coal-fired steam power plants and natural-gas-fired power plants. They burn coal or natural gas to drive a *turbine* to generate electricity. Nuclear power plants work based on principles similar to those in fossil-fueled power plants. The difference is that the heat in nuclear power plants is created by nuclear reactions instead of burning fossil fuels. Hydroelectric power plants harness the power of moving water to drive the turbine to generate electricity. Hydroelectric systems may include a dam or a reservoir to store water. The water can be released to generate electricity when needed.

Fossil fuels generate a significant portion of global greenhouse gas emissions that have influenced global warming [6]. Nuclear power generation also poses risks regarding long-term radioactive waste management as well as accidents and other safety concerns. Hence, there has been growing interest in recent years to shift away from fossil-fuel and nuclear power plants and invest more in renewable power generation. Hydroelectric power is a renewable power resource.

Solar Power, Wind Power, and Other Renewable Resources

Renewable resources, other than hydroelectric power, still provide only a small portion of the electricity supply worldwide. However, there has been a major shift in this area in recent years, in particular due to the increasing number of *solar* and *wind* power generation installations. For example, over the past 10 years, the share of solar and wind power generation in the U.S. state of California has increased from 3% to 23% [36].

Solar power generation is the process of converting the power in sunlight into electricity. It is done either directly by using technologies such as photo-voltaic (PV) cells; or indirectly by using technologies such as *concentrated solar power* (CSP). PV cells are semiconductor devices that generate electricity upon exposure to light [37]. CSP uses mirrors or lenses to concentrate sunlight onto a receiver that contains heat-transfer fluid, such as molten salt. The heat is then used to generate electricity, for example, by using a steam turbine [3].

Regardless of the technology being used, the amount of generated solar power depends highly on *solar irradiance*, which is the power per unit area that is received from the sun. Solar irradiance depends on geographical location and season. It is less during winter and more during summer. Solar irradiance is less on a cloudy day and more on a day with clear sky. In fact, even the movement of clouds during a day with a relatively clear sky can significantly change the amount of solar power generation. As a result, solar power generation can fluctuate during the day. Accordingly, solar power generation is considered *intermittent*.

Wind power generation is the process of converting the kinetic power in wind into electricity. It is done by using *wind turbines*. Wind turbines can be installed on land for *onshore* wind power generation, or on the sea for *offshore* wind power generation. Onshore wind power generation has a lower cost of installation and maintenance, and it is easier to integrate to the electric grid. Offshore power generation is more expensive to install, maintain, and integrate, but it has the advantage that wind speeds are higher off shore compared to on land; therefore, the amount of electric generation is higher per the amount of capacity installed [38].

Wind power generation is *intermittent* because it depends highly on *wind speed* and *wind direction*. Wind speed depends on geographical location and season. It may also change significantly during the day and/or night.

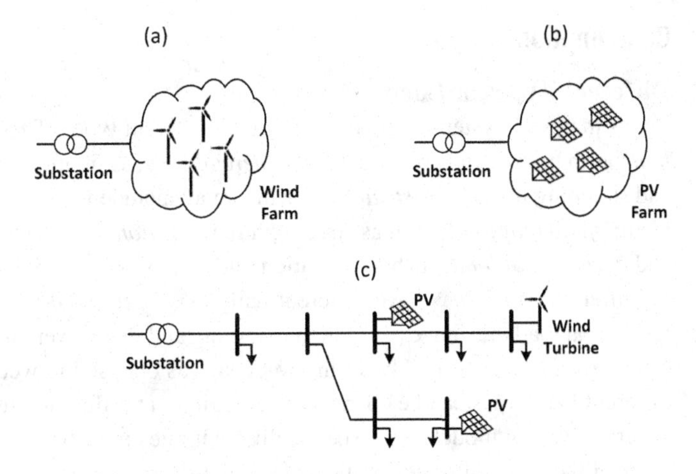

Figure 1.12 Wind and solar power generation: (a) wind farm; (b) solar farm; (c) DERs.

Larger wind power generation and solar power generation facilities are called *wind farms* and *solar farms*, respectively. Based on the current installations, the largest wind farm in the United States is the Alta Wind Energy Center, which is located in California and has a combined capacity of 1550 MW [39]. The largest solar farm in the United States is the Solar Star installation, also located in California, with a combined capacity of 579 MW [40]. Given their large size, wind farms and solar farms typically require an on-site substation to integrate them into the electric grid; see Figures 1.12(a) and (b), respectively.

Some of the other renewable resources that are used for electric power generation include *biomass*, *geothermal*, *tidal*, and *wave*; see [3] for more details.

Distributed Energy Resources

Solar power generation, and to some extent wind power generation, can be done in the form of *distributed energy resources* (DERs) [41–43]. DERs are often at customer locations and have much smaller generation capacity. For example, rooftop PV panels at a house may have a few kilowatts of generation capacity. DERs at commercial and industrial load locations may have a capacity of hundreds of kilowatts. DERs are usually integrated to the electric grid via load transformers and across power distribution feeders; see Figure 1.12(c).

The integration of the DERs into the electric grid can be coordinated by *microgrids*. A microgrid is a mini-version of the electric grid. It includes clusters of locally controlled DERs, in addition to various types of loads, which can operate as a self-sufficient electric grid. Microgrids can operate under *grid-connected* and *grid-disconnected* modes. The latter mode is commonly referred to as the *islanded* operation mode. More details about micro-grids are available in [44–47].

1.3.3 Consumption

Different Types of Loads

Loads in power systems can be divided into different types. These include *residential load*, as in houses and apartments; *commercial load*, as in office buildings, restaurants, and shopping malls; *industrial load*, which may include various machinery and equipment; *municipal load*, such as street lighting; *traction load*, such as electric railways; and *agricultural load*, such as irrigation pumps [48].

Different load types have different daily or weekly load profiles. For example, a residential load may have peaks in the morning and in the evening; or certain commercial loads may occur mainly during weekdays as opposed to weekends. Accordingly, different load types can be charged for electricity at different rates and even based on different pricing models; e.g., see the different rates in [49].

With the increasing penetration of *behind-the-meter* renewable generation installations, such as rooftop solar panels, many customers now have on-site generation capability. Their local power generation can offset part of their load; and they may even inject power into the grid, therefore becoming net generators.

Load Flexibility

Traditionally, the task of balancing supply and demand in an electric grid is the responsibility of the generators; they are the ones that must balance the equation, not the customers. However, this paradigm is gradually changing in the era of the smart grid, due to the development of new technologies that can manage load flexibility, such as in *building energy management systems*, see [50–52], and also the new pricing models that provide *incentives* for customers to utilize their load flexibility to meet the needs in grid operation; e.g., see [53–56].

An emerging type of flexible load is the charging load of *electric vehicles* (EVs). The battery capacities of the current EVs may vary, such as from 17.6 kWh to as high as 100 kWh [57]. EVs are usually plugged in for several hours when they are parked at home or at charging stations, often longer than the time that they need to fully charge. Therefore, they are flexible in terms of the exact timing of charging them, as long as they are charged before they are unplugged.

Example 1.13 Consider an EV that needs four hours to be charged, and suppose it is plugged in from 9:00 AM till 4:00 PM. It is sufficiently flexible that it can be charged during any four hours within the seven hours that it is plugged in. Load flexibility in this example is with respect to the *timing* of the load.

EVs may also *discharge* their batteries into the grid, thus acting as generators in *vehicle-to-grid* (V2G) or *vehicle-to-building* (V2B) configurations [58–60].

1.3.4 Storage

Large-scale energy storage has been traditionally considered to be *not* economically viable. Lack of large-scale energy storage requires the generated electric power to be consumed almost instantly; that is why we always need to balance electric power generation and electric power consumption. However, there have been important advancements in this field in recent years. For example, as of 2018, the three largest investor-owned utilities in California have procured or are seeking approval to procure almost 1500 MW of storage capacity with specific targets for *transmission-connected*, *distribution-connected*, and *customer-side* energy storage systems [61].

Different Technologies

A wide range of energy storage technologies have been considered in recent years for integration into the power grid. Here are a few examples: pumped hydro [62]; compressed air [63]; flywheels [64]; heat thermal [65]; cold thermal [66]; batteries [67–69]; and ultra capacitors [70].

Different Characteristics and Different Applications

Energy storage systems can be characterized by a wide range of parameters. Some basic parameters are *energy rating*, *power rating*, and *efficiency*. Energy rating is the total kWh or MWh of energy that can be stored in the energy storage system. Power rating is the maximum kW or MW power that can be charged to or discharged from the energy storage system. Some energy storage systems may have a different charge rate versus discharge rate. Also, the power rating may depend on not only the energy storage technology but also the way that the energy storage system is integrated into the electric grid. For example, the power rating of a battery system is highly affected by the battery charger-inverter as well as the configuration of connecting the batteries; e.g., see [71, 72]. Efficiency is the percentage ratio of the energy that is charged to the energy storage system versus the energy that is discharged from the energy storage system, considering all losses during the charge and discharge process.

Example 1.14 An energy storage system has an energy rating of 1 MWh, a power rating of 100 kW, and efficiency of 90%. Accordingly, if this energy storage system is fully depleted, and then we charge it at 100 kW for a duration of 10 hours, we can later discharge it at 100 kW for a total duration of 9 hours.

Other factors that characterize an energy storage system may include self-discharging rate, aging and life-cycle, memory, weight, cost per kWh or cost per cycle, temperature tolerance, depth-of-discharge, and response time.

Different energy storage technologies can be suitable for different smart grid applications. Different applications may require different *discharge power*, different *discharge time duration*, and different *discharge frequency*. For example, to offer regulation service, the energy storage should be charged and discharged frequently to help balance generation and consumption. This requires the ability to support a high discharge frequency; and to have a short response time and a high life-cycle. But it does not require large discharge power or long duration of discharge [73]. In contrast, offering bulk power during peak load hours requires the ability to support high discharge power and high duration of discharge; but it does not require the ability to support high discharge frequency, because the energy storage may have to be discharged only once a day or even only once every few days.

1.4 Overview of Chapters 2 to 7

We end this chapter by briefly reviewing the content of Chapters 2 to 7.

Chapter 2: Measuring Voltage and Current

This chapter is about the basics of measuring voltage and current, and the applications of such measurements in the field of smart grids. We start the chapter by talking about different options to deal with the high ranges of voltage and current in the electric grid. We discuss the use of *instrument transformers* to step down voltage and/or current before we can measure them. We also discuss the alternative option of using *non-contact sensors* that measure current and voltage *without* electric contact. Other basic concepts that are discussed in this chapter include *sampling rate*, *reporting rate*, *measurement accuracy*, *measurement aliasing*, and the impact of *averaging filters*, all of which are useful throughout this book. The rest of this chapter is about RMS voltage and current measurements and their characteristics and applications. First, we discuss RMS voltage and current *profiles*, as time series and also in other forms, such as in histograms and scatter plots. Next, we discuss RMS voltage and current *transient responses*, which can be caused by *faults*, *equipment actuation*, *load operation*, etc. Next, we discuss RMS voltage and current *oscillations*, such as in *wide-area* oscillations across the power transmission system, or in *local* transient oscillations at a power distribution system. Some methods are discussed to mathematically characterize oscillations. Next, we discuss *three-phase* RMS voltage and current measurements. We talk about some applications, such as measuring *phase unbalance* and *phase identification*. Next, we discuss the fundamental subject of *events* in smart grid measurements. Events are defined broadly as any change in any component in the power system that is worth studying. Events may include major faults and equipment failures that can affect stability and reliability of the system. While the focus in this chapter is on events in RMS voltage and current measurements, we also discuss methods for *event detection* that are useful throughout this book. Finally, we discuss measuring *frequency* and talk about the *frequency responses* of the electric grid; as well as measuring *frequency oscillations*.

Chapter 3: Measuring Phasors and Synchrophasors

This chapter focuses on measuring voltage and current *phasors* and *synchrophasors*, and the many applications of such measurements in the filed of smart grids. We start the chapter by talking about some basic concepts, such as: calculating phasors from the *fundamental components* of the voltage and current waveforms; *time synchronization*, e.g., by using satellite-synchronized clocks; as well as *data concentration*. We also discuss different types of *phasor measurement units*. We emphasize what is unique about phasor measurement units, which is their ability to measure not only the magnitude but also the *phase angle* of voltage and current. Hence, many of the discussions in this chapter are focused on measuring phase angles and the applications of phase angle measurements. We discuss the impact of the changes in the frequency of the power system on measuring phase angle. We also discuss measuring the *rate of change of frequency* and the applications of *synchronized frequency measurements*. Next, we introduce two fundamental quantities with respect to phasor measurements: *relative phase angle difference* and *phasor differential*. The applications of these quantities are discussed, such as in the analysis of wide-area oscillations and in *identifying the location* of events, respectively. Next, we discuss the *events* in phasor measurements. We extend the discussions about events in Chapter 2 to the important subject of *event classification*. In this regard, we talk about *feature selection* and *supervised and unsupervised classification*. Next, we discuss three-phase and unbalanced phasor measurements, including unbalanced phasor events, phase identification, and the application of *symmetrical components* in the analysis of unbalanced phasor measurements. Next, we discuss *state estimation* and *parameter estimation* in power systems using phasor measurements, as well as the subject of *topology identification*. Finally, we discuss the accuracy in phasor measurements, including different *performance classes* of phasor measurement units, steady-state performance, and dynamic performance.

Chapter 4: Measuring Waveforms and Power Quality

This chapter is about measuring voltage and current *waveforms* as well as *power quality*, and the traditional and emerging applications of such measurements. Waveform sensors can show the wave shape of voltage and current, as opposed to only their RMS values, as in Chapter 2; or only the phasor representation of their fundamental components, as in Chapter 3. Thus, they have a huge amount of data to report. In practice, a typical waveform sensor provides two types of data: a *continuous* reporting of power quality and steady-state *waveform distortion metrics*; and an *event-triggered* reporting of the signal waveform itself. Both aspects and their applications are covered in detail in this chapter. We start by talking about measuring different kinds of distortions in voltage and current waves, including *harmonics*, *inter-harmonics*, and *notching*; as well as metrics such as total harmonic distortion and crest factor. Next, we discuss several methods to detect and capture events in voltage and current waveform measurements. We also discuss how the captured event waveforms can help us understand different types of faults in the system, in particular *incipient faults*, which may indicate a potential catastrophic failure in the future, i.e., a major failure that is still in its *early stages*. In this regard, we discuss faults in underground power cables, overhead power lines, transformers, capacitor banks, and DERs. Next, we extend the discussions about *events* in Chapters 2 and 3 and talk about the statistical analysis of events in voltage and current waveform measurements. In this regard, we define various *quantitative features* that can characterize each event and allow conducting signature evaluation, event classification, and pattern recognition. In each case, we provide examples of how each feature may help us make conclusions with respect to the captured events. Next, we talk about *harmonic synchrophasors* and their applications in harmonic state estimation and topology identification. After that, we discuss *synchronized waveform measurements* and their emerging applications. Finally, we discuss the accuracy in waveform measurements, including the impact of noise and interference.

Chapter 5: Measuring Power and Energy

This chapter focuses on measuring *instantaneous power*, *active power*, *reactive power*, the *power factor*, and *energy*, and the wide range of the applications of such measurements in the field of smart grids. We start the chapter by talking about characterizing different types of power system components, based on their *power profile*, with respect to active power, reactive power, and the power factor. Next, we discuss measuring energy based on *fixed intervals* and *fixed increments*, as well as *net* energy metering and *feed-in* energy metering. Next, we talk about smart meters in length. We discuss the role of smart meters in facilitating price-based and incentive-based *demand response* programs in smart grid development. We discuss the use of smart meter measurements in *baseline calculation* in demand response applications. We also discuss other applications of smart meter measurements, such as in *load profiling* and *load classification*. Next, we discuss *load disaggregation*, *net load disaggregation*, and *sub-metering*. We then discuss *load modeling*, which covers both *static* load models and *dynamic* load models. Next, we discuss *state estimation* by using power measurements, where we discuss both nonlinear and linearized formulations. We then turn to three-phase and unbalanced power measurements and their applications. Finally, we discuss the accuracy in power and energy measurements, including different *accuracy classes* and meter accuracy versus system accuracy, as well as different factors that affect accuracy.

Chapter 6: Probing

This chapter covers the use of probing techniques in smart grid sensing. Probing is the broad technique of *perturbing* the power system for the purpose of enhancing our monitoring capabilities. Rather than only *passively* collecting measurements, probing methods make use of various grid components in order to *actively* create opportunities to learn more about the power system and its unknowns. We start this chapter by talking about conducting state and parameter estimation with the means of probing. We explain how probing can *enhance observability and redundancy*. Next, we discuss the application of probing in topology identification and phase identification, as well as the use of probing in model-free control and modal analysis. We talk about different examples of creating the probing signal, such as by using resistive brakes or intermittent wave modulation. Next, we focus on the application of *power line communications* as a probing tool. By observing the power line communications signal that is sent through the power grid, we can learn about the status of the power grid itself, such as how to diagnose incipient faults in power line cables or to identify the location of a fault. We also discuss the probing application of power line communications in topology identification and phase identification.

Chapter 7: Other Sensors and Off-Domain Measurements

This chapter focuses on *other types* of smart grid sensors as well as *off-domain* measurements. While the fundamentals of smart grid sensors are covered in Chapters 2 through 6, there are other sensors and measurements that can have specific applications in the field of smart grids. Off-domain measurements are the kind of measurements that are *not* primarily intended for the power sector; yet they can help with certain smart grid applications. These additional data and measurements can be used as *stand-alone* data, or in *cross examination* with some of the measurements that we discussed in the previous chapters. We start this chapter by talking about device and asset sensors, which may provide different types of electrical, mechanical, and chemical measurements. Different types of assets are discussed, including transformers, capacitor banks, line conductors, wind turbines, solar panels, and batteries. Next, we look at the recent advancements in *building sensors* that can support smart grid applications at customer locations. We discuss different types of *occupancy* sensors, *temperature and illuminance* sensors, and measurements related to *electric vehicles*. Next, we discuss the applications of *financial data*, including pricing and billing data, and data from electricity markets. Next, we discuss the applications of images, laser images, and drones in smart grid monitoring. Finally, we briefly discuss some other types of off-domain measurements, such as weather data, data from the national lightning detection network, traffic data, and data from social media.

Exercises

1.1 Consider the AC circuit in Figure 1.13, which includes a voltage source:

$$v_{\text{source}} = 120\sqrt{2}\cos(2\pi \times 60 \times t). \tag{1.69}$$

(a) Obtain the current $i(t)$ as a phasor and as a sinusoidal function of time.

(b) Obtain the voltage $v(t)$ as a phasor and as a sinusoidal function of time.

1.2 Consider the circuit in Example 1.3, which consists of two AC voltage sources with the same magnitude at 120 V but different phase angles θ_1 and θ_2. Suppose $\theta_1 = 0°$ is fixed. Suppose θ_2 can vary between 0° and 90°.

(a) Express the current phasor $I\angle\phi$ in terms of parameter θ_2.

(b) Express $I\angle\phi$ in time domain in form of a sinusoidal function $i(t)$.

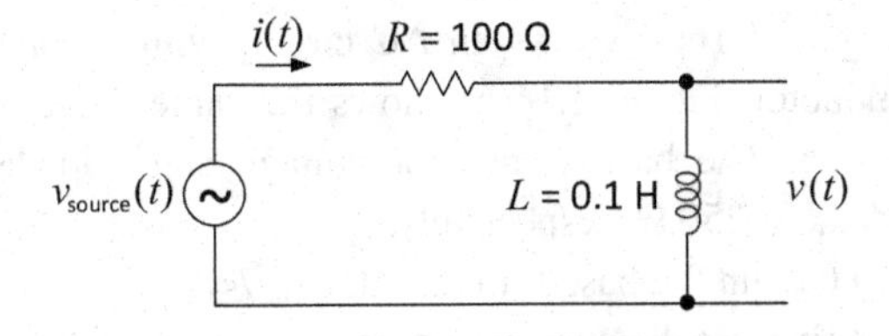

Figure 1.13 The AC circuit in Exercise 1.1.

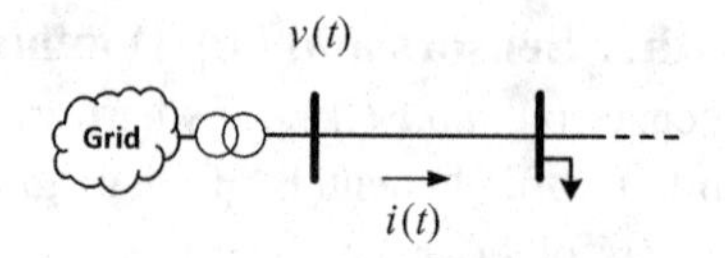

Figure 1.14 The power distribution feeder in Exercise 1.3.

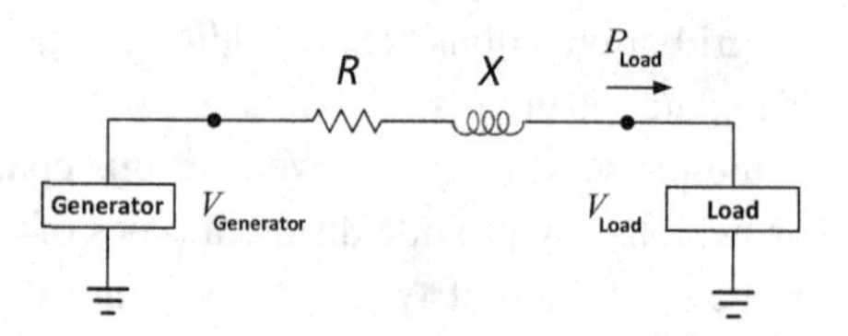

Figure 1.15 The circuit and its parameters in Exercise 1.5.

(c) Plot the instantaneous power $p(t)$ that is dissipated at the $R = 12\Omega$ line resistor. Calculate the average power P that is dissipated at this line resistor for $\theta_2 = 30°, 45°$, and $60°$. Explain your observations.

1.3 Consider a portion of a power distribution system as shown in Figure 1.14, where $v(t) = 7200\sqrt{2}\cos(\omega t)$ and $i(t) = 50\sqrt{2}\cos(\omega t) + 28\sqrt{2}\cos(3\omega t)$.
(a) Plot $v(t)$ and $i(t)$ over a period of 100 msec.
(b) Use (1.4) to numerically obtain the RMS value for $v(t)$ and $i(t)$.

1.4 In Exercise 1.3, obtain and plot the instantaneous power $p(t) = v(t)i(t)$ that is drawn from the substation over a period of 100 msec. Also obtain the apparent power $S = V_{\text{rms}} I_{\text{rms}}$ that is drawn from the substation.

1.5 A small generator is trying to deliver $P_{\text{Load}} = 15$ kW of real power through a 120 V (RMS) single-phase power line to a load that has a lagging power factor of 0.9. The circuit is shown in Figure 1.15. The power line has resistance $R = 0.05\ \Omega$ and inductance $X = 0.1\ \Omega$. What RMS voltage should the generator provide on its end of the power line?

1.6 Consider a balanced three-phase 1.5 MVA load with a lagging power factor of 0.75. The line-to-line voltage is 480 V. What is the size of the balanced three-phase capacitor bank that needs to be added to the load to improve the power factor to 0.9? Give your answer in Farads per phase.

1.7 Consider the definition of neutral current in three-phase systems in (1.39). Use the expressions in (1.35) to show that if the three-phase current is balanced, then the neutral current is zero.

1.8 Figure 1.16(a) shows an AC circuit with a voltage source, a resistor, and an inductor. Figure 1.16(b) shows the same AC circuit but in per unit representation. The base values for voltage and impedance are $V_{\text{base}} = 500$ kV and $Z_{\text{base}} = 250\ \Omega$, respectively.
(a) Obtain the base for current, i.e., I_{base}.
(b) Express the current I in per unit.

1.9 A load transformer rated at 1000 kVA is operating near capacity as it supplies a load that draws 900 kVA with a power factor of 0.7 [3].

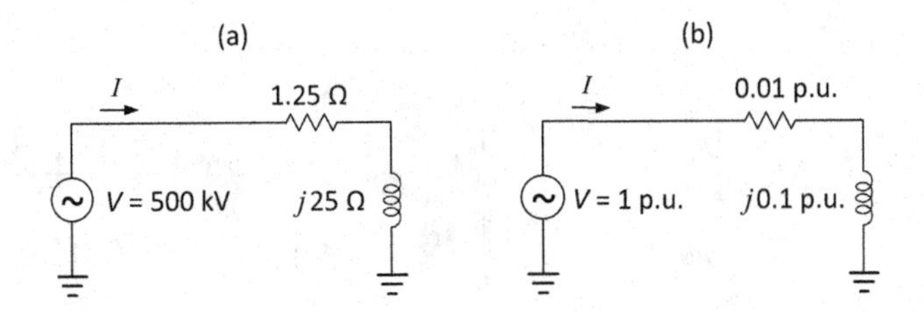

Figure 1.16 The parameters in Exercise 1.8: (a) regular representation; (b) per unit.

(a) How much active power does the load draw?

(b) How much reactive power does the load draw?

(c) How much additional active power can the load draw from this load transformer if the load's power factor remains at 0.7?

(d) How much *additional* active power can the load draw from this load transformer if the load's power factor is corrected to 0.9?

1.10 Consider the three-phase power system in Example 1.8.

(a) Plot $i_A(t)$, $i_B(t)$, and $i_C(t)$ over six cycles.

(b) Obtain and plot the neutral current over six cycles.

1.11 Consider a three-phase load transformer with turn ratio 26:1. The transformer is ideal, i.e., it has no internal power loss or leakage, and it does not cause any shift in the phase angle. Suppose the voltage phasors on the secondary side, i.e., the low-voltage side, of the transformer are as follows:

$$\begin{aligned} V_A \angle \theta_A &= 275.3\angle 14.8^\circ, \\ V_B \angle \theta_B &= 281.6\angle -105.1^\circ, \\ V_C \angle \theta_C &= 277.8\angle 134.9^\circ. \end{aligned} \tag{1.70}$$

(a) Obtain the line-to-neutral voltage phasors on the primary side.

(b) Obtain the line-to-line voltage phasors on the primary side.

1.12 Figure 1.17(a) shows the power generation profile of a PV unit on a cloudy day. Three points are marked on the figure. Point ① is at early morning when solar power generation has not yet started. Point ② is at around 10 AM when solar power generation is picking up. Point ③ is at around 1 PM when the maximum solar power generation occurs on this day. This PV unit is connected to bus 2 at a 3-bus power distribution system, as shown in Figure 1.17(b). In addition to supplying active power, the PV unit also supplies reactive power at a fixed rate of 127 kVAR. There is a fixed load of 245 kW and 184 kVAR at bus 3. The voltage at bus 1 is fixed at $V_1 = 7200\angle 0$. The impedance of each distribution line is $Z_{\text{Line}} = 0.0412 + j0.0625\ \Omega$. Obtain the voltage phasor at bus 2 and the voltage phasor at bus 3 at each of the following cases:

(a) At point ① where solar power generation is $P_{\text{PV}} = 0$.

(b) At point ② where solar power generation is $P_{\text{PV}} = 158$ kW.

(c) At point ③ where solar power generation is $P_{\text{PV}} = 340$ kW.

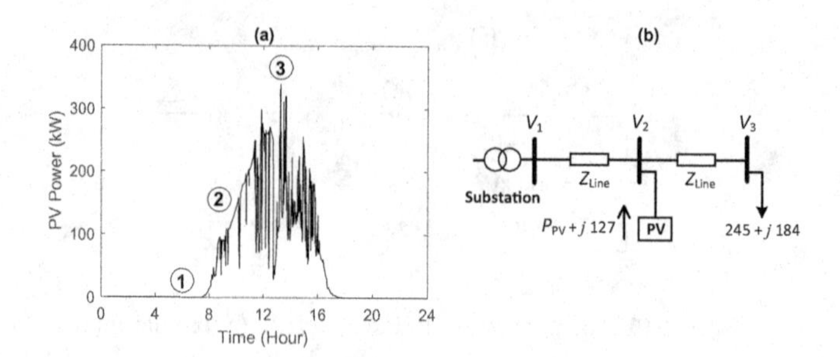

Figure 1.17 The power system in Exercise 1.12: (a) PV power generation; (b) the power distribution network with the PV unit at bus 2 and a fixed load at bus 3.

1.13 Figure 1.18 shows the square wave approximation of the sinusoidal wave in an inverter. Obtain the RMS value of this square waveform.

1.14 Consider the charge-discharge power profile of a 15 kWh energy storage system over a period of 12 hours, as shown in Figure 1.19. A positive power value indicates a charge action, and a negative power value indicates a discharge action. The energy storage system is ideal, meaning that it has 100% charge efficiency and 100% discharge efficiency. The initial state-of-charge (SoC) at the beginning of this time frame, i.e., at time zero, is 20%. Plot the SoC curve in percentage over this time period.

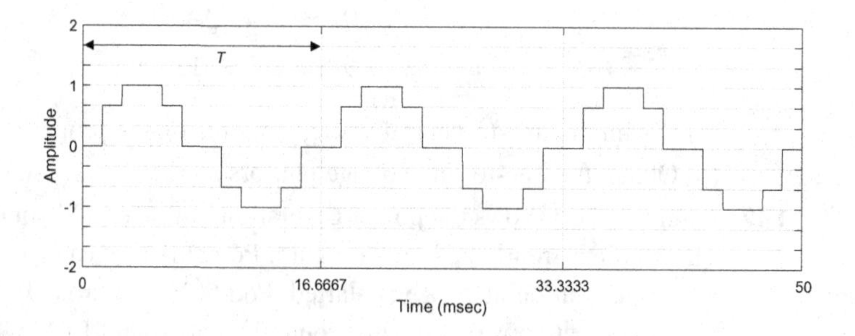

Figure 1.18 The square wave approximation of the sinusoidal wave in Exercise 1.13.

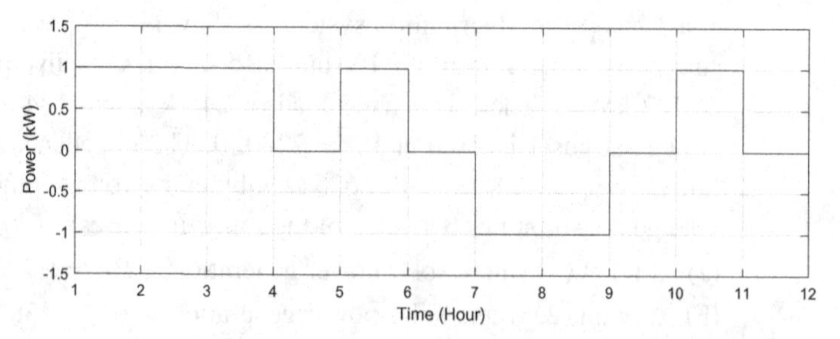

Figure 1.19 The charge-discharge power profile of the storage unit in Exercise 1.14.

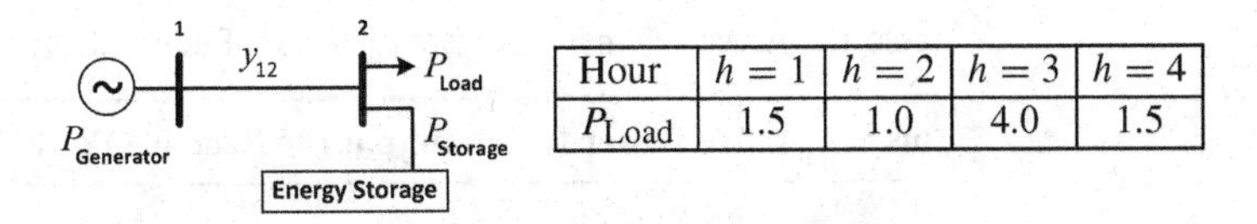

Hour	$h = 1$	$h = 2$	$h = 3$	$h = 4$
P_{Load}	1.5	1.0	4.0	1.5

Figure 1.20 The power network and the hourly load (in p.u.) in Exercise 1.15.

1.15 Consider the power grid with two buses as shown in Figure 1.20.

(a) The load factor (LF) for a load profile is defined as the ratio of the average amount of the load to the maximum load. What is the load factor for the load profile P_{Load} based on the table that is provided in this figure?

(b) We define the *net load* at bus 2 as $P_{\text{Load}} + P_{\text{Storage}}$, where P_{Storage} denotes the power exchange between the energy storage unit and bus 2. A positive P_{Storage} indicates charging, and a negative P_{Storage} indicates discharging. Suppose the storage unit is *ideal*, i.e., it experiences no loss during the charge and discharge cycles. The initial charge level of the storage unit is zero. Suppose the final charge level (i.e., at the end of hour $h = 4$) of the storage unit is also zero. What is the *minimum power rating* and the *minimum energy rating* needed to achieve a load factor of 0.8 for the *net load*? Assume that the entire charge capacity of the storage unit is usable.

(c) Obtain the required charge and discharge schedule for the storage unit to achieve 0.8 load factor for the net load in Part (b).

(d) Suppose the storage unit has 4/5 = 80% charge efficiency and 3/4 = 75% discharge efficiency. Answer the question in Part (b) in this case.

(e) Obtain the required charge and discharge schedule for the storage unit to achieve a 0.8 load factor for the net load in Part (d).

1.16 Consider the 5-bus transmission network in Figure 1.21. The admittance for each transmission line is $0.5 - j10$ p.u. The load at bus 3 is $3.0 + j1.2$ p.u. The load at bus 4 is $2.0 + j0.8$ p.u. There is no shunt capacitor at any bus. The voltage phasors at all buses are also given in the figure.

(a) How much is the active and reactive power injection by each generator?

(b) How much is the total active power loss on transmission lines?

(c) How much is the total reactive power loss on transmission lines?

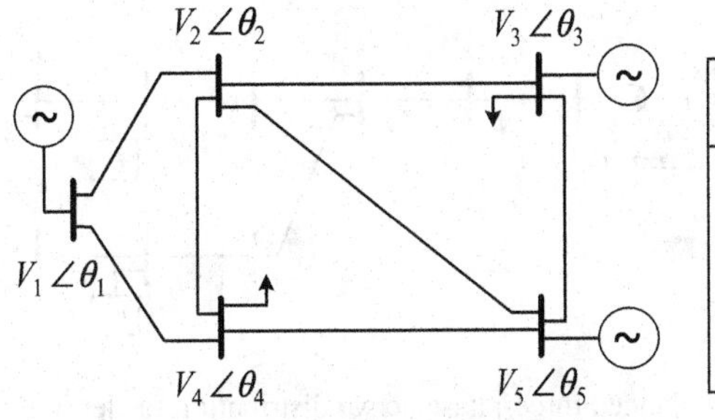

Bus #	Voltage Phasor (p.u.)
1	$1.047261\angle 19.3593°$
2	$1.005200\angle 16.2548°$
3	$0.966485\angle 11.8524°$
4	$0.991939\angle 14.5205°$
5	$1.021064\angle 18.9208°$

Figure 1.21 The power network and the voltage phasors in Exercise 1.16.

Table 1.1 Active and reactive power injection in Exercise 1.17.

Bus #	Active Power Injection (p.u.)	Reactive Power Injection (p.u.)
1	1.2810	1.1240
3	−1.5680	−0.6460
4	−2.1420	−0.8740
5	2.4506	0.8277

1.17 Consider the 5-bus transmission network in Exercise 1.16. Suppose we know only the voltage phasor at bus 1, which is $1.0104\angle 32.4381^\circ$. We do *not* know the voltage phasors at the rest of the buses. However, we do know the active and reactive power injection at each bus; as shown in Table 1.1.
(a) Calculate the active power flow on each transmission line.
(b) Calculate the reactive power flow on each transmission line.
(c) Calculate the total power loss on all transmission lines.

1.18 Consider the balanced three-phase power distribution system in Figure 1.22. The per-phase impedance of each distribution line is $0.0412 + j0.0625\ \Omega$. The voltage phasors at all buses are shown in Table 1.2.
(a) How much is the active and reactive power injection at each bus?
(b) How much is the active and reactive power flow on each line?

1.19 Consider the power distribution network in Exercise 1.18. We know that the voltage phasor at bus 1 is $7200\angle 0^\circ$. Bus 1 is the reference bus. We do *not* know the voltage phasors at the rest of the buses. However, we do know the active and reactive power injection at all buses, as shown in Table 1.3.

Table 1.2 Voltage phasors at all buses in Exercise 1.18.

Bus #	Voltage Phasor (V)	Bus #	Voltage Phasor (V)
1	$7200.0\angle 27.8214^\circ$	6	$7055.7\angle 27.3628^\circ$
2	$7143.9\angle 27.6195^\circ$	7	$7079.4\angle 27.4349^\circ$
3	$7097.4\angle 27.4856^\circ$	8	$7072.9\angle 27.4784^\circ$
4	$7076.8\angle 27.4269^\circ$	9	$7063.9\angle 27.4148^\circ$
5	$7066.9\angle 27.4413^\circ$		

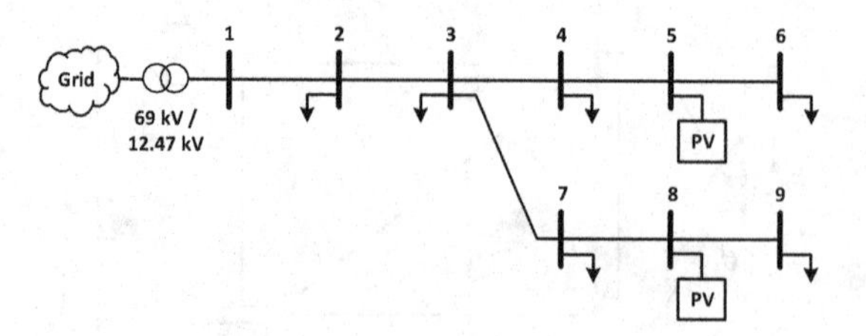

Figure 1.22 The balanced three-phase active distribution feeder in Exercise 1.18.

Table 1.3 Active and reactive power injection in Exercise 1.19.

Bus #	Active Power Injection (kW)	Reactive Power Injection (kVAR)
2	−1104.7	−253.9
3	−592.4	−278.7
4	−1079.2	−342.6
5	993.5	−193.2
6	−1312.0	−316.0
7	−1304.3	−259.1
8	1292.2	−474.2
9	−987.8	−245.8

(a) Calculate the active power flow on each distribution line.
(b) Calculate the reactive power flow on each distribution line.
(c) Calculate the total power loss on all distribution lines.

1.20 Consider the 2-bus power transmission network with two parallel transmission lines, as shown in Figure 1.23. Admittance $y = -j10$ p.u. is fixed. The controllable admittance y_{control} can be adjusted between 0 and $-j20$ p.u. Select y_{control} such that one-fourth of the power transfer from bus 1 to bus 2 flows through the top transmission line and three-fourth of the power from bus 1 to bus 2 flows through the bottom transmission line.

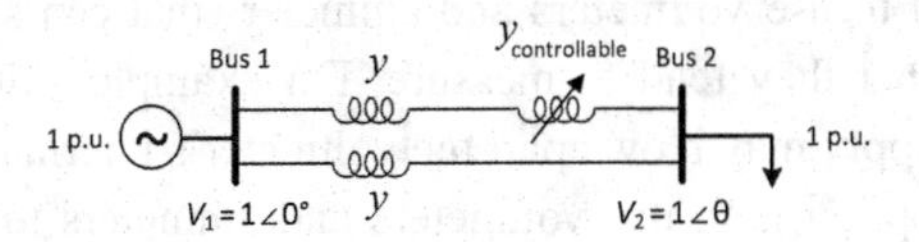

Figure 1.23 The network with controllable admittance in Exercise 1.20.

2 Voltage and Current Measurements and Their Applications

Measuring voltage and current is the foundation of monitoring power systems. A *voltmeter* is the basic instrument to measure voltage; see Figure 2.1(a). It is wired in parallel with the two points of the circuit at which the voltage difference is intended to be measured. An ideal voltmeter would have an infinite impedance such that it would take no current. An *ammeter* is the basic instrument to measure current; see Figure 2.1(b). It is wired in series with the circuit that carries the current that is intended to be measured. An ideal ammeter would have zero impedance such that it would create no voltage drop. Detailed discussions on digital voltmeters and ammeters and their circuits are available in [74].

2.1 Instrument Transformers

It is vital to use voltmeters and ammeters that can support the range of voltage and current that they tend to measure. For example, connecting a 30 V voltmeter to a 230 V supply may blow apart the voltmeter's internal circuitry.

One option is to use voltmeters and ammeters that can support higher ranges of voltage and current. However, this option often comes at a higher cost. A common alternative option is to use instrument transformers to first *step down* voltage or current and then measure it using a voltmeter or an ammeter, respectively.

Two basic types of instrument transformers, namely the *current transformer* (CT) and the *potential transformer* (PT), are shown in Figure 2.2. They are used to measure the primary current i_1 and the primary voltage v_1, respectively, by providing the secondary current i_2 and the secondary voltage v_2 that are *proportional* to those of the electric circuit but at significantly reduced magnitudes.

2.1.1 Turn Ratio

For an ideal CT, we have

$$\frac{i_2}{i_1} = \frac{N_1}{N_2}, \tag{2.1}$$

where N_1 and N_2 indicate the number of turns in the primary and the secondary windings, respectively. Some typical turn ratios are 50:5, 100:5, 400:5, and 4000:5,

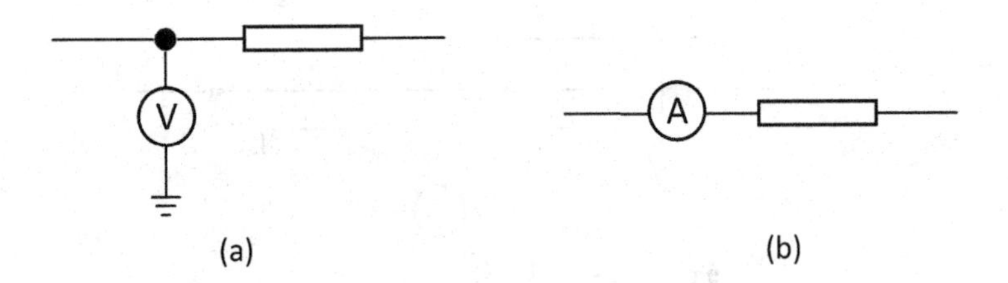

Figure 2.1 Basic measurement instruments: (a) voltmeter; (b) ammeter.

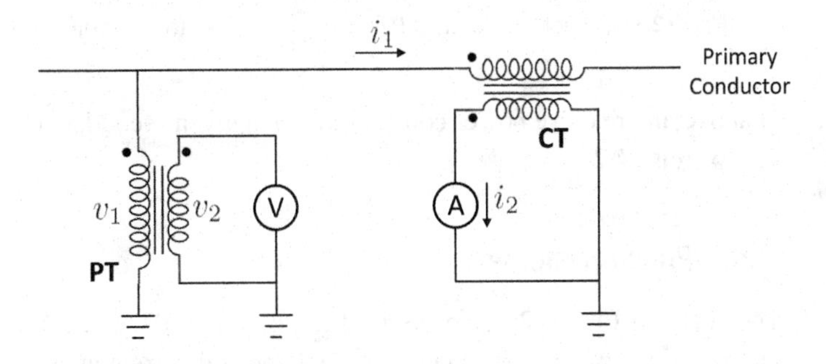

Figure 2.2 Instrument transformers, PT and CT, to measure voltage and current.

where 5 on the right-hand side indicates 5 A, which is the typical reduced current level on the secondary windings. Also, for an ideal PT, we have

$$\frac{v_2}{v_1} = \frac{N_2}{N_1}. \tag{2.2}$$

Some typical turn ratios are 25:1, 60:1, 100:1, and 4500:1. The actual reduced voltage level on the secondary windings may vary for different instruments.

Example 2.1 Suppose a CT with a turn ratio of 100:5 is used to measure current. If the ammeter shows 3.48 A, then the actual current is measured as

$$3.48 \times (100/5) = 69.6 \text{ A}. \tag{2.3}$$

2.1.2 Load Rating and Burden

Instrument transformers often have nameplate parameters for *load rating* and *burden.* The load rating, which is defined mostly for CTs, indicates the maximum load current that can be applied to the primary windings of a CT. That is, for a CT, the load rating is a limit on i_1. In contrast, the burden is associated with the secondary windings of a CT or PT. It indicates the amount of impedance made by the elements of the metering circuit, which may be connected to the secondary windings of the transformer without causing a metering error greater than that specified by its accuracy classification. The burden for each metering device may be indicated in terms of impedance or in terms

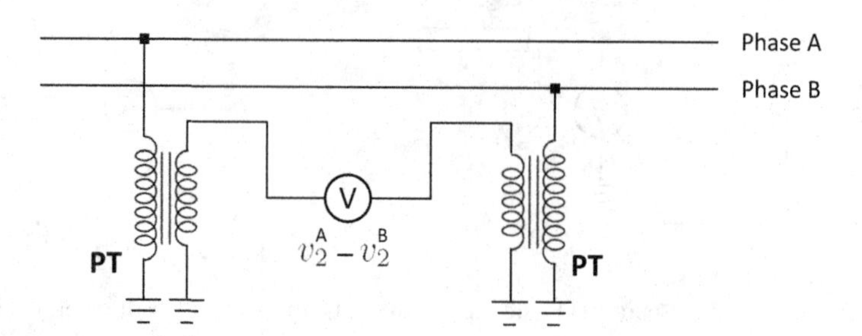

Figure 2.3 Using two identical PTs to measure line-to-line voltage across two phases.

of active and reactive power consumption at a given secondary current and frequency; see Exercise 2.2.

2.1.3 Three-Phase Systems

The setup in Figure 2.2 can be used to connect a PT and a CT to each phase of a three-phase power system to measure voltage and current at each phase.

Voltage in three-phase systems also can be measured across phases. For example, one can use two identical PTs to measure the voltage difference across Phase A and Phase B, as shown in Figure 2.3. Here, the voltmeter measures the *phase-to-phase* voltage, also known as the *line-to-line* voltage. Other possible configurations to use PTs to measure phase-to-phase voltages are explained in [75].

Example 2.2 Suppose two PTs are used to measure line-to-line voltage on a three-phase power system. The turn ratio is 60:1 for both PTs. If the voltmeter shows 206.33 V, then the actual current line-to-line voltage is measured as

$$206.33 \times 60 = 12.38 \text{ kV}. \tag{2.4}$$

2.2 Non-Contact Voltage and Current Sensors

Conventional ammeters and voltmeters are installed as integrated parts of the electric circuit. That is why they need CTs and PTs in order to measure large current and voltage levels. An alternative, more recent approach is to use sensors that measure current and voltage *without* electric contact. Specifically, one can measure the *magnetic field* surrounding a wire to estimate the current that flows through the wire. One can also measure the *electric field* surrounding a wire to estimate the voltage at the wire. The relationships are proportional, between the magnetic field and current and between the electric field and voltage. Non-contact current and voltage sensors are installed on or around *overhead* power lines. A relatively small 42 cm × 17 cm × 22 cm line-mounted overhead transmission line sensor, also known as a *bird on wire* sensor, may estimate conductor current up to 1500 A and conductor voltage up to 750 kV [76].

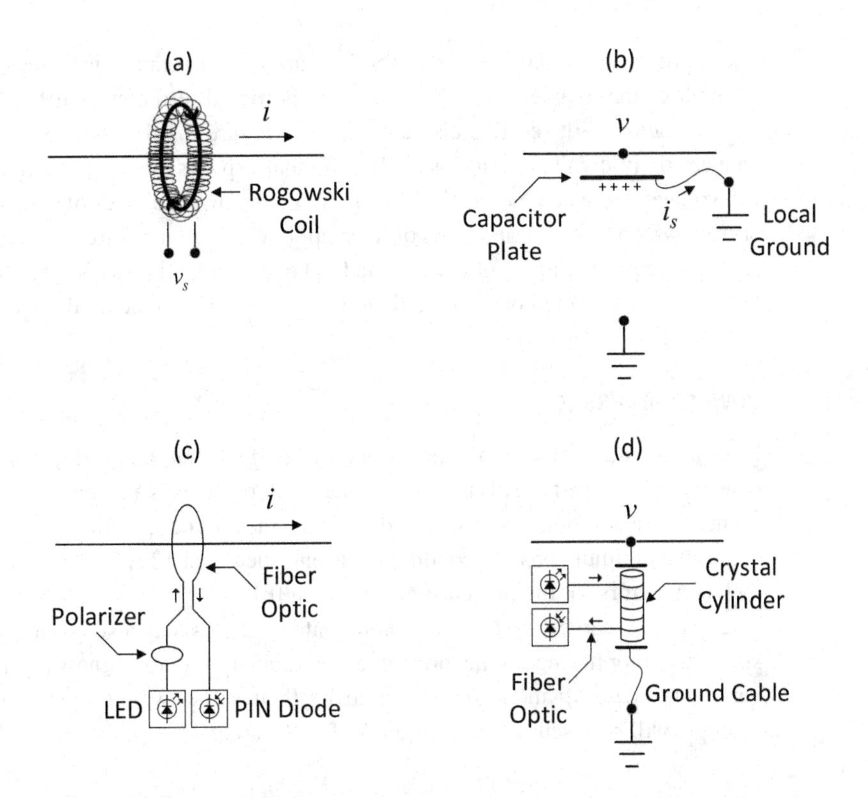

Figure 2.4 Different non-contact overhead line sensor technologies: (a) Rogowski coil to measure current; (b) capacitor plate to measure voltage; (c) magneto-optic transducer to measure current; (d) electro-optic transducer to measure voltage.

2.2.1 Different Working Principles

Different non-contact sensors may use different principles to measure the magnetic and electric fields. For example, in Figure 2.4(a), the current that goes through a conductor is measured by a Rogowski coil that captures the magnetic field surrounding the wire and generates magnetic-field-induced voltage v_s, which is proportional to the rate of change of current. Once this voltage is integrated over time, it provides a signal that is proportional to the current i that goes through the conductor [77]; see Exercise 2.5. In Figure 2.4(b), the voltage between the conductor and the ground is measured by measuring the current i_s between an isolated capacitor plate and a local ground, where the isolated capacitor plate is charged by the electric field surrounding the wire. The measured current is proportional to the voltage v between the conductor and the ground [78].

Another generation of non-contact voltage and current sensors works on principles in *optics*. For example, in Figure 2.4(c), current is measured by a magneto-optic current transducer (MOCT) that exploits the Faraday effect. MOCT detects the impact of the magnetic field surrounding the conductor on rotating the polarization of plane-polarized light that is transmitted by a light-emitting diode (LED) through a single fiber-optic pass around the conductor. The amount of rotation is proportional to the strength of the magnetic field; thus it is proportional to the current flowing through the conductor [79]; see Exercise 2.4. Also, in Figure 2.4(d), voltage is measured by an electro-optic voltage transducer (EOVT) that exploits the Pockles effect. Here, the full voltage of the conductor is applied between the two end faces of a cylinder-shaped crystal, which has several rounds of fiber-optic winding on its circumferential surface. EOVT detects the impact of the electric field on the optical phase shift in the transmitted light. The detected phase modulation is proportional to the total voltage [80].

2.2.2 Power Harvesting

Most non-contact overhead line sensors are *self-powered*, harvesting power from the conductor's magnetic fields in a non-contact fashion. This is a useful feature because it reduces the need for maintenance and therefore the cost of operation. Power harvesting starts at a minimum *pick-up* conductor current, such as at 12 A on 0.375-inch to 1.030-inch conductors for the line current sensors in [81].

The power-harvesting feature of non-contact sensors may also introduce some new smart grid monitoring applications due to the binary nature of the *sleep* mode (zero) versus the *wake-up* mode (one) caused by their self-powered operation. One such example will be discussed in Section 7.1.2 in Chapter 7.

2.3 Sampling Rate, Reporting Rate, and Accuracy

When addressing power grid modernization, it is natural to focus exclusively on digital sensors, as opposed to the traditional electromechanical meters. Digital sensors are more reliable and provide high measurement accuracy. Two important specifications of digital sensors are the *sampling rate* and the *reporting rate*.

2.3.1 Sampling Rate

The sampling rate of a sensor is indicated in samples per second or samples per cycle of the AC signal. Even if a measured voltage or current signal is purely sinusoidal, its reconstruction requires a minimum sampling rate, namely *twice the frequency of the signal*, according to the Nyquist-Shannon sampling theorem [82]. Proper signal filtering is also needed in order to avoid measurement *aliasing*.

Example 2.3 Consider a sensor that takes eight samples per second. As shown in Figure 2.5, this sensor cannot distinguish a 9 Hz sinusoidal signal from a 1 Hz sinusoidal signal. Hence, a 9 Hz signal may appear to be, i.e., may be aliased to, a 1 Hz signal. From the Nyquist-Shannon sampling theorem, the 8 Hz sampling rate for this sensor allows properly measuring periodic signals with up to $8/2 = 4$ Hz frequency. Therefore, to avoid measurement aliasing, an *anti-aliasing filter*, which is essentially

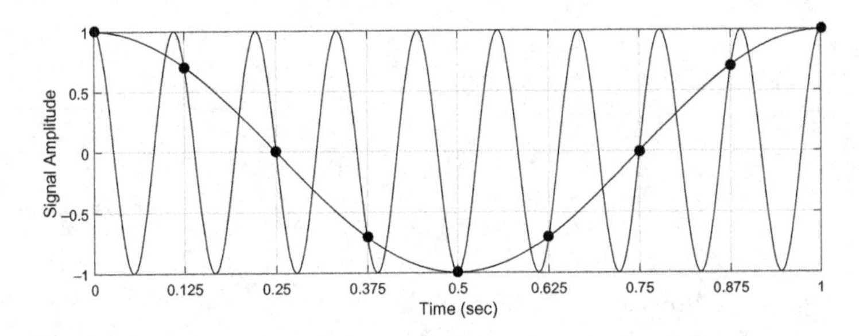

Figure 2.5 An example for aliasing a high-frequency signal to a low-frequency signal.

a low-pass filter, must be placed *before* the signal sampler in order to filter out any measurement signal with a frequency higher than 4 Hz.

More details about anti-aliasing filters are available in [83–86].

2.3.2 Reporting Rate

The reporting rate is the rate at which the sensor reports its measured data. It is often a multiple of the sampling rate to allow pre-processing of the measured data within each reporting interval. As shown in Example 2.4, the reporting rate of a sensor directly affects the type of applications that it can support.

Example 2.4 Consider the RMS readings of the voltage at the secondary of a transformer at a commercial building – one phase only, as shown in Figure 2.6(a). Two voltage sensors are used. One sensor reports 60 readings per second, i.e., it reports one RMS value per AC cycle. The other sensor reports one reading per minute, i.e., it reports the average RMS value of 3600 AC cycles. The data from the low-resolution sensor is sufficient to identify a voltage regulator *tap-changing event* at time 8:37 AM. However, with the use of the higher-resolution sensor, we can zoom in at time 8:45 AM and also identify the *transient response* of a thermostatic load when it switches on, as shown in Figures 2.6(b) and (c). Here, the load switching has caused momentary voltage sag and momentary inrush current. The transient responses take about 100 msec before new steady-state conditions are reached. Such transient responses are *not* visible to the lower-resolution sensor.

Further increasing the reporting rate will introduce another class of sensors that capture the *waveform* of the voltage and current signals; see Chapter 4.

For those sensors where the sampling rate is much higher than the reporting rate, the reported data may include not only the average but also the *minimum* and *maximum* values of the measured quantity. In Example 2.4, the low-resolution sensor may report not only the average RMS value of the 3600 AC cycles, but also the minimum and maximum RMS values across those 3600 AC cycles.

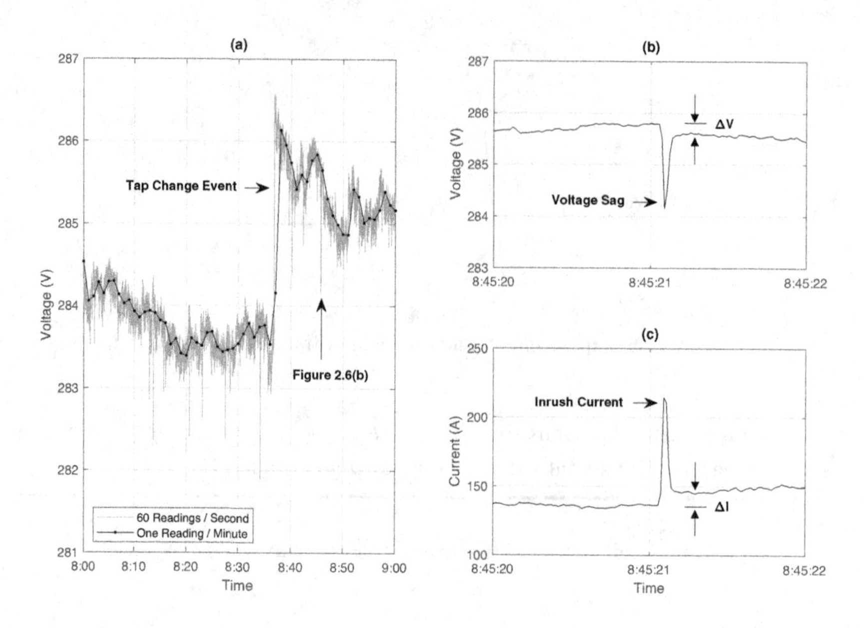

Figure 2.6 (a) An example to compare measuring voltage at two different reading rates; (b) momentary voltage sag; (c) momentary current surge.

2.3.3 Accuracy

The accuracy of a digital voltmeter or ammeter is often given in the form of a percentage of the reading (`rdg`) and/or a percentage of the full scale (`FS`).

Example 2.5 Consider a digital voltmeter with a measurement range of 0 V to 345 V. The accuracy of this voltmeter is ±0.05% `rdg` ±0.05% `FS`. If the RMS reading of the voltage is 275 V, then the accuracy of this reading is

$$\pm 0.0005 \times 275 \pm 0.0005 \times 345 = \pm 0.31 \text{ V}. \tag{2.5}$$

If a PT or CT is used, then the accuracy of the measurement depends not only on the accuracy of the measurement device but also on the accuracy of the instrument transformer. For example, the accuracy of a CT is defined based on its rated primary current. For an IEC 60044-1 standard grade, i.e., Class 1.0 CT, accuracy is ±1.0% at 100% of the rated primary current, ±1.5% at 20% of the rated primary current, and ±3.0% at 5% of the rated primary current. Also, a *ratio-correction factor* (RCF) is often defined for CTs as the fraction of the true turn ratio over the nameplate turn ratio. RCF is often provided as a curve plotted against multiples of the rated primary or secondary current for a given constant burden.

Example 2.6 If RCF is 1.01 at a certain primary current for a CT with a nameplate ratio of 100:5, then the true turn ratio is $1.01 \times 100/5 = 20.2$, as opposed to $100/5 = 20$. In other words, the true turn ratio is 100:4.95.

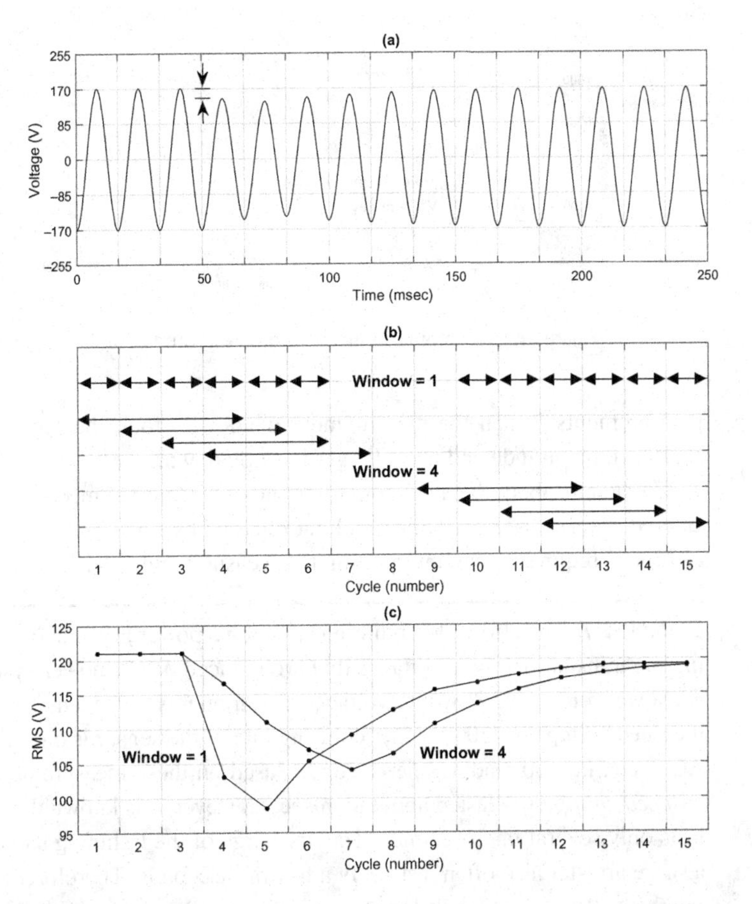

Figure 2.7 The impact of averaging filters: (a) the voltage signal to be measured; (b) measurement windows of lengths one and four cycles; (c) reported RMS voltages.

2.3.4 Impact of Averaging Filters

Most voltage and current sensors use some sort of *averaging filters*; these may include, among others, the low-pass anti-aliasing filter that we discussed in Section 2.3.1. As a result, each individual measurement that is reported by the sensor is the average or weighted average of multiple raw measurements over a finite time interval, known as the *measurement window*. The length of the measurement window varies among sensors and affects their class and performance.

The basic concept of average filtering is illustrated in Figure 2.7. The voltage waveform to be measured is shown in Figure 2.7(a). Two choices of the measurement window are considered: one and four, as shown in Figure 2.7(b). The reporting rate is the *same* in both cases. If the window size is one, then the RMS value of the most recently measured voltage cycle is reported. If the window is four, then the average of the RMS values of the *four most recent* voltage cycles is reported. Accordingly, the transient resolution reported by the sensor is different in each case, as shown in Figure 2.7(c). We can see that the use of averaging filters results in *smoothing* the

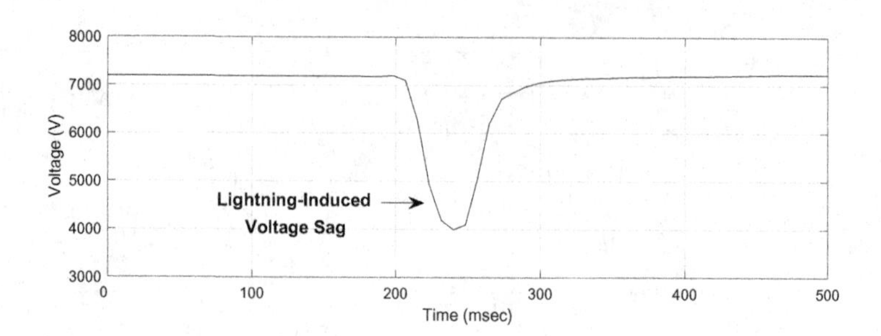

Figure 2.8 A lightning-induced voltage sag, studied in Example 2.7 [87].

measurements. When the measurement window is four, the reported voltage sag is shorter in magnitude and takes longer to settle down.

The details about the averaging filter and other internal mechanisms are often considered proprietary information and not released by sensor manufacturers. Therefore, caution is required when interpreting the measurement data.

Example 2.7 The effect of a lightning strike on a power grid can be traced by monitoring the transient impulse voltages that are induced on the power system. An example is shown in Figure 2.8, where voltage measurements are taken at a power line where the rated voltage is 7200 V. The reporting rate of the sensor is one RMS value per one AC cycle, i.e., 60 readings per second. Based on these measurements, the *lightning-induced voltage sag* lasted about 80 msec. However, it is known that a lightning strike lasts only several microseconds. The discharge of the lightning surge current through a surge arrester also often lasts only a few milliseconds. Therefore, it is likely that the relatively slow response in Figure 2.8 is due to filtering and other internal dynamics of the sensor [87].

The impact of averaging filters is negligible in a steady-state analysis, and even in a dynamic analysis if the transient response is relatively slow compared to the sensor's reporting rate. However, if one tends to analyze very fast events, such as a lightning strike, one may have to use *waveform* sensors; see Chapter 4.

2.4 RMS Voltage and Current Profiles

Traditionally, the voltage and current measurements in the power industry have been mostly in terms of the RMS values of the AC voltage and current waves. This has been particularly the case in the traditional *supervisory control and data acquisition* (SCADA) systems. Therefore, it is important to learn the kinds of analysis that one can do with the RMS voltage and current measurements. For the rest of this chapter, we will focus on the RMS voltage and current measurements. We will discuss measuring voltage and current phasors in Chapter 3 and measuring voltage and current waveforms in Chapter 4, as they require using more advanced sensors.

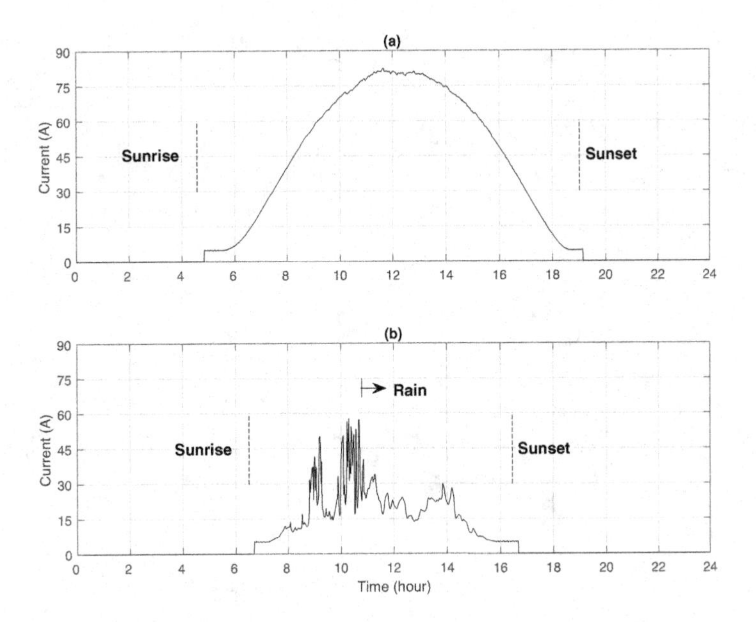

Figure 2.9 Two daily RMS current profiles at a PV inverter's output: (a) a sunny day in summer; (b) a cloudy and rainy day in winter [88].

2.4.1 Daily Profiles

It is often informative to examine the daily RMS profiles of voltage and current, e.g., at a substation, a PV inverter, or a load. In most cases, there is no need for a sensor with a very high reporting rate. A one-minute reporting interval is often sufficient to see the overall trends and to identify the major voltage and current events.

Example 2.8 Two daily profiles of the RMS current measurements that are taken at the output terminals of a solar PV inverter are shown in Figure 2.9. The profile in Figure 2.9(a) is measured on a *sunny* day in *summer*. The profile in Figure 2.9(b) is measured on a *cloudy* and *rainy* day in *winter*. The generation output highly depends on the time of day, season, and weather conditions.

The changes in the current profiles in Example 2.8 are solely due to the changes in solar irradiance at the location of the PV unit. However, when it comes to voltage profiles, the changes and fluctuations can be due to a wide range of causes, as we will see in the next example and its follow-up discussions.

Example 2.9 Figure 2.10 shows the daily RMS profiles of voltage and current measurements at a power distribution feeder [89, 90]. The reporting rate is once per minute. The rated voltage is 7200 V. The RMS voltage profile is relatively stable, remaining between 7050 V, i.e., 0.979 p.u., and 7350 V, i.e., 1.021 p.u. The RMS current profile is generally higher during the day and lower at night. Of interest are

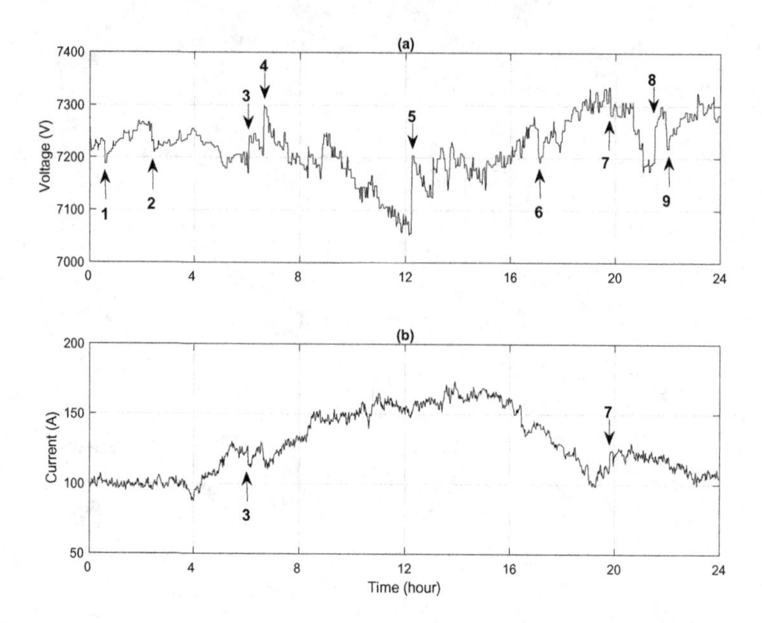

Figure 2.10 Examples of daily RMS profiles – one phase only – at a power distribution feeder: (a) daily voltage profile; (b) daily current profile.

the sudden changes in the RMS voltage profile. A subset of those sudden changes are marked with numbers 1–9. Events 1, 2, 6, 7, and 9 demonstrate voltage sags. Events 3, 4, 5, and 8 demonstrate voltage swells.

The voltage sags and swells that are identified in Example 2.9 are caused by different sources. They could be due to a *local* issue, such as a sudden change in this feeder's load, or the operation of a capacitor bank, a voltage regulator, or another device on this feeder. They could also be due to a *non-local* issue, i.e., a voltage sag or swell that occurred at the substation, sub-transmission system, or transmission system, which also showed up on this feeder's voltage profile.

Events 1, 2, 4, 5, 6, 8, and 9 on the voltage profile are less likely to be due to sudden changes in the feeder's load, because they do *not* coincide with a major change in the current profile. For instance, the current profile is almost flat from midnight until around 3:45 AM, yet there are major changes in the voltage profile in this period, including Events 1 and 2. As for Events 3 and 7, they *do* coincide with some noticeable changes in the current profile, which would be more visible if we zoom in on the figure. Thus, Events 3 and 7 could be *load induced* or caused by other local issues. These two events will be further discussed in Section 5.2 in Chapter 5.

Next, suppose we also obtain the daily RMS voltage profile of a *neighboring* distribution feeder, but under the *same* substation, and on the *same* day. This is shown in Figure 2.11. The reporting rate is again once per minute. The timings of the same nine events that we identified in Example 2.9 are marked and enumerated also in this figure. Events 1, 2, 4, 5, 6, and 9 also showed up quite noticeably in this figure. From

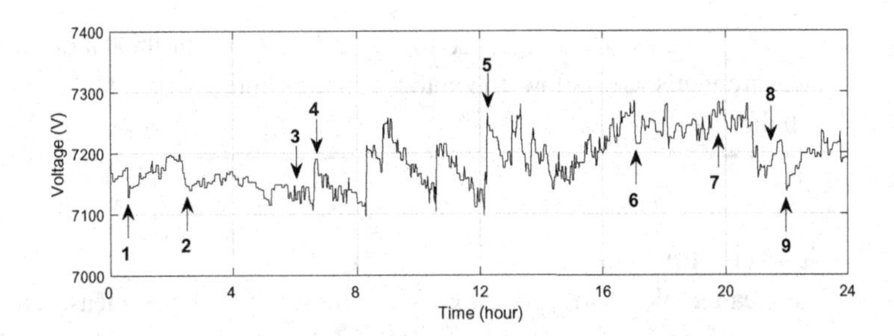

Figure 2.11 Daily RMS voltage profile at a neighboring feeder; compare with Figure 2.10(a). Events 1, 2, 4, 5, 6, and 9 also appear in this feeder's RMS voltage profile.

this, together with the analysis in the previous paragraph, we can conclude that all these events very likely have *root causes* at the substation, sub-transmission system, or transmission system, i.e., they are very likely *not* caused by local issues on either of the two feeders. Similarly, we can now conclude that events 3, 7, and possibly 8 *are* caused by some local issues in the first feeder.

While the above analysis is intuitive and manual, one can use techniques from *statistics* and *machine learning* to extract the above patterns automatically and from large sets of measurement data; see Section 2.7 as well as Section 3.7 in Chapter 3.

2.4.2 Histograms and Scatter Plots

The time period of the daily RMS voltage and current profiles can be extended to several days, weeks, or months. However, for such longer periods, it is often beneficial to present the measurements in a way that can highlight some of their *statistical characteristics*. In this section, we discuss two common options.

Histograms

Histograms can be used to provide an approximate representation of the *distribution* of a large set of measurements. To construct a histogram, we divide the entire range of the measurements into a series of intervals, called *bins*, and we count the number of measurements whose values fall into each interval [91].

Example 2.10 Suppose RMS voltage measurements are available on a minute-by-minute basis at two power distribution feeders for a period of one month. The histogram representation of the measurements at each of these two feeders is shown in Figure 2.12. The range of the measurements is from 6900 V to 7400 V. This range is divided into 25 bins. The first bin includes all the measurements between 6900 V and 6920 V, the second bin includes all the measurements between 6920 V and 6940 V, and so on. The histogram corresponding to Feeder 1 is almost evenly distributed around the rated voltage at 7200 V; see Figure 2.12(a). About 50.5% of the reported voltage measurements are below the rated voltage, and about 49.5% are above the rated voltage. The histogram corresponding to Feeder 2 is wider and placed mostly to the left of the rated voltage; see Figure 2.12(b). About 85.8% of the reported voltage measurements are below the rated voltage, and about 14.2% are above the rated voltage.

Scatter Plots

The scatter plot corresponding to the minutely voltage measurements in Example 2.10 is shown in Figure 2.13. Here, the voltage measurements at Feeder 2 are plotted *against* their *corresponding* voltage measurements at Feeder 1. Each point represents the measurements at one minute. A total of $60 \times 24 \times 30 = 43,200$ points are shown on this figure. Two key observations are marked on the figure.

First, consider the dashed diagonal line, which is marked as ①. The points *below* this line are associated with the minutes during which the voltage at Feeder 1 was *higher* than the voltage at Feeder 2. This accounts for 90.6% of the points. Conversely,

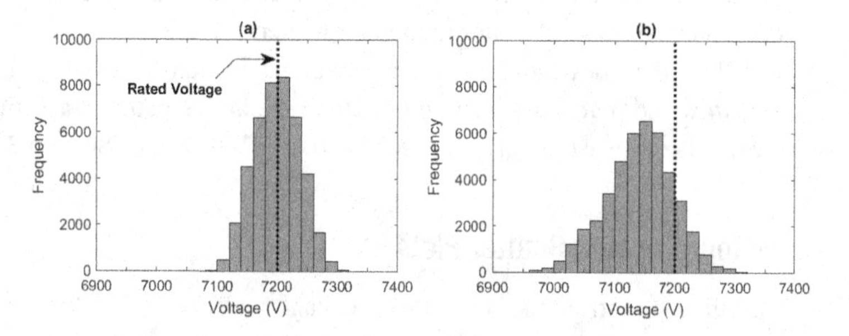

Figure 2.12 The histogram representation of the minutely RMS voltage profiles of two power distribution feeders over one month: (a) Feeder 1; (b) Feeder 2.

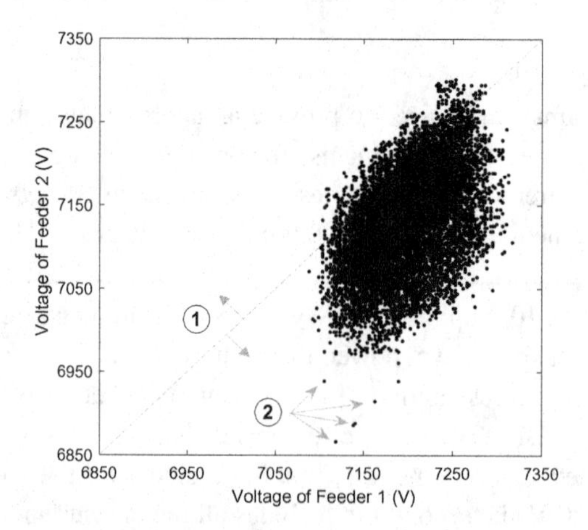

Figure 2.13 Scatter plot corresponding to the voltage measurements in Example 2.10. The diagonal line helps with making comparisons, and the arrows point at the outliers.

the points *above* this line are associated with the minutes during which the voltage at Feeder 1 was *lower* than the voltage at Feeder 2. That accounts for 8.4% of the points. Note that 1.1% of the points are *on* the diagonal line. These are the cases where the voltages were *equal* on both feeders to the extent of the reading resolution of the sensors.

Second, consider the four arrows, which are collectively marked as ②. They point at a few *outliers*. During the minutes associated with these outliers, Feeder 2 experienced unusually major voltage sags that were likely only *local* to Feeder 2 because Feeder 1 did *not* similarly experience such major voltage sags.

Clustering

Scatter plots can sometimes reveal *clusters* of points that can be considered for *classification*, i.e., grouping of measurements that have *similar features*.

Example 2.11 Recall from Example 2.4 in Section 2.3.2 that switching on major loads can cause *inrush current*, which can be captured in the current measurements if the reporting rate of the sensor is sufficiently high. An example is shown in Figure 2.14(a). Here, the inrush current is denoted by ΔI_{inrush}. The change in the *steady-state current* that is due to the increased load is also denoted by ΔI_{steady}. Figure 2.14(b) shows the scatter plot of these two quantities based on all the inrush current cases that were observed at a commercial facility during one day. One immediate observation is related to the outlier case that is marked as ①. More importantly, we can see that the majority of the points in this scatter plot are *clustered* into two groups, which are marked as ② and ③. Note that the points in ③ experience much larger inrush current than the points in ② when compared at the same amount of change in the steady-state current.

We will discuss clustering and classification further in Section 3.7 in Chapter 3, Section 4.3 in Chapter 4, and Section 5.4 in Chapter 5.

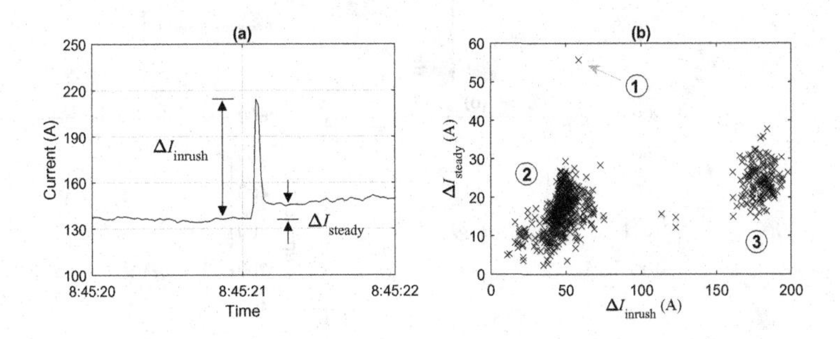

Figure 2.14 Scatter plot for inrush current events: (a) definition of ΔI_{inrush} and ΔI_{steady}; (b) two separate clusters and a few outlier points.

2.5 RMS Voltage and Current Transient Responses

So far we have seen that an event as simple as a load switching, such as the one in Example 2.4, can create a transient response in the power system, at least locally, i.e., at the location where the load switching occurs. Such load-induced transient responses comprise the majority of the (rather minor) transient responses that one can see in voltage and current measurements on an ongoing basis.

However, there are also many other, often more major and more important, causes of transient responses, such as faults and equipment actuations. Their impact usually can be seen more broadly across the power system.

Analysis of transient responses, whether at the load level, the distribution level, or the transmission level, is useful in many power system applications, such as in control, protection, stability analysis, and reliability assessment. The time resolution that is needed depends on the time scale of the transient event being monitored.

2.5.1 Transient Responses Caused by Faults

Example 2.12 The voltage and current transients during an animal-caused fault event at a distribution feeder are shown in Figure 2.15. The fault occurred on a lateral not too far from the substation. Measurements are done on the secondary side of a load transformer at a commercial load location on *another* lateral [92]. Three transient stages are marked by numbers ①, ②, and ③. Stage ① shows the momentary drops in voltage and current during the initial animal-caused short-circuit fault. Stage ②

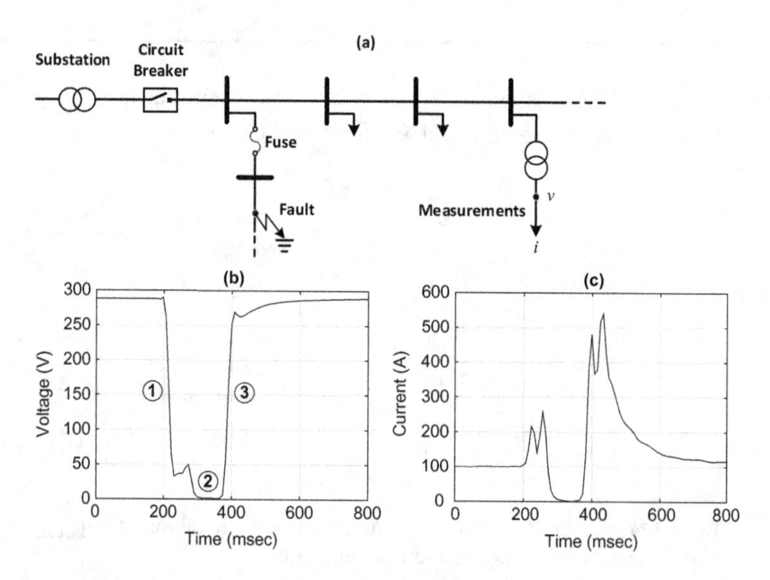

Figure 2.15 (a) One-line diagram of a distribution feeder during an animal-caused fault; (b) voltage readings during the fault; (c) current readings during the fault [92].

shows the momentary power outage (at the location of the sensor) due to the operation of the circuit breaker. Sometime during stages ① and ② the fuse on the faulted lateral burns, isolating the fault. Stage ③ shows the transient in voltage and current after the circuit breaker recloses and restores service.

Impact at Other Locations

Given the significance of the transient response in Example 2.12, one may expect that the fault may have also caused transient responses in other locations on the power grid. That is indeed the case, as we can see in Figure 2.16. The voltage measurements in Figure 2.16(a) are taken at the distribution substation at the *same* distribution feeder. The voltage measurements in Figure 2.16(b) are taken at another substation that is *several miles away* from the faulted location. As we move away from the faulted location, the transient response becomes less significant.

While the animal-induced fault in Example 2.12 caused transient responses in a region that may span tens of miles; there are also other types of more significant disturbances in power systems that may cause transient responses all across the interconnected power grid; that may span hundreds of miles. We will see such system-wide transient responses in Sections 2.6.1 and 2.9.2.

2.5.2 Transient Responses Caused by Equipment Actuations

Example 2.13 A PV farm is interconnected to a power distribution system at a distribution substation. Figure 2.17(a) shows the RMS voltage that is measured at the point of inter-connection. Only one phase is shown here. A major step change is visible, caused by a step-up transformer tap change event in the power system. This voltage event resulted in a *transient response* in the RMS current of the PV farm, as shown in Figure 2.17(b). The pattern of the changes in the RMS current reveals how the inverters at the PV farm respond to the step-up disturbance in voltage; see [93].

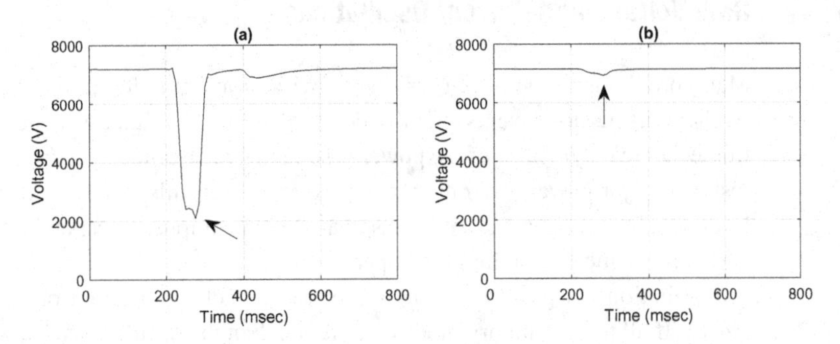

Figure 2.16 Transient responses that are caused by the fault in Figure 2.15 at two other locations: (a) the substation at the same feeder; (b) another substation [92].

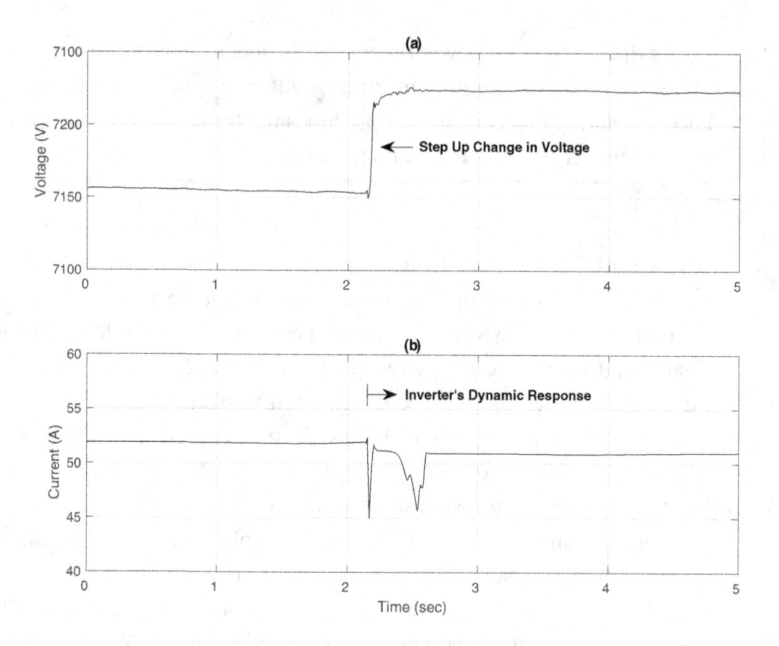

Figure 2.17 Dynamic step response of a PV farm's inverter to a step up voltage event due to a transformer tap change: (a) RMS voltage profile; (b) RMS current profile [93].

The transient response in Figure 2.17(b) takes only about one second. It includes major *undershoots* and a few small *overshoots*. The transient signature in this figure is for the most part a reflection of the dynamics of the inverter control system.

Monitoring the transient responses that are caused by equipment operation can inform us about the *state of the health* of the equipment and the possible signs of *incipient faults* (i.e., early-stage faults) that could become problematic in the future, as well as the overall response of the system under potential contingencies. We will discuss incipient faults in details in Section 4.3 in Chapter 4.

2.6 RMS Voltage and Current Oscillations

Many problems in the power grid begin with or are manifested through *oscillations*. At the transmission level, system-wide oscillations are often associated with the electromechanical dynamics of the power system that are excited by a *disturbance*, such as losing a major power generation unit or losing a transmission line. At the distribution level, local oscillations are often associated with equipment responses to disturbance, circuit resonance, or certain load operations.

Oscillations in power systems are often characterized based on their *oscillatory modes*. If all the oscillatory modes are *stable*, then the oscillations decay and diminish over time. However, if one or more oscillatory modes are *unstable*, then the oscillations grow in magnitude until corrective actions are taken.

In this section, we discuss examples of system-wide and local oscillations that can be seen in voltage and current measurements. We also discuss how to obtain the oscillatory modes of the system from voltage and current measurements.

2.6.1 Wide-Area Oscillations in Power Transmission Systems

System-wide oscillations, also known as *wide-area oscillations*, are common phenomena in power transmission systems. They are caused by the electromechanical oscillations of rotational generators in response to faults in the system, transmission line switching, a sudden change in the output of generators, or a sudden change in major loads. Wide-area oscillations may affect the magnitude, phase angle, and frequency of voltage across the entire interconnected power system.

Some of the common classes of wide-area oscillations include: *local plant mode oscillations*, where one generator swings against the rest of the power system; *inter-area mode oscillations*, where two coherent groups of generators swing against each other, causing excessive power transfers across the network; and *control mode oscillations*, where some generators suffer from poorly tuned exciters, governors, or other generator controllers. These different oscillation modes can sometimes be identified based on the frequency of the oscillations [94]:

- Control mode oscillations: 0.01 Hz to 0.15 Hz;
- Inter-area mode oscillations: 0.15 Hz to 1.0 Hz;
- Local plant mode oscillations: 1.0 Hz to 2.0 Hz.

The above frequency ranges are approximate.

Example 2.14 Figure 2.18 shows the transient inter-area oscillations that were observed in voltage measurements at a 500 kV transmission line after a generation disturbance. The initial disturbance, which is due to losing a major generator unit in the system, resulted in a voltage sag followed by *ring-down* oscillations, i.e., decaying oscillations that ultimately disappear in ambient noise after a few seconds. The dominant mode of oscillations had a frequency of 0.7 Hz.

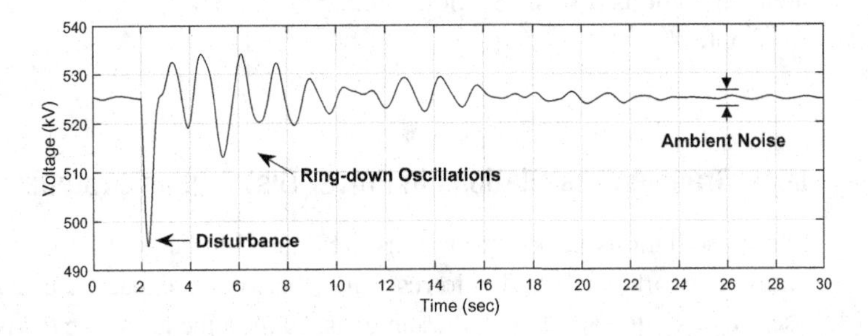

Figure 2.18 Oscillations in voltage at a 500 kV transmission line due to a generation disturbance [95]. The oscillations gradually decay and ultimately disappear in ambient noise.

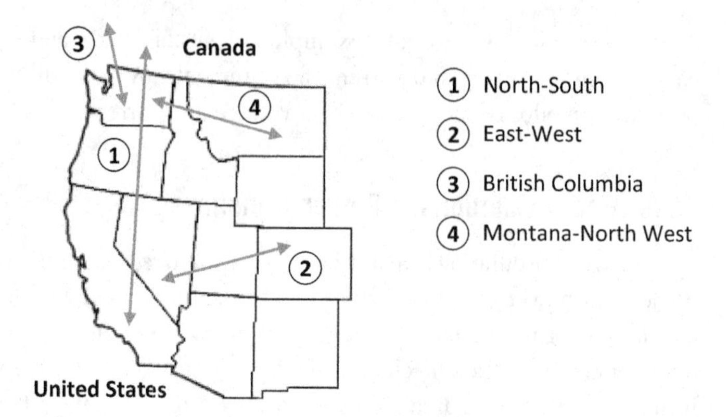

Figure 2.19 The primary modes of inter-area oscillation in the Western Interconnection in the United States. Each mode involves two regions that are connected with an arrow.

Inter-Area Oscillations

Among different types of wide-area oscillations, understanding *inter-area oscillations* is particularly important due to their impact on power system stability and the fact that they involve many parts of the power system in distant regions with complex dynamic behavior. Inter-area oscillations are used in the analysis of power system dynamics because of their relation to small-signal stability [94]. Inter-area oscillations can arise for a variety of reasons, such as excessive power transfers, inefficient damping controls at some generators, or unfavorable load characteristics.

Figure 2.19 shows some of the primary modes of inter-area oscillations in the Western Interconnection in the United States [96]. Each mode involves two geographical regions. For example, the North-South mode involves generators in Canada and the Pacific Northwest region in the United States oscillating against generators in the Desert Southwest and Southern California, causing excessive power swings between Northern California and Oregon as well as between Northern and Southern California. Studies have shown that damping this inter-area oscillation mode can improve system stability and increase the transfer capability of the California Oregon Intertie (COI) corridor on the Western Interconnection; e.g., see [97, 98]. We will discuss inter-area oscillations also in Section 2.9.2, Section 3.4.3 in Chapter 3, and Section 6.4 in Chapter 6.

2.6.2 Local Transient Oscillations in Power Distribution Systems

Some oscillations in power systems are caused by a local issue in the power distribution network, such as due to resonance between a capacitor and an inductor. Such oscillations often affect only the same distribution feeder where the issue occurs. Also, they are likely to be visible only in current measurements, compared to the voltage measurements. As we will see in Chapter 4, some local transient oscillations may

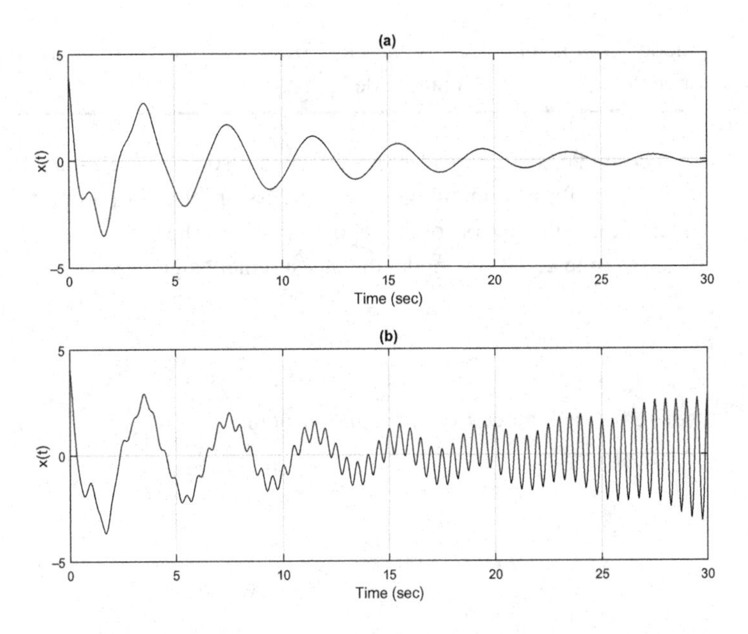

Figure 2.20 Examples of oscillatory modes: (a) stable; (b) unstable.

have very high frequencies at the order of kHz and may last for only a fraction of an AC power system cycle.

2.6.3 Modal Analysis of Oscillations

An oscillatory signal $x(t)$ can be mathematically modeled as

$$x(t) = \sum_{i=1}^{m} A_i e^{\sigma_i t} \cos(w_i t + \varphi_i), \tag{2.6}$$

where m is the number of *oscillatory modes*. For each mode i, notations A_i, φ_i, w_i, and σ_i indicate the *amplitude*, *phase angle*, *angular frequency*, and *damping factor*, respectively. Note that if signal $x(t)$ includes a DC term, then the DC term can be modeled as a mode with zero frequency, a zero damping factor, and a zero phase angle. Signal $x(t)$ can be any *time series*, such as the RMS readings of voltage or current.

Example 2.15 Two oscillatory signals $x_1(t)$ and $x_2(t)$ are shown in Figures 2.20(a) and (b), respectively. Signal $x_1(t)$ has $m = 2$ modes, where $A_1 = 3.5$, $\varphi_1 = \pi/4$, $w_1 = 2\pi \times 0.25$, $\sigma_1 = -0.1$, $A_2 = 1.75$, $\varphi_2 = \pi/6$, $w_2 = 2\pi \times 0.8$, and $\sigma_2 = -0.5$. Both modes are stable and decay over time. Signal $x_2(t)$ has $m = 3$ modes. The first and the second modes are stable and identical to those of signal $x_1(t)$. The parameters of the third mode are $A_3 = 0.15$, $\varphi_3 = 0$, $w_3 = 2\pi \times 2$, and $\sigma_1 = 0.1$. The third mode is *not* stable, because $\sigma_3 > 0$. As a result, the oscillations corresponding to this mode quickly grow in magnitude.

The model in (2.6) can be represented also in *discrete-time* domain by applying the Z-Transform from digital signal processing [99, 100]. Let Δt denote the reporting interval, i.e., the inverse of the reporting rate, of the sensor; see Section 2.3.2. At each discrete time k, we can model the measurement signal as

$$x(t = \Delta t \times k) = \sum_{i=1}^{m} {z_i}^k R_i, \tag{2.7}$$

where for each mode $i = 1, \ldots, m$, we define:

$$z_i = e^{(\sigma_i + jw_i)\,\Delta t} \tag{2.8}$$

and

$$R_i = A_i e^{j\varphi_i}. \tag{2.9}$$

Since our focus is on digital sensors that report measurements in discrete time, for the rest of this section we use the model in (2.7) instead of (2.6). Also, for notational simplicity, we use $x(k)$ to denote the measurement at discrete time k, instead of the more precise but cumbersome notation $x(t = \Delta t \times k)$.

Prony Method

A popular modal analysis technique is the *Prony method* [101]. It can estimate all four parameters, w_i, σ_i, A_i, and φ_i, for a given measurement signal $x(t)$ based on the discrete-time model in (2.7)–(2.9). This is done in three steps.

First, suppose a total of n measurement points are available, where $n \gg m$. A discrete linear auto-regressive (AR) predictor model is used to capture the dynamics of $x(t)$. In this regard, $x(t)$ at discrete time m is approximated as

$$x(m) = a_1 x(m-1) + a_2 x(m-2) + \ldots + a_m x(0). \tag{2.10}$$

Here we approximate the measurement at discrete time m as a linear combination of all the *previous* measurements at discrete times $m-1, m-2, \ldots, 0$. We can use (2.10) and similarly approximate $x(m+1), x(m+2), \ldots, x(n-1)$; and derive:

$$\begin{bmatrix} x(m) \\ x(m+1) \\ x(m+2) \\ \vdots \\ x(n-1) \end{bmatrix} = \begin{bmatrix} x(m-1) & \cdots & x(0) \\ x(m) & \cdots & x(1) \\ x(m+1) & \cdots & x(2) \\ \vdots & & \vdots \\ x(n-2) & \cdots & x(n-m-1) \end{bmatrix} \begin{bmatrix} a_1 \\ \vdots \\ a_m \end{bmatrix}. \tag{2.11}$$

Let $\boldsymbol{\Psi}$ denote the $(n-m) \times 1$ vector on the left and $\mathbf{X}$ denote the $(n-m) \times m$ matrix in the middle. Both $\boldsymbol{\Psi}$ and $\mathbf{X}$ depend solely on measurements. Also, let $\mathbf{a}$ denote the vector of unknown coefficients $a_1, \ldots, a_m$. Since $n \gg m$, we can formulate the following *least-squares* (LS) optimization problem based on the relationship in (2.11) in order to obtain the unknown coefficients:

$$\min_{\mathbf{a}} \ \|\mathbf{\Psi} - \mathbf{X}\mathbf{a}\|_2 . \tag{2.12}$$

We can solve the above LS problem by using an LS solver; such as `lsqlin` in MATLAB [102]. Alternatively, we can obtain the solution in closed-form as [103]:

$$\mathbf{a} = (\mathbf{X}^T\mathbf{X})^{-1}\mathbf{X}^T\mathbf{\Psi}. \tag{2.13}$$

Second, we use the coefficients $a_1, \ldots, a_m$ that we obtained in the previous step in order to solve the following *discrete-time characteristics polynomial* of degree m, which is associated with the difference equation in (2.10):

$$1 - \sum_{i=1}^{m} a_i z^{-i} = 0. \tag{2.14}$$

After reordering the terms, we have:

$$z^m - \sum_{i=1}^{m} a_i z^{m-i} = \begin{bmatrix} 1 & -\mathbf{a}^T \end{bmatrix} \begin{bmatrix} z^m \\ z^{m-1} \\ z^{m-2} \\ \vdots \\ z \\ 1 \end{bmatrix} = 0. \tag{2.15}$$

We can solve the above equation over unknown z by using a solver such as `roots` in MATLAB [104]. The *roots* (solutions) of the discrete-time characteristics polynomial provide the *poles* corresponding to the oscillatory modes of the measured signal. Such poles are in the discrete-time domain. From (2.8), we can derive:

$$\sigma_i + jw_i = \ln(z_i)/\Delta t, \quad i = 1, \ldots, m. \tag{2.16}$$

By applying the definition of complex logarithm, we can rewrite (2.16) as

$$\sigma_i = \ln\left(|z_i|\right)/\Delta t, \ w_i = \angle z_i/\Delta t, \quad i = 1, \ldots, m. \tag{2.17}$$

Third, once z_i is known for all $i = 1, \ldots, m$, we can use (2.7) to construct:

$$\begin{bmatrix} x(0) \\ x(1) \\ x(2) \\ \vdots \\ x(n-1) \end{bmatrix} = \begin{bmatrix} 1 & \cdots & 1 \\ z_1 & \cdots & z_m \\ z_1^2 & \cdots & z_m^2 \\ \vdots & & \vdots \\ z_1^{n-1} & \cdots & z_m^{n-1} \end{bmatrix} \begin{bmatrix} R_1 \\ \vdots \\ R_m \end{bmatrix}. \tag{2.18}$$

Let $\mathbf{\Phi}$ denote the $n \times 1$ vector on the left and $\mathbf{Z}$ denote the $n \times m$ matrix in the middle. Vector $\mathbf{\Phi}$ depends solely on measurements. Matrix $\mathbf{Z}$ is the solution of the discrete-time characteristics polynomial in (2.15) that we obtained in the second step. Therefore, both $\mathbf{\Phi}$ and $\mathbf{Z}$ are known. Also, let $\mathbf{R}$ denote the vector of unknown residues

Table 2.1 Oscillation modes obtained in Example 2.16

Mode	Frequency (Hz)	Damping (Hz)	Amplitude (kV)	Phase Angle (°)
1	± 1.6056	−0.9081	5.0083	±45.419
2	± 1.5002	−0.2515	1.1597	∓74.344
3	± 1.2395	−0.3032	2.2171	±55.979
4	± 1.0872	−0.4831	2.2250	±45.605
5	± 0.7241	−0.0767	2.3696	±30.464
6	± 0.6540	−0.1128	3.0296	±105.92
7	± 0.2890	−0.3598	4.6563	±162.87

$R_1, \ldots, R_m$. We can formulate an LS optimization problem similar to the one in (2.12) and obtain its solution as

$$\mathbf{R} = (\mathbf{Z}^T\mathbf{Z})^{-1}\mathbf{Z}^T\boldsymbol{\Phi}. \tag{2.19}$$

Given the above solution for **R**, we can use (2.9) to obtain:

$$A_i = |R_i|, \quad \varphi_i = \angle R_i, \quad i = 1, \ldots, m. \tag{2.20}$$

Once all the m oscillatory modes are calculated, one may still need to do some post-processing before the results can be used. For example, it is known that the Prony method usually results in *aliased modes* [105]. In fact, based on an analysis similar to that in Section 2.3.1, we can show that any oscillatory mode with a frequency that is higher than half the reporting rate of the sensor is prone to aliasing. Such oscillatory modes are not reliable and should be discarded from the results.

Example 2.16 Again, consider the ring-down oscillations in the voltage measurements in Figure 2.18 in Example 2.14. The reporting interval for the measurements is $\Delta t =$ 0.1 second. By choosing $m = 101$, the Prony method results in calculating 101 oscillatory modes. A total of 86 modes have frequencies higher than $0.5/0.1 = 5$ Hz. Thus, they are prone to aliasing and are discarded. The remaining 15 modes include one DC mode with an amplitude of 524.19 kV and seven pairs of complex conjugate oscillatory modes, as listed in Table 2.1.

The modes that we identified in Example 2.16 are shown in Figure 2.21. The DC mode is shown in Figure 2.21(a). The seven oscillatory modes are shown in Figures 2.21(b)–(h). All modes are plotted starting at time $t = 2$ seconds, which is the moment when the disturbance occurs, which is the start of the oscillations

When combined, that is, when they are added together, the above 15 modes provide a Prony estimation for the original measurements in Figure 2.18. The Prony estimation and the original measurements are compared in Figure 2.22. We can see that the Prony estimation is particularly accurate during the first few seconds and then gradually drifts from the measurements. Accuracy of the Prony estimation may increase by increasing the parameter m; see Exercise 2.17.

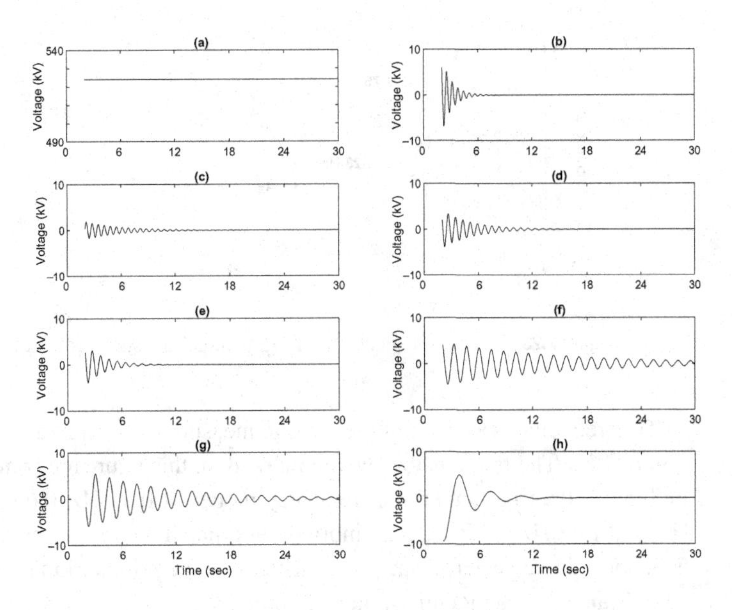

Figure 2.21 All the modes that were obtained by using the Prony method in Example 2.16: (a) DC mode; (b)–(h) complex conjugate oscillatory modes as listed in Table 2.1.

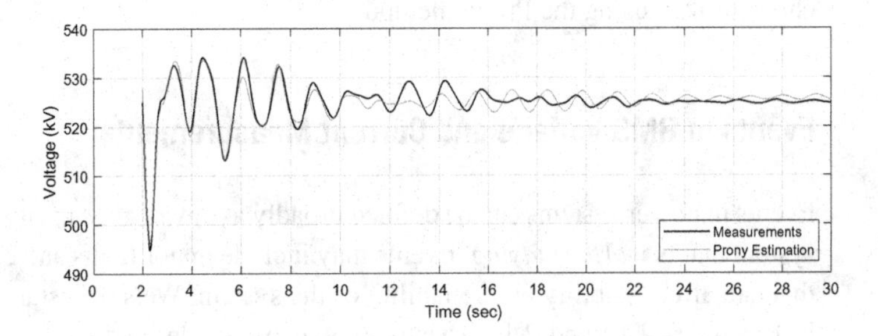

Figure 2.22 Comparing the Prony estimation with the original measurements in Example 2.16. The Prony estimation is the summation of all the modes in Figure 2.21.

The Prony method has been widely studied for the analysis of oscillatory modes in different types of measurements in power systems. Adjustments, improvements, and alternative methods have been proposed in the literature, such as in [106–113].

Fourier Method

Prony analysis is an extension of the well-known *Fourier analysis*. Prony analysis forms a sum of damped sinusoidal terms. In contrast, Fourier analysis forms a sum of sustained sinusoidal terms, where the damping factor is zero at all frequencies. In this regard, we can still use the Fourier Transform (FT) or Fast Fourier Transform (FFT) as a quick method to estimate only the frequency, not the damping factor, of the oscillatory modes in the measurements. The FFT of the measurements can be obtained by using the command `fft` in MATLAB [114].

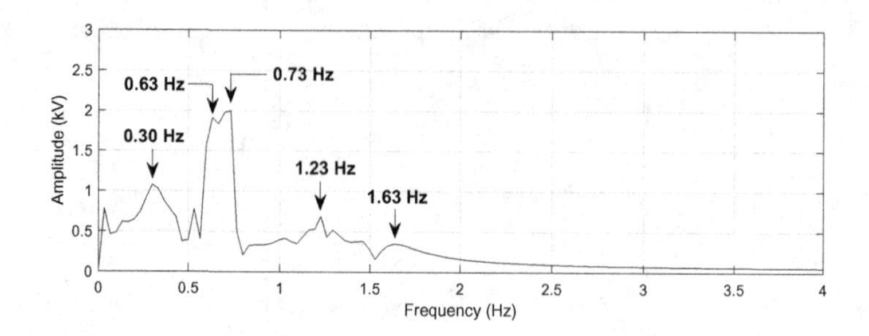

Figure 2.23 Frequency spectrum of the voltage measurements in Example 2.16.

The frequency spectrum of the voltage measurements in Example 2.16 is shown in Figure 2.23. The frequencies that are marked on this figure are comparable with those in Table 2.1. In particular, the frequency modes at 0.30 Hz, 0.63 Hz, 0.73 Hz, 1.23, Hz, and 1.63 Hz in Figure 2.23 more or less match the frequency modes in rows 7, 6, 5, 3, and 1 in Table 2.1, respectively. These are the primary modes of oscillations that an FFT analysis can identify relatively quickly.

Since FFT is computationally less complex than the Prony analysis, it is often used to *detect* oscillations; e.g., see Example 2.35. Once detected, the oscillation can be characterized using the Prony method.

2.7 Events in RMS Voltage and Current Measurements

Events in power systems can be defined broadly as any *change* in any component in the system that is *worth studying*. Events may include major faults and equipment failures that can affect stability and reliability of the system. We saw instances of such events in Examples 2.12 and 2.14. Events may also include load switching or transformer tap changing actions that are *benign* yet may reveal some useful information about the power system and its components. We saw instances of such events in Examples 2.11 and 2.13.

Of course, whether or not an event is "worth studying" depends on what we are trying to achieve in our study. An event could be of great importance for one smart grid monitoring purpose but useless for another smart grid monitoring purpose. For example, when our goal is to study inter-area oscillations at the Western Interconnection, the oscillations in the frequency of the voltage measurements are a critical event to study, as we will see in Section 2.9.2; but an animal-caused fault at a distribution line, such as what we saw in Figure 2.15, is just a small noise in the analysis of system-wide oscillations. In contrast, when our goal is to study the operation of the protection system at a distribution feeder, the animal-caused fault is a key event; but the fluctuations in the voltage frequency due to inter-area oscillations are of no interest.

The definition of events in a study may also depend on the type of sensors that are used and their characteristics. For example, some events could last for only a very short period of time; therefore, they may not be captured by our sensor, depending on

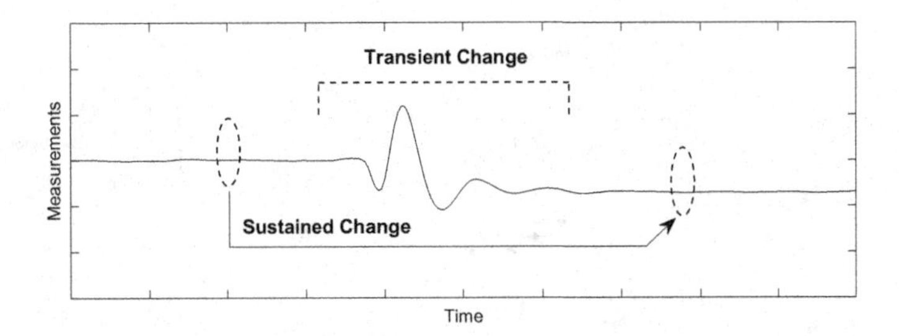

Figure 2.24 An event can show up in the voltage and/or current measurements in form of major transient changes, a major sustained change, or both.

the sampling rate and the reporting rate of the sensor. Furthermore, some events may not manifest in the typical RMS voltage and RMS current measurements, but they do show up if we have access to the phase angle measurements of the voltage and current phasors (see Chapter 3), or the actual waveform measurements of voltage and current (see Chapter 4).

2.7.1 Transient Events versus Sustained Events

Despite the complexities around the definition of events, we can still define two types of events that can help us with the analysis of events throughout this book, namely *transient* events and *sustained* events. The basis for this classification is shown in Figure 2.24. If the event causes a significant transient change in the measurements, then the event is a transient event, and the transient change is likely worth studying. If the event causes a significant sustained change, then the event is a sustained event, and the sustained change is likely worth studying. Note that if an event causes *both* significant transient changes and significant sustained changes, then it could be considered both a transient event and a sustained event, depending on the purpose of our analysis.

The generation disturbance in Example 2.14 caused a major voltage sag and several cycles of oscillations in the voltage measurements, as we saw in Figure 2.18; but as soon as the transient response disappeared, the voltage returned to the same level as it was prior to the event. Therefore, we can consider this event as a transient event as far as the voltage measurements in Figure 2.18 are concerned.

The transformer tap changing event in Example 2.13 in Section 2.5.2 caused a significant sustained change in the voltage measurements in Figure 2.17(a); however, it did *not* cause any major transient response in the voltage measurements in this figure. Therefore, we can consider this event as a sustained event as far as the voltage measurements in Figure 2.17(a) are concerned. As for the same event that is also seen in the current measurements in Figure 2.17(b), it has caused considerable sustained changes as well as significant transient changes. Therefore, we might consider this event as both a sustained event and a transient event as far as the current measurements in Figure 2.17(b) are concerned.

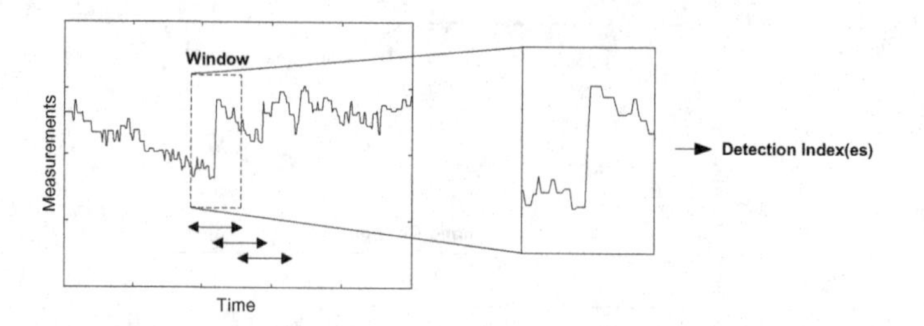

Figure 2.25 Event detection based on calculating a detection index over a window of measurements. The window moves forward in time to detect more events.

2.7.2 Event Detection Methods

In this section we discuss some basic methods to detect events in voltage and current measurements. All these methods include the following three steps:

1. A window of measurements is considered;
2. One or more indexes are calculated based on the window of measurements;
3. An event is detected if the index(es) exceed certain threshold(s).

The window then moves forward in time, and the above three steps are repeated in order to continue detecting more events; see Figure 2.25.

Min-Max

A simple detection index is the difference between the maximum and the minimum values within the window of measurements [113]. Let us define

$$\text{Min-Max Index} = \max\{x_w(t)\} - \min\{x_w(t)\}. \tag{2.21}$$

where $x_w(t)$ denotes the measurements within a given window w. An event is detected in window w if the Min-Max Index exceeds a predetermined threshold.

Example 2.17 Consider the daily voltage profile in Example 2.9 in Section 2.4. Recall that the measurements are reported on a minutely basis. Suppose we set the window size to 20 minutes, i.e., to include 20 measurement points. Thus, the day is divided into $24 \times 60/20 = 72$ windows. Note that, in this example, we assume that the measurement windows do *not* overlap. Figure 2.26 shows the Min-Max Index that is calculated for each of the 72 windows. Suppose we set the detection threshold to 70 kV. Therefore, a window of measurements is deemed to contain an event if the difference between the maximum reported voltage and the minimum reported voltage within the window is at least 70 kV. This results in detecting eight events. Five of them are among the events that we had previously marked in Example 2.9, labeled as 3, 4, 5, 8, and 9. They are positioned *above* the threshold curve. Importantly, four of the events that we had marked in Example 2.9 are *not* detected here, labeled as 1, 2, 6, and 7. They are positioned *below* the threshold curve.

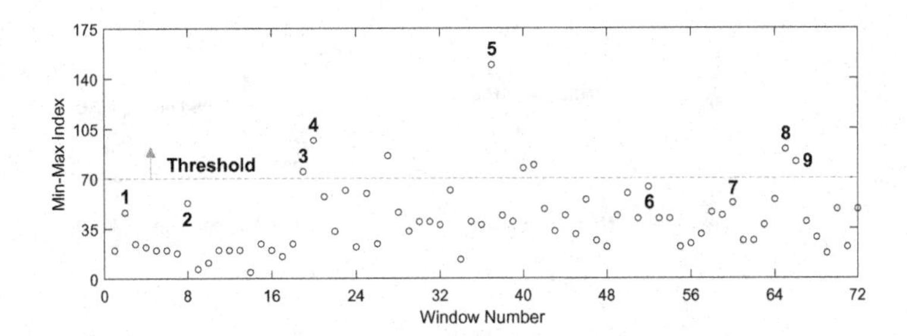

Figure 2.26 Using the Min-Max method to detect the events in voltage measurements in Example 2.17. The numbers refer to the labels of the events in Example 2.9.

The results in Example 2.17 are sensitive to the choice of the threshold. For instance, if we reduce the threshold from 70 kW to 40 kW, the total number of windows that are detected to contain events would increase from 8 to 31; and all of the nine events that we had previously marked in Example 2.9 would be detected. The caveat is that more windows with normal operation would fall in the event category.

Median Absolute Deviation

Detection indexes also can be defined based on principles in statistics. Here, we discuss one such index. Let use define the *median absolute deviation* as

$$\text{MAD} = \gamma \text{ Median}\left\{\left|x_w(t) - \text{Median}\{x_w(t)\}\right|\right\}. \tag{2.22}$$

A typical value for coefficient γ is 1.4826 [115]. An event is detected in window w if *any* of the measurements within window w is *lower* than

$$\text{Median}\{x_w(t)\} - \zeta\text{MAD}; \tag{2.23}$$

or if *any* of the measurements within window w is *higher* than

$$\text{Median}\{x_w(t)\} + \zeta\text{MAD}. \tag{2.24}$$

A typical value for coefficient ζ is 3. Next, we define:

$$\text{MAD-Low Index} = \min\{x_w(t)\} - (\text{Median}\{x_w(t)\} - \zeta\text{MAD}) \tag{2.25}$$

and

$$\text{MAD-High Index} = \max\{x_w(t)\} - (\text{Median}\{x_w(t)\} + \zeta\text{MAD}). \tag{2.26}$$

An event is detected in window w if the MAD-Low Index is less than zero *or* the MAD-High Index is greater than zero. The former indicates the presence of an *undershoot* among the measurements in window w, while the latter indicates the presence of an *overshoot* among the measurements in window w.

Example 2.18 Again, consider the daily voltage profile in Example 2.9 in Section 2.4. As in Example 2.17, suppose the window size is 20. Figure 2.27(a) shows a window of measurements where an event is detected, because part of the measurements appears

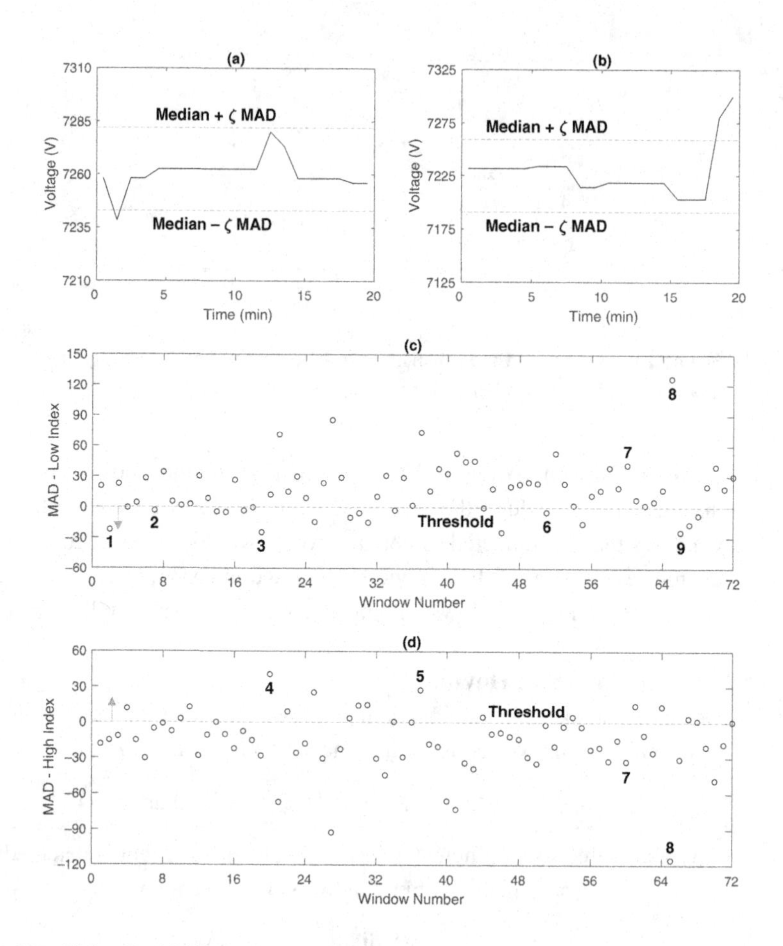

Figure 2.27 Using the MAD method to detect the events in voltage measurements in Example 2.18: (a) and (b) windows with events that are detected due to an undershoot and an overshoot, respectively; (c) and (d) MAD-Low Index and MAD-High Index across all 72 windows, respectively. The numbers refer to the labels of the events in Example 2.9.

below the threshold in (2.23). Figure 2.27(b) shows a window of measurements where an event is detected, because part of the measurements appears *above* the threshold in (2.24). Figures 2.27(c) and (d) show the MAD-Low Index and the MAD-High Index calculated for each of the 72 windows, respectively. Seven of the events that we had previously marked in Example 2.9 are detected, labeled as 1, 2, 3, 4, 5, 6, and 9. Event 8 is not detected; however, this event *can* be detected once we use *partially overlapping windows*, as we will discuss at the end of this section. Event 7 cannot be detected, even with partially overlapping windows, unless we reduce parameter ζ, which may result in detecting too many cases as events. To understand why the situation with Event 7 is different, recall from Example 2.9 that Event 7 is likely caused by a local issue in the distribution network. Therefore, the voltage measurements may not be the best indicator if our goal is to detect local events such as Event 7. In such cases, it is better also to check the current measurements, as we saw in Figure 2.10(b), in order to search for potential locally events.

Fast Fourier Transform

The Min-Max and MAD methods are usually effective in detecting major sustained events as well as transient events that involve major impulses. However, when it comes to transient events that involve oscillations, it might be better to use *frequency spectral* methods to detect such events. A common frequency spectral method is FFT. In this method, FFT is applied to each window of voltage or current measurements in order to obtain the frequency spectrum associated with the measurements in that window. An event is detected if the amplitude of any frequency component in the frequency spectrum exceeds a predetermined threshold.

Figure 2.28(a) shows the voltage measurements during the oscillatory event that we previously saw in Example 2.14 in Section 2.6.1. Figure 2.28(b) shows the corresponding frequency spectrum that is obtained by applying FFT to the window of measurements in Figure 2.28(a). If we set the event detection threshold to 1.5 kV, then this oscillatory event is detected because the frequency spectrum exceeds the threshold around the 0.63 Hz–0.73 Hz frequency range.

Other Detection Methods

There exist several other event detection methods that can be used to detect events in voltage and current measurements. Some of these methods can be seen as variations or extensions of the methods that we discussed above. For example, one can use *mean* instead of *median* in the MAD method. There are also many options among the frequency spectral event detection methods, such as the Yule-Walker method [113] and the Wavelet method [116, 117].

Some recent methods also use machine learning in both *supervised* and *unsupervised* settings. In supervised event detection, prior knowledge is used to identify and characterize the common patterns for the kind of events that we are interested in detecting. The event detection algorithm then looks for those specific patterns in order to detect the intended events; see the methods in [118–122]. In unsupervised event

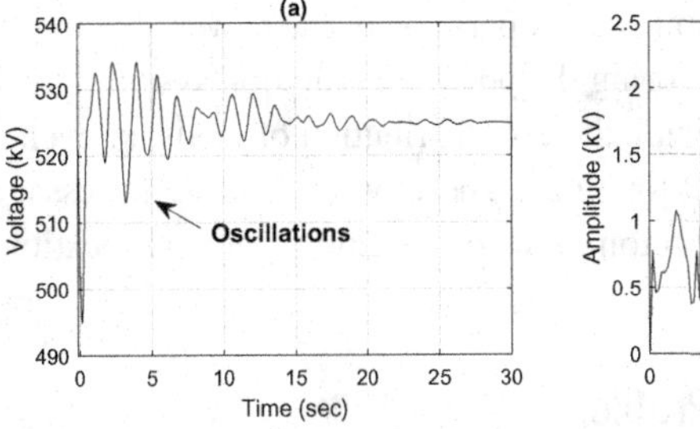

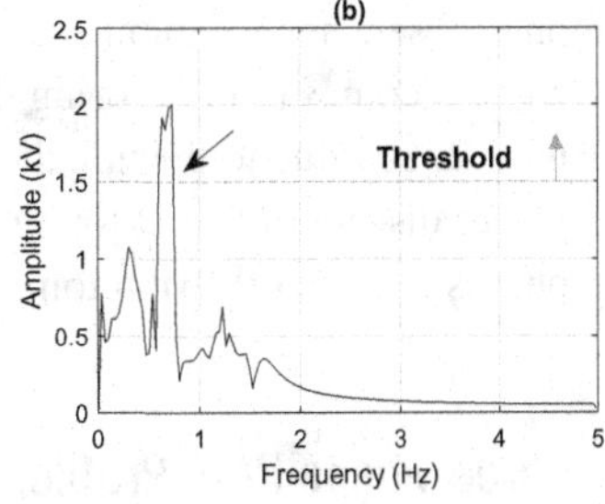

Figure 2.28 Using the FFT method to detect an oscillatory event: (a) a window of voltage measurements; (b) the corresponding frequency spectrum.

detection, no prior knowledge is used. Instead, the goal is to learn the *normal* trends in measurements and then detect an event whenever there is an *abnormality*, i.e., a major deviation from the identified normal trends; see the methods in [123–126].

Overlapping Windows and Dynamic Windows

It is very common in event detection to use windows that have some *overlap* with each other. Such overlap can help avoid missing the events that may fall on the borderline of two adjacent but non-overlapping windows. For instance, in Example 2.18, we did not detect Event 8 because it appeared partially in one window and partially in another window. However, once we use a sequence of windows with 50% overlap, i.e., each window overlaps with half of the previous window and also with half of the next window, we are able to detect this event.

Sometimes it may help to also use *dynamic* windows. For example, if no event is detected in a window of measurements, then we can try *increasing* or *decreasing* the window size before we conclude that there is indeed no event in the considered window of measurements. Using dynamic windows might be a necessity if we expect the events to have different lengths in time; e.g., see [127].

2.7.3 Events in Other Types of Measurements

The focus in this Chapter has been on events in RMS voltage and RMS current measurements. However, events can be of interest in any type of smart grid measurements. We will discuss events in voltage and current phasor measurements in Section 3.7 in Chapter 3, events in voltage and current waveform measurements in Section 4.4 in Chapter 4, and events in active and reactive power measurements and power factor measurements in Section 5.2.1 in Chapter 5.

2.8 Three-Phase Voltage and Current Measurements

A power grid is a three-phase power system. Since the three phases are often *balanced* at the transmission level, it might be sufficient to measure voltage and current on only one phase when we monitor the operation of the power transmission system. However, when it comes to monitoring the power distribution system, the three phases are often *unbalanced* due to the unbalanced distribution of loads, or even due to an unbalanced power distribution grid topology. Therefore, it is often necessary to monitor all three phases in power distribution networks, including at load locations.

2.8.1 Three-Phase RMS Profiles

Figures 2.29(a) and (b) show the RMS voltage and RMS current profiles at each phase of a three-phase load. We can see that the power system is *not balanced*, as the RMS

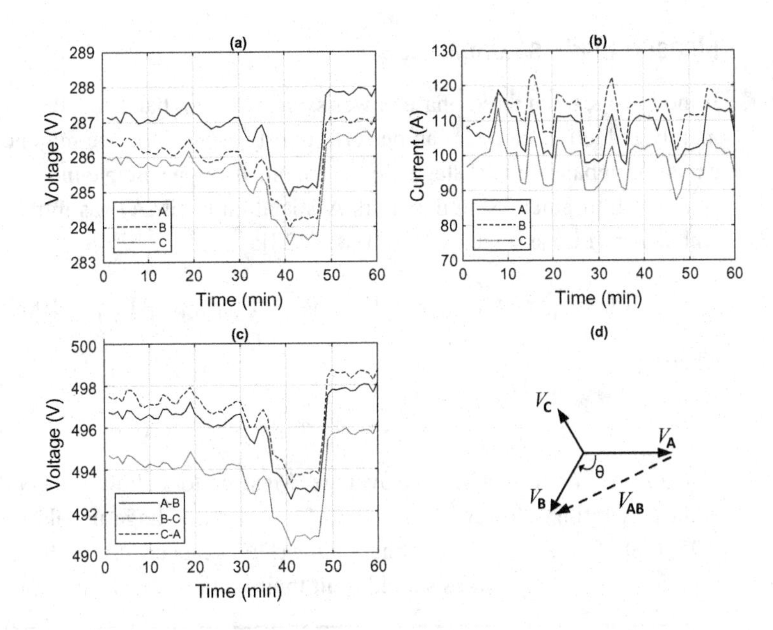

Figure 2.29 Three-phase measurements: (a) voltage; (b) current; (c) line-to-line voltage; (d) relationship between line-to-line voltage and phase voltages.

voltage and current profiles are not identical across the three phases. However, the RMS profiles do generally resemble each other in this example.

Figure 2.29(c) also shows the RMS profile for the line-to-line voltage that is measured at the same three-phase load. By comparing the line-to-line voltage measurements in Figure 2.29(c) with the phase voltage measurements in Figure 2.29(a), we can see that the RMS voltage profiles generally look similar; however, there are advantages in measuring both phase voltages and line-to-line voltages. For example, as shown in Figure 2.29(d), we can use the measurements for line-to-line voltage V_{AB} and the measurements for phase voltages V_A and V_B to obtain the *angle* between voltage *phasors* at Phase A and Phase B, as follows:

$$\cos(\theta) = \frac{{V_A}^2 + {V_B}^2 - {V_{AB}}^2}{2\, V_A\, V_B}. \tag{2.27}$$

Example 2.19 Suppose RMS voltage at Phases A and B is measured as

$$\begin{aligned} V_A &= 286.63\ V \\ V_B &= 287.26\ V. \end{aligned} \tag{2.28}$$

And the RMS line-to-line voltage across Phases A and B is measured as

$$V_{AB} = 497.70\ V. \tag{2.29}$$

From (2.27), we obtain $\cos(\theta) = -0.5042$. Therefore, we have $\theta = -120.28°$.

We will discuss voltage and current phasor measurements in Chapter 3.

2.8.2 Measuring Phase Unbalance

If the voltages in a three-phase power system are not balanced, then some equipment, mainly induction motors, can perform poorly. Therefore, we may need to assess the extent of unbalance in voltage measurements using a suitable index. For instance, the National Equipment Manufacturers Association (NEMA) has introduced Percentage Unbalance (PU) as a metric as follows [128]:

$$PU = \frac{1}{\Gamma} \max \Big\{ \big|V_{AB} - \Gamma\big|, \big|V_{BC} - \Gamma\big|, \big|V_{CA} - \Gamma\big| \Big\} \times 100\%, \tag{2.30}$$

where

$$\Gamma = \frac{1}{3}(V_{AB} + V_{BC} + V_{CA}) \tag{2.31}$$

denotes the line-to-line voltage average. Most motors allow a 1% voltage unbalance without derating. However, 2%, 3%, 4%, and 5% voltage unbalance can result in a 0.95, 0.88, 0.82, and 0.5 Derating Factor (DF), respectively [129]. The ANSI C84.1 standard states that utilities should limit their voltage unbalance to 3%.

Example 2.20 Suppose line-to-line voltages are measured at a load as

$$\begin{aligned} V_{AB} &= 497.70\ V \\ V_{BC} &= 494.87\ V \\ V_{CA} &= 496.98\ V. \end{aligned} \tag{2.32}$$

From (2.30) and (2.31), we can obtain

$$PU = \frac{1.6467}{496.52} \times 100\% = 0.33\%. \tag{2.33}$$

Since PU < 1%, such voltage unbalance does not cause derating in motor loads.

2.8.3 Phase Identification

Electric utilities often do not have reliable records about how the three phases of each distribution feeder are connected to the loads. Even if phase connections are initially recorded correctly, they may change over time due to service restoration, topology reconfiguration, or repairs. Wrong phase labeling is a major source of error in the analysis of power distribution systems. Therefore, it is critical to use available sensor data to correctly identify the loads on each phase.

The basic idea in the phase identification problem is illustrated in Figure 2.30. The feeder serves several loads, including a single-phase load at load point 1 and a three-phase load at load points 2, 3, and 4, as shown in Figure 2.30(a). There are three possibilities for the connection of the single-phase load, as shown in Figure 2.30(b). There are also six possibilities for the connection of the three-phase load, as shown in Figure 2.30(c). The phase identification problem is the problem of identifying the correct phase connection configuration at each load.

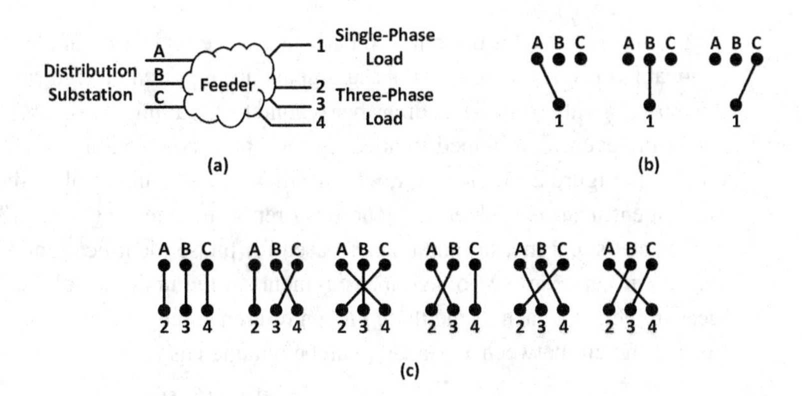

Figure 2.30 The phase identification problem: (a) phase connections are unknown at load points 1, 2, 3, 4; (b) there are three possibilities for the connection of the single-phase load; (c) there are six possibilities for the connection of the three-phase load.

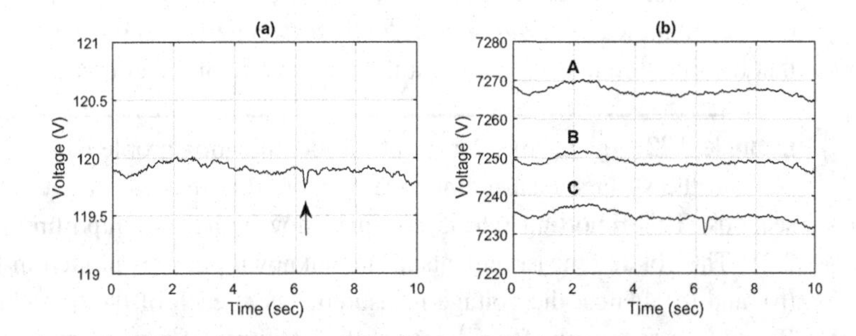

Figure 2.31 Phase identification by comparing voltage profiles: (a) voltage profile at a single-phase load; (b) voltage profiles on three phases at the feeder head at the substation.

Example 2.21 Consider the voltage measurements at a single-phase load as shown in Figure 2.31(a). Suppose the phase connection is unknown for this load. The voltage measurements on three phases at the substation are shown in Figure 2.31(b). They are labeled as Phases A, B, and C. By comparing the voltage profile in Figure 2.31(a) with each of the three voltage profiles in Figure 2.31(b), it is clear that the single-phase load is connected to Phase C. Specifically, the *sudden momentary voltage sag* that is marked in Figure 2.31(a) is seen *only* in Phase C in Figure 2.31(b). All the measurements in this example are reported at 30 readings per second.

The phase identification method in Example 2.21 can also be discussed in the context of the event analysis in Section 2.7. That is, we identified the phase connection by detecting an event on the single-phase measurements in Figure 2.31(a) and comparing it with an event that we detected at the same window on Phase C of the three-phase measurements in Figure 2.31(b), while considering the fact that no similar event was detected at this window on Phase A and Phase B.

Another note about the phase identification method in Example 2.21 is that it is applicable only if the voltage measurements are reported at high rates. If we reduce the reporting rate from 30 readings per second to 1 reading per second, then we cannot detect the event that helped us identify the phase connections based on the voltage profiles in Figure 2.31. However, we can still use some statistical methods to solve the phase identification problem even at lower reporting rates; e.g., see [130, 131].

One statistical measure that can be used for phase identification is the *correlation coefficient* between the voltage measurements at the unknown phase and the voltage measurements at each of the three known reference phases. For example, the correlation coefficient between V_1 and V_A can be obtained as

$$\text{Corr}\,(V_1, V_A) = \frac{\text{Cov}\,(V_1, V_A)}{\sqrt{\text{Var}\,(V_1)\,\text{Var}\,(V_A)}}, \tag{2.34}$$

where Cov(·, ·) and Var(·) denote the co-variance and variance operators, respectively [91]. We can similarly obtain $\text{Corr}\,(V_1, V_B)$ and $\text{Corr}\,(V_1, V_C)$. A higher correlation coefficient indicates *stronger correlation* between the two voltage profiles, suggesting that the measurements are done at the same phase of the circuit.

Example 2.22 Again, consider the phase identification problem in Example 2.21. Suppose all the voltage measurements are reported at the rate of one reading every two seconds. This reporting rate is 60 times slower than the reporting rate in Example 2.21. The voltage measurements at the unknown phase is plotted in Figures 2.32(a), (b), and (c) against the voltage measurements at each of the three known phases A, B, and C, respectively, for a duration of 15 minutes of measurements. The correlation coefficients are obtained as

$$\begin{aligned} \text{Corr}\,(V_1, V_A) &= 0.9289, \\ \text{Corr}\,(V_1, V_B) &= 0.9286, \\ \text{Corr}\,(V_1, V_C) &= 0.9751. \end{aligned} \tag{2.35}$$

Therefore, we can conclude that the single-phase load is connected to Phase C. This result is consistent with the outcome of the analysis in Example 2.21.

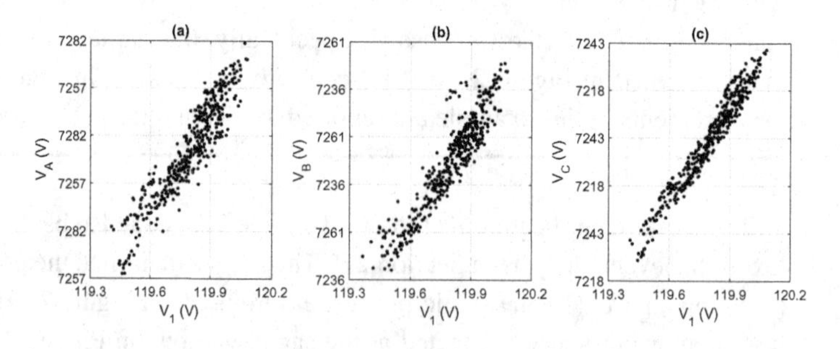

Figure 2.32 Voltage measurements with an unknown phase plotted against the voltage measurements with known phases in Example 2.22: (a) Phase A; (b) Phase B; (c) Phase C.

There are several other ways of solving the phase identification problem depending on the type of smart grid sensor technologies that are available. We will discuss phase identification further in Section 3.6.4 in Chapter 3, Section 5.9.2 in Chapter 5, and Sections 6.2 and 6.7 in Chapter 6.

2.9 Measuring Frequency

Another fundamental characteristic of AC voltage and current signals is frequency. The instrument to measure frequency is the *frequency meter*. Common frequency metering technologies include resonance-type devices, which often have low resolution in the order of 0.25 Hz, as well as time-measurement-based devices, which measure the time interval between two consecutive zero crossings of the voltage waveform. More recently, synchronized phasor measurements are also used to measure power system frequency; see Section 3.3 in Chapter 3. A frequency measurement is more accurate if the signal is sinusoidal with no or little distortion. Therefore, frequency is often measured based on voltage signals, as opposed to current signals that are prone to distortion due to nonlinear loads; see Chapter 4.

An example for frequency measurements during normal grid operating conditions is shown in Figure 2.33. The measurements in this figure are averaged in each second. In North America, frequency is typically maintained at 60 ± 0.036 Hz [132]. The frequency measurements in Figure 2.33 are within this range.

2.9.1 Generation-Load Imbalance

Managing the operation of the electric grid includes a constant effort to *balance* electric power generation with electric power consumption. The impact of unbalance between electric power generation and electric power consumption can be explained based on the *electromechanical* operation of turbine generators.

If electric power consumption exceeds electric power generation, then the turbine generators *slow down* slightly, converting some of their mechanical kinetic energy (*inertia*) into extra electric power to help meet the increased load. Since the frequency of the generated power is proportional to the turbine's *rotor speed*, increasing electric

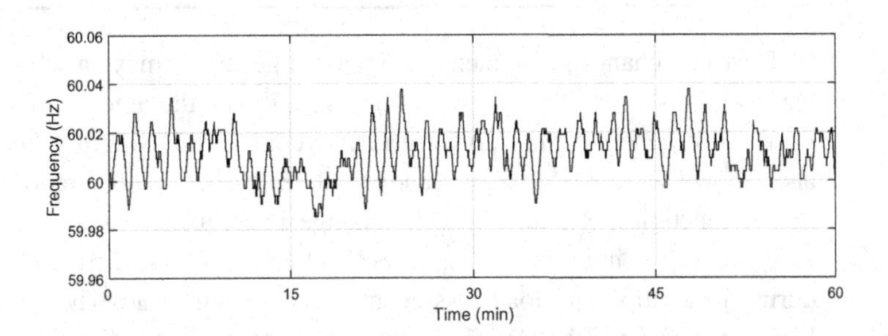

Figure 2.33 Frequency measurements during normal operating conditions.

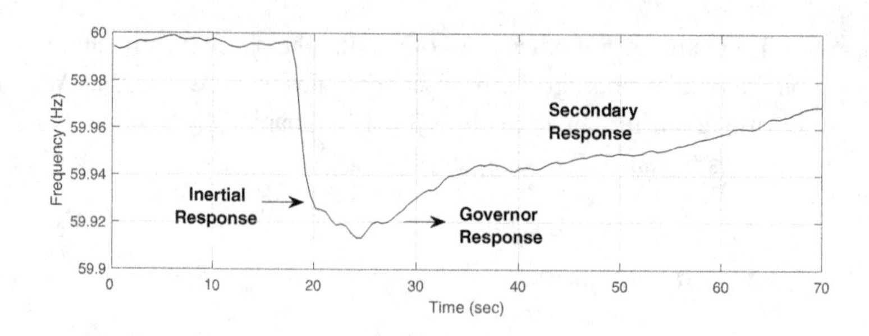

Figure 2.34 Frequency measurements during a major generator loss event.

power consumption results in a drop in the system frequency [3]. It takes a few seconds for a turbine generator to increase its mechanical torque to bring its rotor speed and therefore the system frequency back to normal.

Conversely, if electric power consumption falls below electric power generation, then the turbine generators *speed up* slightly, thus increasing the frequency, before their rotor speed and the system frequency are brought back to normal.

A considerable momentary *under-frequency* operation indicates a considerable momentary generation *deficit*. In contrast, a considerable momentary *over-frequency* operation indicates a considerable momentary generation *surplus*.

Example 2.23 A generator trip can cause a sudden imbalance between generation and load. This can create a sudden drop in frequency, as shown in Figure 2.34, based on a scenario in [133]. Here, a generator trips, and the frequency drops to 59.91 Hz. The rate of frequency drop decays within three to four seconds because of the *inertial response* of the system. For example, when frequency drops, motor loads slow down, which results in less generation-load mismatch. Next, the governor control of generators starts to arrest and then halt the frequency decline within 8–10 seconds. This procedure is known as *primary frequency response*. Finally, the secondary frequency response kicks in by Automatic Generation Control (AGC), which deploys regulating reserves, i.e., the fast-responding generators, to bring the frequency back to the scheduled level.

The exact change in frequency from a lost generator may vary based on the time of day and the season. Nevertheless, one can estimate the amount of frequency decline from a lost generator in a power grid interconnection using the frequency response, also known as the *frequency response characteristic* (FRC), corresponding to that interconnection. FRC indicates the change in frequency as a result of a change in load-generation imbalance. It is calculated using historical frequency measurements during generation and load loss events over several years. The current estimates of FRC for all three North American interconnections are [134]:

- Eastern Interconnection: −2760 MW / 0.1 Hz
- Western Interconnection: −1482 MW / 0.1 Hz
- Texas Interconnection: −650 MW / 0.1 Hz

The negative sign means there is an inverse relationship between generation loss and frequency change. For example, on average, a 1000 MW generation loss in the Western Interconnection causes a frequency change in the order of $-1000 \times 0.1/1482 = -0.067$ Hz. As another example, on average, a 1000 MW load loss in the Eastern Interconnection causes a frequency change in the order of $1000 \times 0.1/2760 = 0.036$ Hz. Conversely, FRC can be used to estimate the amount of generator loss from frequency measurements. For example, the size of the lost generator in Example 2.23 is around $-0.085 \times -1482/0.1 = 1260$ MW.

Inertia of Renewable Energy Resources

Frequency measurements can also be used to estimate system inertia [135], examine the impact of renewable generation on frequency response [136], and evaluate the performance of an interconnection to bring frequency back to normal within a five-minute period after a generation loss or a load loss event [133].

2.9.2 Frequency Oscillations

Recall from Section 2.6.1 that wide-area oscillations can affect the magnitude, phase angle, and frequency of voltage. We already saw the impact of wide-area oscillations on voltage magnitude in Example 2.14. Next, we will see the impact of wide-area oscillations on voltage frequency. We will see the impact of wide-area oscillations on voltage phase angle in Section 3.4.3 in Chapter 3.

Example 2.24 Figure 2.35(a) shows a damped oscillation in frequency measurements. The duration of the oscillation is about six seconds. The largest peak-to-peak amplitude during the event is 10 mHz. The frequency of the most dominant oscillation mode is 1.23 Hz. This oscillation event and its frequency could be detected using FFT

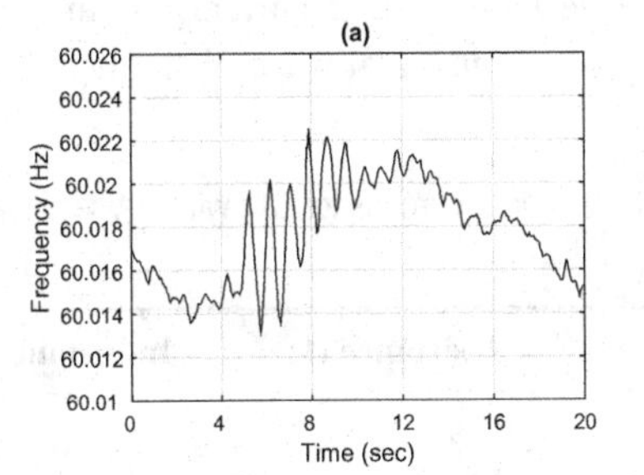

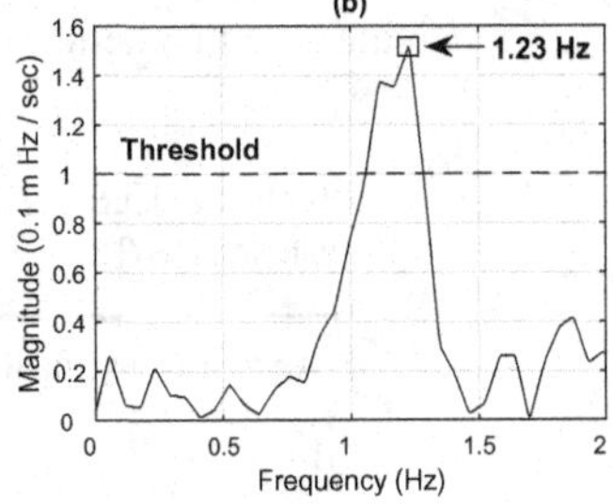

Figure 2.35 Transient oscillations in system frequency: (a) measurements [113]; (b) Fourier analysis. This oscillatory event can be detected by using the FFT method.

analysis, as shown in Figure 2.35(b). The FFT is applied to the differential of the frequency in order to remove its DC offset. Here, the event detection is triggered due to an increase in the FFT magnitude beyond a pre-determined *threshold*. The peak of the FFT magnitude is at 1.23 Hz.

Exercises

2.1 Consider the measurement of RMS current in Example 2.1. Suppose the accuracy of the ammeter that is connected to the secondary side of the CT is $\pm 0.5\%$ `rdg`. What is the possible range of the true current at the primary side of the CT under the following two different scenarios?
(a) The CT is ideal.
(b) The CT is *not* ideal, and its accuracy is $\pm 1\%$.

2.2 The secondary side of a 400:5 CT is connected to an ammeter with burden of 1.47 W and 0.92 VARs, a meter protection relay with a burden of 0.82 W and 0.80 VARs, and a wiring with a burden of 3.60 W and 0.01 VARs. All these elements are connected in series. For all elements, the burden is given at a 5 A–60 Hz secondary current.
(a) How much is the total burden?
(b) Based on Table 2.2, select the burden designation class for this CT.

2.3 The relationship between current I and magnetic field strength B that is measured by a non-contact current sensor in Figure 2.36(a) is expressed as $B = \mu_0 I \cos(\delta)/2\pi r$. Suppose the same sensor is installed on the ground [138], underneath a *balanced* three-phase transmission line in two configurations, as shown in Figures 2.36(b) and (c). Let B_1 and B_2 denote the strength of the magnetic field that is measured by the sensor in each case.
(a) Obtain an expression for B_1 and an expression for B_2.
(b) Express the height of the conductor h as a function of B_1, B_2, and d.

2.4 The working principle of an MOCT is shown in Figure 2.37. The angle of rotation in the polarization is denoted by $\beta = vBd$, where v is a constant and d is the length of the optic tube [79]. If current $i(t)$ in the power cable that creates the magnetic field is a sinusoidal wave, then magnetic field $B(t)$ is also

Table 2.2 IEEE C57.13 Standard burdens for CTs with 5 A secondary winding [137]

Burden Designation	Resistance (Ω)	Inductance (mH)
B-0.1	0.09	0.116
B-0.2	0.18	0.232
B-0.5	0.45	0.580
B-0.9	0.81	1.040
B-1.8	1.62	2.080

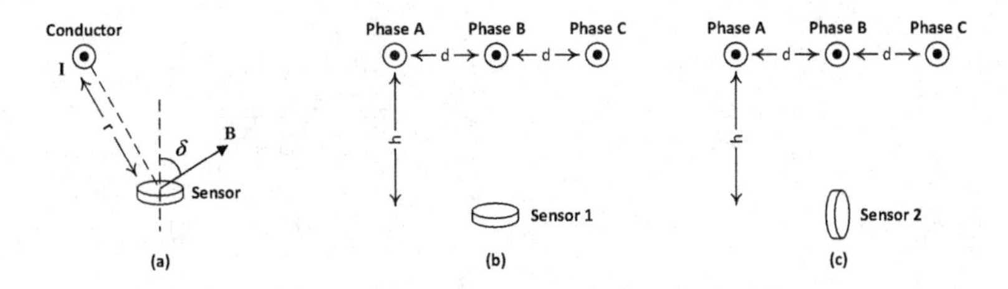

Figure 2.36 Measuring magnetic field strength of conductors in Exercise 2.3.

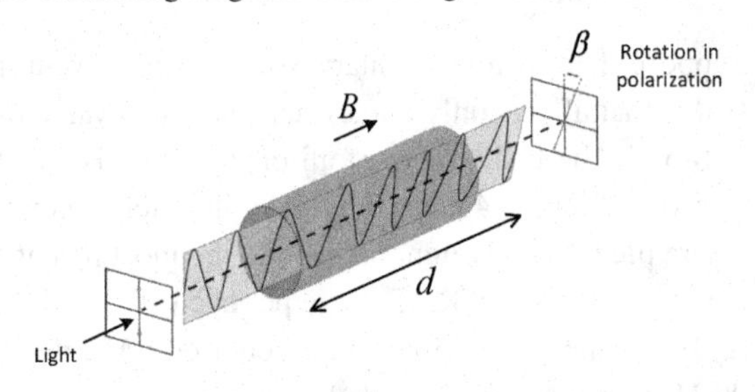

Figure 2.37 The relationship between magnetic field strength B and the amount of rotation in polarization of light that goes through an optic tube; see Exercise 2.4.

a sinusoidal wave; accordingly, $\beta(t)$ is also a sinusoidal wave. Express the RMS value of $\beta(t)$ in terms of the RMS value of $i(t)$.

2.5 The relationship between the magnetic-field-induced voltage $v_s(t)$ in a Rogowski coil in a non-contact current sensor and the current $i(t)$ that flows through the conductor is expressed as follows:

$$v_s(t) = -\frac{AN\mu_0}{l}\frac{di(t)}{dt}, \tag{2.36}$$

where N is the number of turns in the Rogowski coil, A is the area of each turn, l is the length of the winding, and μ_0 is a constant. Suppose $i(t)$ is purely sinusoidal. Express the RMS value of $v_s(t)$ in terms of the RMS value of $i(t)$.

2.6 In Example 2.3, suppose an anti-aliasing filter is designed to filter out any measurement with a frequency above 6 Hz, instead of above 4 Hz.

(a) Is this sufficient to prevent the aliasing scenario in Figure 2.5?

(b) What frequencies may still alias in presence of this anti-aliasing filter?

2.7 From the measurements in Figure 2.6, we have $\Delta V = 285.8 - 285.6 = 0.2$ V and $\Delta I = 145.5 - 136.6 = 8.9$ A. How much is the increase in apparent power load on the one phase that is shown in this figure?

2.8 File `E2-8.csv` contains voltage measurements over a period of 12 hours. The reporting rate of the measurements is *almost* one reading per minute. The first few readings are shown in Figure 2.38. Suppose τ denotes the time difference (in seconds) between any two consecutive readings of voltage.

(a) Calculate τ for all the measurements in the file.

(b) Plot the histogram for τ. Explain your observation.

(c) Is there any *missing* measurement? How many?

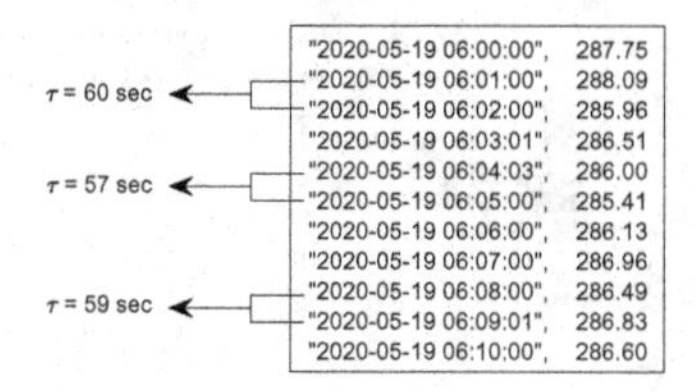

Figure 2.38 Some of the measurements in Exercise 2.8 and examples for calculating τ.

2.9 Suppose there is a momentary voltage spike event, possibly due to a lightning strike, that affects only one cycle. The RMS value of the affected voltage cycle is 145 V. The RMS value of all other cycles is 120 V. Suppose a measurement window of length 4 is used by the voltmeter with *weights* 0.4, 0.3, 0.2, and 0.1, where the highest weight is given to the most recent cycle. The reporting rate of the voltmeter is one RMS value per cycle.
(a) How much is the *peak* of the reported RMS values?
(b) How many of the reported RMS values are affected by the event?

2.10 Two voltage swell events at a distribution feeder are marked with numbers 1 and 2 in Figure 2.39(a). The current on the *same* feeder and the voltage on a *neighboring* feeder are shown in Figures 2.39(b) and (c), respectively. Specify whether each event is a local event or a nonlocal event.

2.11 Consider two neighboring power distribution feeders, and suppose we cross-examine all RMS voltage events above 0.5% rated magnitude. The results based on one month of minute-by-minute RMS voltage measurements are shown in Figure 2.40(a). Each point indicates one event. The location on the x-axis indicates the *change* in voltage on Feeder 1. The location on the y-axis indicates the *change* in voltage on Feeder 2. Suppose we *classify* the event points into eight groups, as shown in Figure 2.40(b). The number of events in each of these eight groups is 107, 102, 67, 0, 106, 104, 65, and 0, respectively.
(a) What percentage of the voltage sag events that are detected on Feeder 1 are caused by the loads and equipment on Feeder 1?
(b) What percentage of the voltage sag events that are detected on Feeder 2 are caused by the loads and equipment on Feeder 2? Compare the results with Part (a).

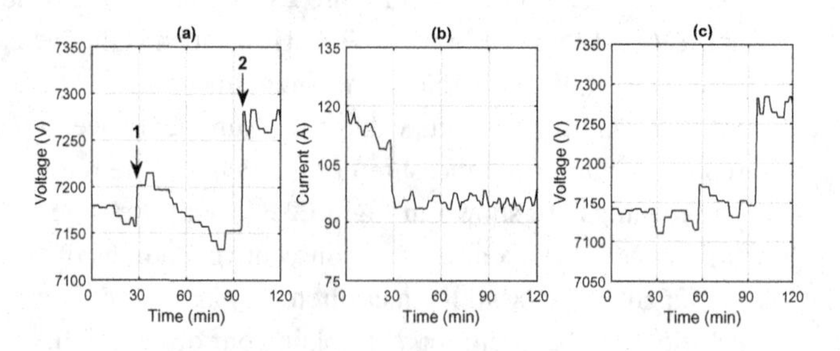

Figure 2.39 Minutely RMS voltage and current measurements in Exercise 2.10.

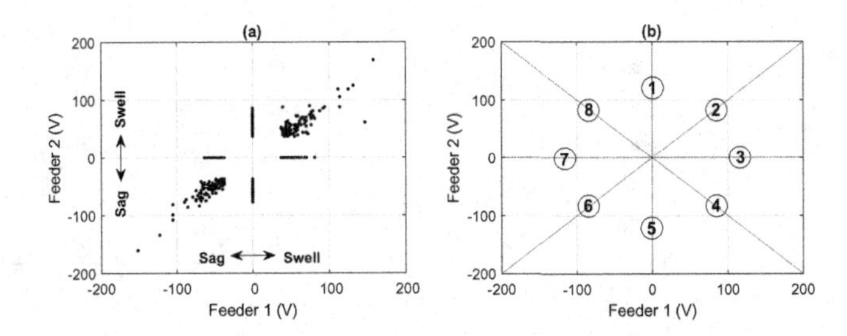

Figure 2.40 Cross-examining voltage events between two feeders in Exercise 2.11 [139].

2.12 File `E2-12.csv` contains the voltage and current measurements, averaged across three phases, at the AC side of a utility-scale PV inverter over a period of one week at one reading per five minutes.

(a) Plot voltage versus current in a scatter plot.

(b) Repeat Part (a) for the measurements between 8 AM and 6 PM. That is, *exclude* the measurements that are obtained *before* 8 AM or *after* 6 PM.

2.13 Consider $n \gg 1$ noise-contaminated samples from signal $x(t)$. Suppose the fixed component of the samples, i.e., their average, is removed. We define:

$$\mathbf{Y} = \begin{bmatrix} x(0) & x(1) & \cdots & x(l) \\ x(1) & x(2) & \cdots & x(l+1) \\ \vdots & \vdots & & \vdots \\ x(n-l-1) & x(n-l) & \cdots & x(n-1) \end{bmatrix}, \tag{2.37}$$

where $l = \lfloor n/2 \rfloor$. The Singular Value Decomposition (SVD) of this matrix is obtained as $\mathbf{Y} = \mathbf{U}\mathbf{\Sigma}\mathbf{V}^T$, where $\mathbf{\Sigma}$ is a diagonal matrix containing the singular values of matrix $\mathbf{Y}$, denoted by $\varsigma_1, \ldots, \varsigma_{l+1}$. The diagonal entries in $\mathbf{\Sigma}$ are sorted in a descending order, i.e., $\varsigma_1 \geq \ldots \geq \varsigma_{l+1}$. A simple noise reduction method works based on setting all singular values that are smaller than ϵ percentage of ς_1 to zero, and then reconstructing samples $x(0), \ldots, x(n)$ from the revised matrix $\mathbf{Y}$. Apply this noise reduction method to the RMS voltage measurements in file `E2-13.csv`. Try $\epsilon = 1\%$ and $\epsilon = 5\%$.

2.14 Consider the *single-mode* oscillation in Figure 2.41. Starting from time $t = 0$, the signal takes the following form: $x(t) = Ae^{\sigma t}\cos(wt + \varphi)$. Obtain the amplitude A, phase angle φ, frequency w, and damping factor σ. You can estimate the frequency by examining the oscillation interval; such as from one positive peak to another positive peak. Other parameters can be estimated similarly by examining various aspects of the signal in the figure.

2.15 In practice, it is common to quantify the damping factor of each oscillatory mode in terms of its *damping ratio*, in percentage, which is obtained as

$$\zeta_i = \frac{-\sigma_i}{\sqrt{\sigma_i^2 + w_i^2}} \times 100\%. \tag{2.38}$$

Obtain the damping ratio for all the oscillatory modes in Table 2.1.

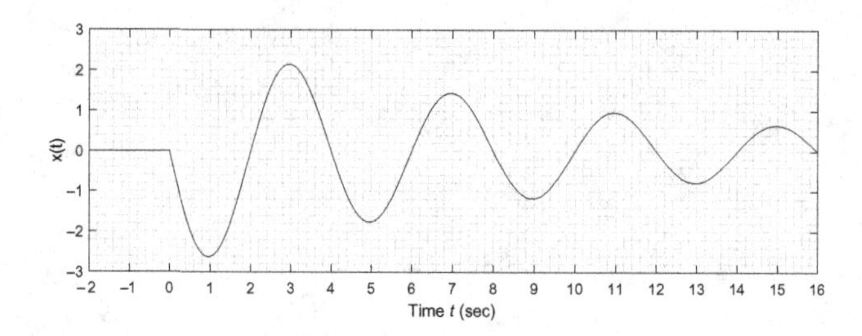

Figure 2.41 Single-mode oscillation in Exercise 2.14. Oscillation starts at time $t = 0$.

2.16 File `E2-16.csv` contains the voltage measurements for the transient oscillations in Figure 2.18. We want to go through the steps to obtain the modes that are shown in Table 2.1. Recall that in Example 2.16 we have $m = 101$.

(a) Obtain vector $\boldsymbol{\Psi}$ and matrix $\mathbf{X}$.

(b) Solve the LS problem in (2.12) to obtain $a_1, \ldots, a_m$.

(c) Solve the characteristics polynomial in (2.15) to obtain $z_1, \ldots, z_m$.

(d) Use (2.17) to obtain $w_1, \ldots, w_m$ and $\sigma_1, \ldots, \sigma_m$.

(e) Obtain vector $\boldsymbol{\Phi}$ and matrix $\mathbf{Z}$.

(f) Use (2.19) to obtain $R_1, \ldots, R_m$.

(g) Use (2.20) to obtain $A_1, \ldots, A_m$ and $\varphi_1, \ldots, \varphi_m$.

2.17 Repeat Exercise 2.16, but this time set $m = 201$. Present the oscillation modes in a table similar to Table 2.1.

2.18 The accuracy of the Prony method for any given number of modes m can be evaluated by calculating the *root mean square error* (RMSE) between the original measurements and the Prony estimation of the measurements:

$$\text{RMSE} = \sqrt{\frac{1}{n} \sum_{\tau=1}^{n} \left(x(\tau) - \sum_{i=1}^{m} A_i e^{\sigma_i \tau} \cos(w_i \tau + \varphi_i) \right)^2}. \tag{2.39}$$

(a) Obtain the RMSE for the results in Exercise 2.16.

(b) Obtain the RMSE for the results in Exercise 2.17.

2.19 File `E2-19.csv` contains a voltage profile at a 120 V single-phase load. We want to identify the events in this voltage profile by using the Min-Max method, with a threshold of 2 V. The detection window size is 10.

(a) Use non-overlapping detection windows. List the events that you detect.

(b) Use detection windows with 50% overlap. List the events that you detect.

2.20 Repeat Exercise 2.19, but use the MAD method with $\gamma = 1.4826$ and $\zeta = 3$.

2.21 File `E2-21.csv` contains the minute-by-minute measurements for phase voltages V_A and V_B and line-to-line voltage V_{AB} for a duration of one hour. Plot θ, i.e., the angle between the voltage phasors at Phase A and Phase B, while it changes during the hour; see Eq. (2.27).

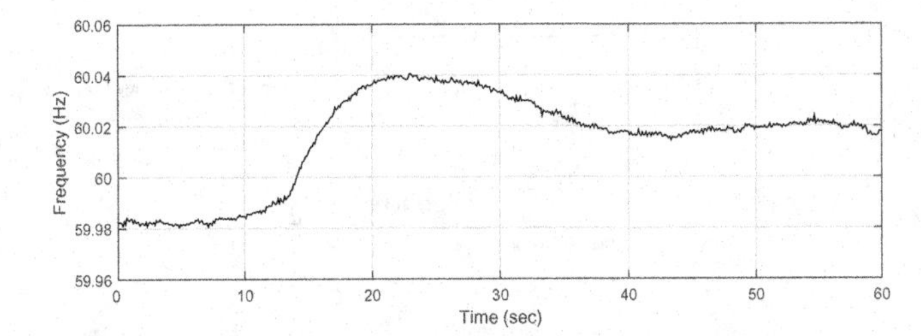

Figure 2.42 Measurements during a system-wide frequency event in Exercise 2.26 [133].

2.22 File `E2-22.csv` contains two sets of second-by-second three-phase voltage measurements that are taken at two close-by locations on the same circuit. The phases for the first set of measurements are labeled as A, B, and C. The phases for the second set of measurements are labeled as a, b, and c.

(a) Obtain the correlation coefficients across all phases, i.e., between Phase A and Phase a, between Phase A and Phase b, between Phase A and Phase c, etc. A total of *nine* correlation coefficients should be calculated.

(b) Solve the phase identification problem. That is, take phases A, B, and C as reference and identify the phase connectivity for Phases a, b, and c.

2.23 File `E2-23.csv` contains minute-by-minute three-phase voltage measurements at a load location over a period of one week.

(a) Plot the histogram of the PU during this week.

(b) Does PU exceed 1% at any time?

2.24 File `E2-24.csv` contains the frequency measurements for duration of one day at one reading per second. Nominal frequency is 60 Hz.

(a) Plot the histogram for these measurements at 0.01 Hz resolution.

(b) What percentage of the measurements are outside the 60 ± 0.036 range?

(c) Plot the frequency during the event in which it drops below 59.95 Hz. How long does the frequency stay below $60 - 0.036$ Hz during this event?

2.25 Based on their FRC values, the system frequency in which North American interconnection is likely to be *least* affected by losing a 500 MW generator?

2.26 Consider the frequency measurements in Figure 2.42.

(a) Suppose the measurements are done in California. Explain the reason for the surge in frequency during the 10th and the 20th seconds. In particular, estimate the amount of change (in MW) of generation or consumption.

(b) Suppose the measurements are done in Texas. Repeat Part (a).

2.27 Consider an energy storage unit with 1.5 MWh energy rating and 250 kW power rating. Suppose this storage unit is used to regulate frequency. The storage unit is discharged whenever the frequency drops below $60 - 0.036 = 59.964$ Hz; see Figure 2.43(a). The storage unit is charged whenever the frequency exceeds above $60+0.036 = 60.036$ Hz; see Figure 2.43(b). Suppose the storage

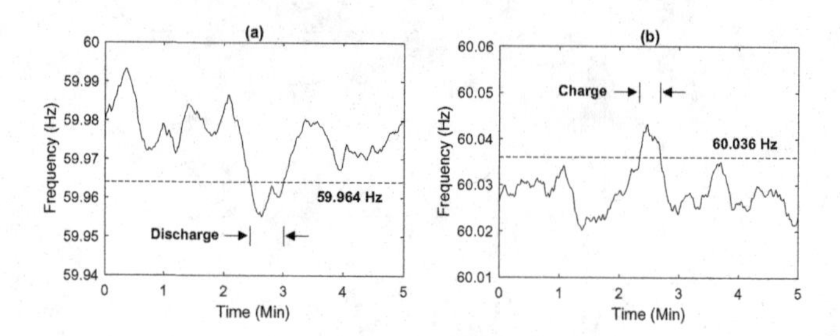

Figure 2.43 Operation of the energy storage unit in response to changes in frequency in Exercise 2.27: (a) charge the storage unit when the frequency drops below a threshold; (b) discharge the storage unit when the frequency exceeds above a threshold.

unit responds to the second-by-second frequency measurements in file `E2-27.csv`. The initial state of charge (SoC) is 50%.

(a) Identify all the charge intervals and all the discharge intervals.

(b) Suppose the storage unit is ideal. Plot the SoC curve versus time.

3 Phasor and Synchrophasor Measurements and Their Applications

The instrument to measure phasors is the *phasor measurement unit* (PMU). PMUs have received great attention over the past two decades; and they have been deployed widely throughout the world to support a variety of applications.

In this chapter we discuss the fundamental characteristics of PMUs and the important issues in working with phasor measurements. We will also discuss a wide range of traditional and emerging applications for PMU measurements.

3.1 Measuring Voltage and Current Phasors

Recall from Section 1.2.2 in Chapter 1 that a sinusoidal signal $x(t) = \sqrt{2}X \cos(\omega t + \theta)$ can be represented by phasor $X\angle\theta$, which is a *complex number* with magnitude X and phase angle θ. It is common to represent and analyze both voltage and current waveforms in AC circuits by using their corresponding phasors.

Example 3.1 The voltage at a motor load is measured as $v(t) = 120\sqrt{2}\cos(\omega t)$, and the current is measured as $i(t) = 1.63\sqrt{2}\cos(\omega t - 0.7532)$. The angular frequency is $\omega = 2\pi \times 60$. These voltage and current waveforms are shown in Figure 3.1(a), and their phasor representations are shown in Figure 3.1(b).

If voltage and current waveforms are purely sinusoidal, as in Example 3.1, then the magnitude of the phasor is equal to the RMS value of the waveform. In Example 3.1, the RMS value is 120 V for voltage and 1.63 A for current.

Fundamental Component

In practice, voltage and current may *not* have purely sinusoidal waveform. Therefore, voltage and current phasors are defined based on the *fundamental component* of their respective waveforms. In this regard, Fourier analysis is used to extract the fundamental component from the sampled measurements.

3.1.1 Phasor Calculation Using Discrete Fourier Transform

Phasor measurements are obtained by applying Discrete Fourier Transform (DFT) to sampled data of voltage and current measurements. Consider a window of N samples

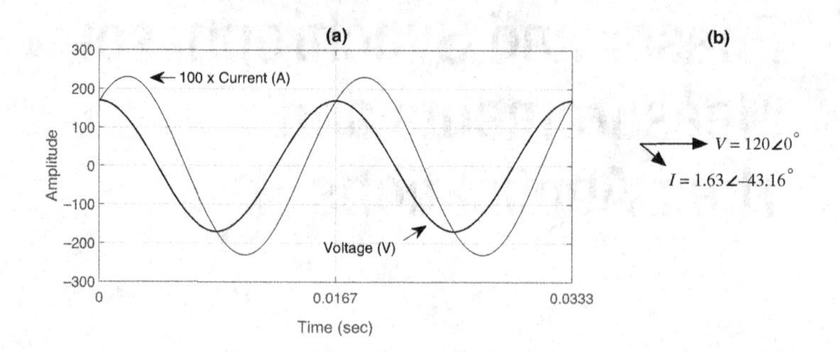

Figure 3.1 Voltage and current in Example 3.1: (a) waveforms; (b) phasors.

taken from one cycle of periodic signal $x(t) = \sqrt{2}X\cos(\omega t + \theta)$. The samples are denoted by $x(0), \ldots, x(N-1)$. The real and the imaginary components of phasor $X\angle\theta$ corresponding to these N samples are obtained as

$$\mathrm{Re}\{X\angle\theta\} = \frac{\sqrt{2}}{N}\sum_{n=0}^{N-1} x(n)\cos(2\pi n/N) \tag{3.1}$$

and

$$\mathrm{Im}\{X\angle\theta\} = -\frac{\sqrt{2}}{N}\sum_{n=0}^{N-1} x(n)\sin(2\pi n/N), \tag{3.2}$$

respectively. One can confirm that (3.1) is equal to $X\cos(\theta)$ and (3.2) is equal to $X\sin(\theta)$; see Exercise 3.1. The next phasor estimation is then obtained by moving the window by one sample and repeating the DFT calculation on samples $x(1), \ldots, x(N)$. An alternative approach is to use *recursive* calculation and obtain the phasor estimation corresponding to samples $x(1), \ldots, x(N)$ in terms of the phasor estimation corresponding to samples $x(0), \ldots, x(N-1)$; see [140].

PMUs must have enough computation power in order to conduct the DFT calculation within *each* reporting interval. Note that, to support 10 phasor readings per second, a PMU may internally sample voltage or current at a *much higher rate*, such as at 48 samples per cycle [2]; also see Section 2.3 in Chapter 2.

3.1.2 Time Reference to Measure Phase Angle

The value of the phase angle in phasor calculation depends on the *time reference*. One can change the time reference and obtain a different phasor representation for the same signal, which has the same magnitude but a different phase angle.

For instance, consider the voltage and current waveforms in Example 3.1. If we shift the time axis to the right by $(\pi/6)/\omega$ seconds, $(\pi/3)/\omega$ seconds, and $(\pi/2)/\omega$ seconds, then the phasor representations change to the following:

$$V = 120\angle 30^\circ,\ I = 1.63\angle{-13.16^\circ},$$
$$V = 120\angle 60^\circ,\ I = 1.63\angle 16.84^\circ, \tag{3.3}$$
$$V = 120\angle 90^\circ,\ I = 1.63\angle 46.84^\circ,$$

respectively. The above phasor representations can be obtained by *rotating* the phasors in Figure 3.1(b) by 30°, 60°, and 90° counterclockwise, respectively.

The choice of the time reference may not be an issue when we work with only *one* PMU. For instance, any of the phasor representations in (3.3) correctly captures the *relative* relationship between the voltage phasor and the current phasor in Example 3.1. However, if we are using *multiple* PMUs, then it is necessary to use the *same* time reference so that the phasor measurements that are obtained by different PMUs can be comparable with each other and thus can be used in the same analysis. We will discuss this issue in greater detail in the next section.

3.2 Time Synchronization and Synchrophasors

PMUs measure what is called *synchrophasors*, i.e., phasors whose phase angles are measured *relative* to a common reference cosine function at the nominal system frequency which is synchronized to the Universal Time Coordinated (UTC) standard time. An example is shown in Figure 3.2. Here, the reference signal $x_{\text{ref}}(t)$ has a maximum at time $t = 0$. Signal $x(t)$ has a positive zero crossing at time $t = 0$. The offset is -90°. Therefore, $\theta = -90^\circ$, or alternatively, $\theta_2 = 270^\circ$.

3.2.1 Precise Time Synchronization

Synchrophasor measurements require a precise time reference. According to the IEEE C37.118 Standard, the clock must be accurate to better than *one microsecond* [141]. This is to ensure that the error in synchrophasor measurements is less than 1%; as we will discuss in detail in Section 3.9.

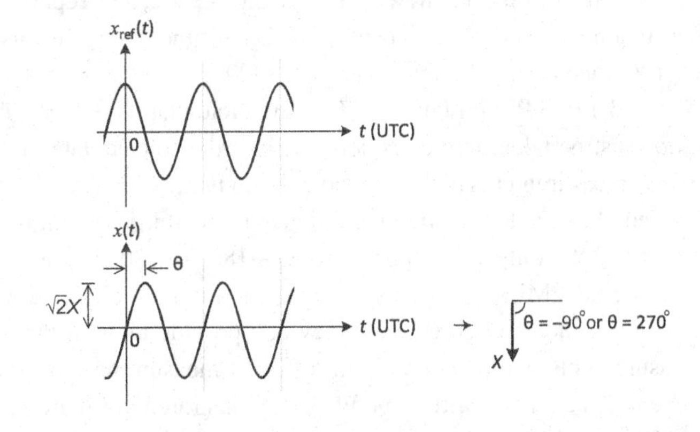

Figure 3.2 Measuring the synchrophasor for signal $x(t)$ in reference to signal $x_{\text{ref}}(t)$.

Table 3.1 An example for voltage synchrophasor measurements recorded by a PMU.

UTC Time (micro-second)	Magnitude (V)	Phase Angle (°)
1579101467916666	39832.582	183.680175
1579101467933333	39830.183	183.729736
1579101467950000	39831.321	183.780303
1579101467966666	39830.669	183.832092
1579101467983333	39831.177	183.887145
1579101468000000	39832.093	183.939575
1579101468016666	39833.123	183.991912
1579101468033333	39832.088	184.046417
1579101468050000	39830.748	184.098785
1579101468066666	39831.515	184.145889
1579101468083333	39833.174	184.197860

There are different methods to achieve such precise time synchronization across multiple PMUs. For example, one can use certain time synchronization protocols among the intended group of PMUs, such as the protocols in [142, 143].

However, currently, the most common method of time synchronization across PMUs is to use the Global Positioning System (GPS) [140]. GPS uses *satellite-synchronized clocks*, which requires the atomic clocks on all GPS satellites to be synchronized. The GPS civilian code has a time-accuracy specification of 340 nanoseconds; however, it typically performs at 35 nanoseconds [144]. These are both well within the time synchronization precision that is required by PMUs.

Each PMU is equipped with a GPS receiver to receive synchronization signals from GPS satellites. All phasors that are measured by a group of GPS-synchronized PMUs are on the same time reference; hence, they are comparable.

Time Stamps

For each synchrophasor, a PMU reports the *magnitude*, the *phase angle*, and the *time stamp*. An example is shown in Table 3.1. Here, the reporting rate is 60 Hz, i.e., one reading every 16.667 msec. The UTC time stamp in this example is given in microseconds. Time stamp 1579101468000000 indicates the start of a second, i.e., the start of 3:17:48 PM on January 15, 2020. Note that if a PT or CT is used, then its turns ratio must be taken into consideration in order to calculate the magnitude. However, for the measurements in the second column in Table 3.1, the PT's turns ratio is already applied. Also, note that the phase angles in the third column in Table 3.1 are given in the 0° to 360° range, as opposed to the −180° to 180° range.

Since all PMUs are time-synchronized and they all provide high-precision time-stamps for their measurements, they can provide us with the desired synchrophasor measurements, which come from different measurement locations across the power system. An example with three PMUs is illustrated in Figure 3.3.

Note that issues such as data aggregation and working with PMUs with different reporting rates may arise; we will discuss this subject in Section 3.2.4.

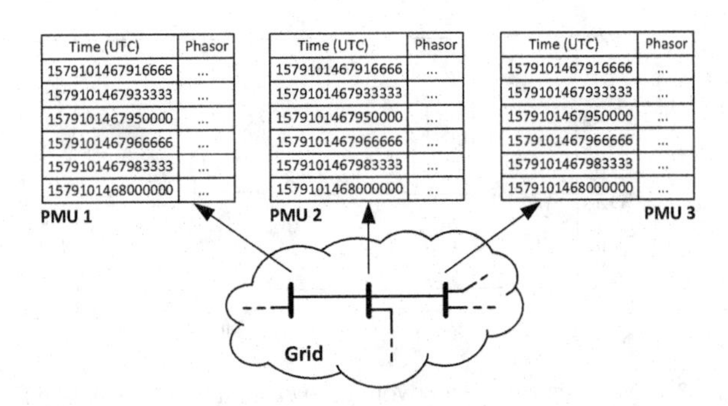

Figure 3.3 Time-stamped measurements from three PMUs at three different locations.

Frame per Second

The reporting rate of PMUs is sometimes stated in *frames per second* (fps), where each frame refers to one reading of the PMU, such as each row in Table 3.1. PMUs are required by the IEEE C37.118 Standard for Synchrophasor Measurements to report at least 10 readings per second, i.e., at 10 fps, for a 60 Hz system [141]. Different types of PMUs have different typical reporting rates; see Section 3.2.3.

3.2.2 Application of Time Synchronization

Apart from the fact that PMUs can measure phase angle; their ability also to measure voltage and current magnitude at *high reporting rates* and with *precise time stamps* is by itself an important advantage for PMUs, when compared with the traditional voltage and current sensors that we discussed in Chapter 2.

Example 3.2 Suppose we seek to extend the analysis in Example 2.9 in Chapter 2 to identify whether a voltage event, lasting only a few milliseconds, is caused by a system-wide issue at the *transmission level* or a local issue at the *distribution level*. We compare voltage measurements at two nearby feeders, as shown in Figure 3.4. Given the small time scale of the analysis, even one or two seconds drift between the clocks of the two sensors may create misleading results. Therefore, PMUs with precise time synchronization are used for this analysis. Four transient voltage events are marked on Figure 3.4. Event ① is visible in Feeder 2 but not in Feeder 1. Events ② and ④ are visible in Feeder 1 but not in Feeder 2. Event ③ is visible in both feeders. Therefore, event ③ is caused in transmission or sub-transmission systems while events ①, ②, and ④ are caused in distribution systems.

3.2.3 Different Types of PMU Technologies

Informally, one can divide the existing phasor measurement technologies into different groups, such as PMUs, Distribution-level PMUs (D-PMUs), and Frequency Disturbance Recorders (FDRs). Traditionally, PMUs have been installed at high

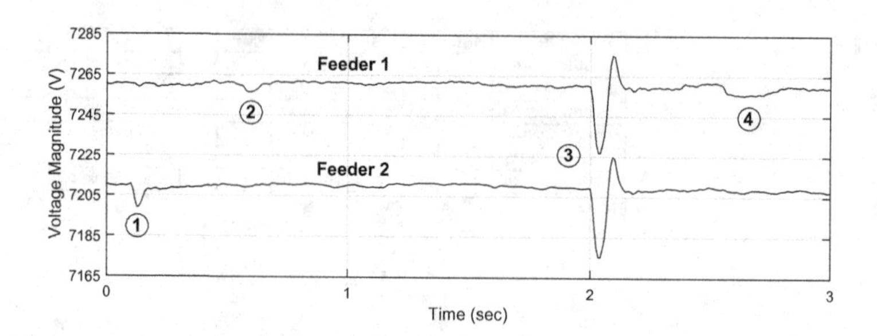

Figure 3.4 Analysis of transient voltage events on two distribution feeders by using two D-PMUs [139]. The four events that are marked are discussed in Example 3.2.

voltage transmission-level substations [145]. Their typical reporting rate is 30 fps or 60 fps [146]. D-PMUs, also known as micro-PMUs, are installed at medium voltage distribution-level substations and low voltage load transformers. Their typical reporting rate is 120 fps [139, 147]. FDRs are installed at ordinary 120 V outlets. They often serve as low-cost synchronized frequency sensors to analyze system-wide frequency dynamics. Their typical reporting rate is 10 fps [133]. Compared with PMUs and FDRs that have been deployed in practice for several years, D-PMUs are relatively new technologies that have been gaining increasing attention recently.

3.2.4 Synchrophasor Data Concentration

Several synchrophasor measurement applications require access to data from multiple PMUs at remote locations. Also, PMUs have limited local data storage capacity. Therefore, in practice, *phasor data concentrator* (PDC) devices are used to gather data from several PMUs, identify and reject bad data, align the time stamps, and create a coherent record of all collected synchrophasor data.

An important function of any PDC is *data aggregation*, where the PDC aligns data that it receives from multiple PMUs or even other PDCs and transmits the combined data. The incoming data may have different reporting rates. If the reporting rate in a PDC output stream is lower than the reporting rate in a PDC input stream, then the PDC must conduct *down-conversion* on input data. If the reporting rate in a PDC output stream is higher than the reporting rate in a PDC input stream, then the PDC must conduct *up-conversion* on input data.

Other functions of PDCs may include latency calculation, data validation, bad data detection and correction, data communications, and cyber-security [148].

3.3 Nominal and Off-Nominal Frequencies

A phasor is defined for a given frequency $\omega = 2\pi f$. Ideally, the frequency should be the same, at its rated value, all across the same power grid. For example, in North America, the frequency in each interconnection is rated at $f = 60$ Hz. However, in

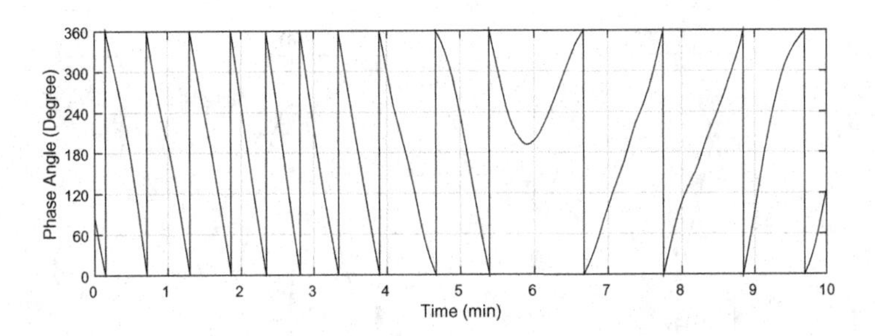

Figure 3.5 Measurements of voltage phase angle made by a PMU.

practice, the frequency constantly changes, as we saw in Section 2.9 in Chapter 2. In this section, we discuss how the changes in frequency affect the way that the phase angles are measured by PMUs. We also discuss how PMU measurements can be used to obtain synchronized frequency measurements.

3.3.1 Impact of Frequency on Measuring Phase Angle

The phase angle measurements in Table 3.1 change at every reading. In fact, phase angles in this table appear to constantly increase at a small pace. These changes are for the most part due to the fact that real-world power systems seldom operate at nominal frequency, which we denote by $f_0 = 60$ Hz. Instead, they may operate at an *off-nominal* frequency $f = f_0 + \Delta f$, where $\Delta f \neq 0$.

Example 3.3 Consider the voltage phase angle measurements for a duration of 10 minutes in Figure 3.5. For the first few minutes, the phase angle decrements in cycles that run from 360° to 0° before *wrapping around* and jumping back to 360°. At around minute six, there is a *reverse* in this pattern, and the phase angle starts increasing. For the last few minutes, the phase angle increments in cycles that run from 0° to 360° before *wrapping around* and jumping back to 0°.

The impact of frequency on phase angle measurements is explained in Figure 3.6. A PMU with reference signal $v_{\text{ref}}(t)$ is used to measure phase angles for signal $v_1(t)$ at frequency $f_1 = 60$ Hz and signal $v_2(t)$ at frequency $f_2 = 61$ Hz:

$$v_1(t) = \sqrt{2}\cos(2\pi f_1 t + 270°) = \sqrt{2}\cos(2\pi f_0 t + 270°) \tag{3.4}$$

$$v_2(t) = \sqrt{2}\cos(2\pi f_2 t + 270°) = \sqrt{2}\cos(2\pi f_0 t + 2\pi\Delta f t + 270°). \tag{3.5}$$

The phase angle that the PMU reports for signal $v_1(t)$ is 270°, which is the same at all reporting instances at $t = 0$, $t = 0.0167$, $t = 0.0333$, $t = 0.0500$, $t = 0.0667$, $t = 0.0833$, and $t = 0.1000$. This is shown in Figure 3.6(b). In contrast, the phase angle that the PMU reports for signal $v_2(t)$ is $2\pi\Delta f\, t + 270°$, which changes over time,

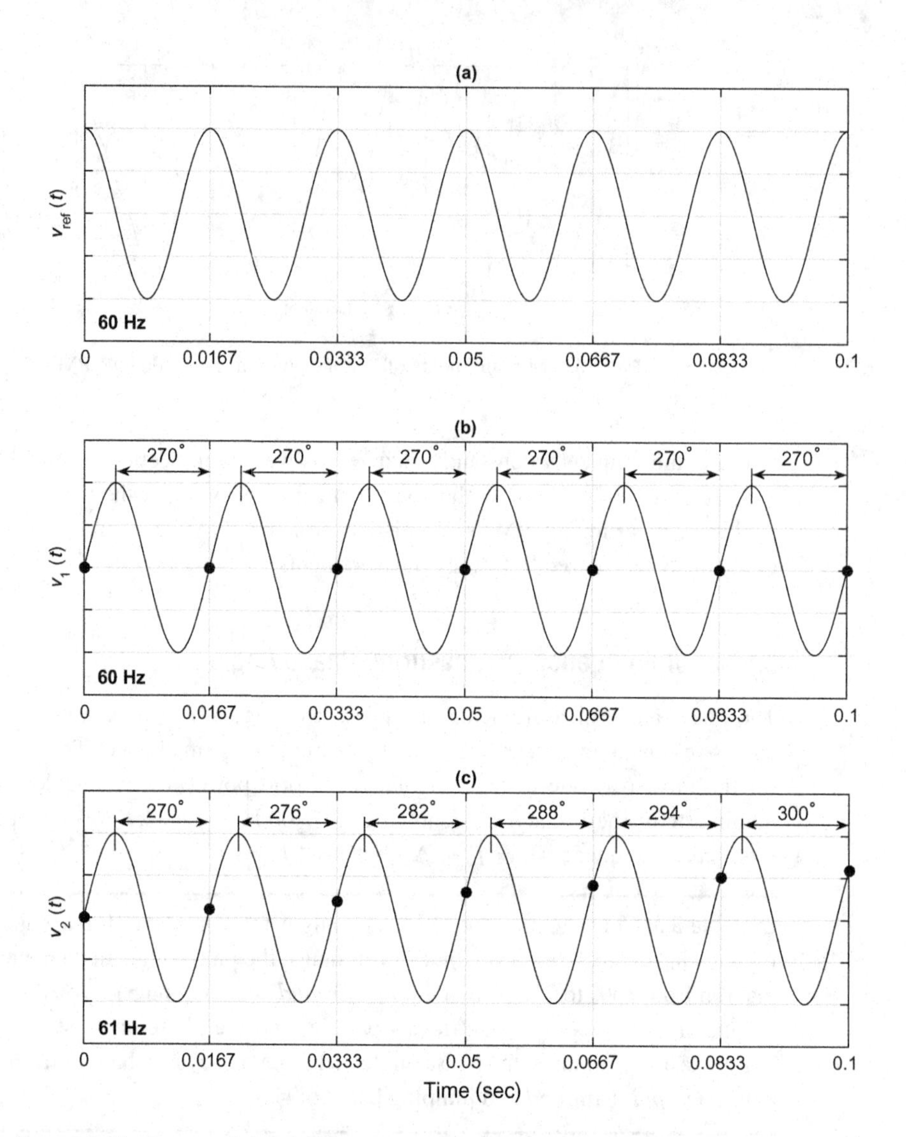

Figure 3.6 The impact of signal frequency on measuring phase angle: (a) reference signal; (b) signal with nominal frequency; (c) signal with off-nominal frequency.

taking the values of 270°, 276°, 282°, 288°, 294°, and 300°, at $t = 0$, $t = 0.0167$, $t = 0.0333$, $t = 0.0500$, $t = 0.0667$, $t = 0.0833$, and $t = 0.1000$, respectively. This is shown in Figure 3.6(c). Note that the increments in the phase angle measurements in this figure are fixed at $360^\circ \times \Delta f \times \Delta t = 360^\circ/60 = 6^\circ$.

From the above analysis, the measurements on phase angles can be used to estimate the frequency. If frequency is fixed, or it changes very slowly, then it can be estimated from the measured phase angles as [140]:

$$f = f_0 + \Delta f = f_0 + \frac{1}{2\pi}\frac{\Delta\theta}{\Delta t}, \tag{3.6}$$

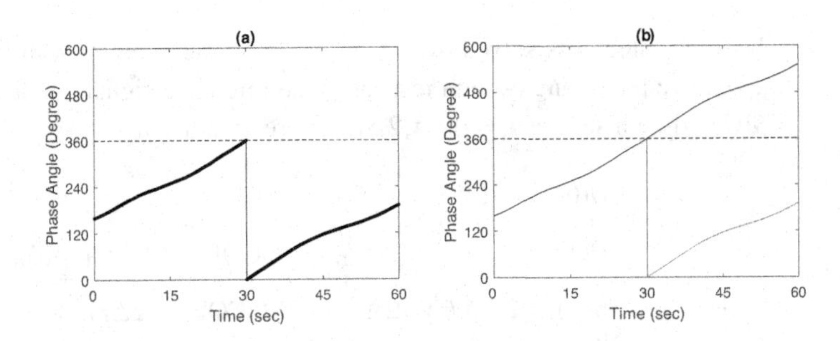

Figure 3.7 Phase angle measurement: (a) wrapped; (b) unwrapped.

where $\Delta\theta$ denotes the difference between two consecutive phase angle readings. If the phase angles are given in degrees, then 2π is replaced with 360°. From (3.6), and after reordering the terms, we derive $\Delta\theta = 360° \times \Delta f \times \Delta t$, which explains the fixed 6° increments in the phase angle measurements in Figure 3.6(c).

Unwrapping Phase Angle Measurements

An important note in calculating frequency from phase angle measurements is that, if $\Delta\theta$ is calculated at a point where the phase angle measurement wraps around, then the phase angle measurement needs to be unwrapped before it is used in (3.6). An example is shown in Figure 3.7. Here, the original phase angle measurements that are reported by the PMU are wrapped around from 360° to 0° at time $t = 30$ sec. The phase angle measurements that are reported by the PMU are unwrapped by *adding* 360° to all measurements at time $t = 30$ sec and beyond. For the case where the original phase angle measurements are wrapped around from 0° to 360°, they are unwrapped by *subtracting* 360°.

3.3.2 Rate of Change of Frequency

If frequency is not fixed, then (3.6) may not provide an accurate estimation of frequency. Next, suppose frequency itself is a function of time, denoted by $f(t)$. Using the first two terms in the Taylor series approximation, we can write:

$$f(t) = f_0 + \Delta f + \frac{df}{dt}t. \tag{3.7}$$

The derivative in the third term is referred to as the *rate of change of frequency* (ROCOF). Consider measuring the phase angle of a sinusoidal signal $x(t) = \sqrt{2}X\cos(\vartheta(t))$. Suppose the signal has a time-varying frequency $f(t)$ as in (3.7). From the analysis of power system dynamics, we know that $\omega(t) = d\vartheta(t)/dt$, where $\omega(t) = 2\pi f(t)$ is the *instantaneous* angular frequency [149]. Thus, we have:

$$\begin{aligned} \vartheta(t) &= 2\pi \int_0^t f(\tau)d\tau + \vartheta(0) \\ &= 2\pi f_0 t + \left(2\pi\Delta f t + \pi \text{ROCOF} t^2 + \vartheta(0)\right), \end{aligned} \tag{3.8}$$

where the integral is solved over $f(\tau)$ based on the expression in (3.7). From (3.8), the relative phase angles with respect to the reference signal that are reported by the PMU at time instances $t = 0, \Delta t, 2\Delta t, \ldots$ are obtained as

$$\begin{aligned}
\theta(0) &= \vartheta(0), \\
\theta(1) &= 2\pi\Delta f \times (\Delta t) + \pi\text{ROCOF} \times (\Delta t)^2 + \vartheta(0), \\
\theta(2) &= 2\pi\Delta f \times (2\Delta t) + \pi\text{ROCOF} \times (2\Delta t)^2 + \vartheta(0), \\
&\vdots
\end{aligned} \tag{3.9}$$

If ROCOF $= 0$, then we can use (3.9) to obtain $\Delta\theta = 2\pi\Delta f \Delta t$, which after reordering the terms would result in (3.6). However, in general, ROCOF is likely *not* zero and *not* known in advance. Therefore, the system of equations in (3.9) can be used to simultaneously estimate Δf and ROCOF. This is done by using a window of N phase angle measurements. From (3.9), we can derive:

$$\begin{bmatrix} \theta(1) - \theta(0) \\ \theta(2) - \theta(0) \\ \theta(3) - \theta(0) \\ \vdots \\ \theta(N-1) - \theta(0) \end{bmatrix} = \begin{bmatrix} 2\pi\Delta t & \pi\Delta t^2 \\ 4\pi\Delta t & 4\pi\Delta t^2 \\ 6\pi\Delta t & 9\pi\Delta t^2 \\ \vdots & \vdots \\ 2(N-1)\pi\Delta t & (N-1)^2\pi\Delta t^2 \end{bmatrix} \begin{bmatrix} \Delta f \\ ROCOF \end{bmatrix}. \tag{3.10}$$

The unknowns can now be estimated using the LS method:

$$\begin{bmatrix} \Delta f \\ ROCOF \end{bmatrix} = (\mathbf{A}^T\mathbf{A})^{-1}\mathbf{A}^T\mathbf{\Phi}, \tag{3.11}$$

where $\mathbf{\Phi}$ denotes the $N - 1 \times 1$ vector of measured phase angle differences, and $\mathbf{A}$ denotes the $N - 1 \times 2$ matrix of coefficients. Note that, matrix $\mathbf{A}$ depends only on Δt. Therefore, it can be pre-calculated and stored for use in real time. Once Δf and ROCOF are estimated, the frequency itself is estimated using (3.7).

Example 3.4 One can use the phase angle measurements in the first $N = 10$ rows in Table 3.1 and solve (3.11) to estimate $\Delta f \approx 0.008528$ and ROCOF ≈ 0.002193. By substituting these results in (3.7), frequency is estimated as

$$f \approx 60 + 0.008528 + 0.002193 \times \frac{1}{60} \approx 60.0085645. \tag{3.12}$$

Here, $\mathbf{\Phi}$ is a 9×1 vector and $\mathbf{A}$ is a 9×2 matrix.

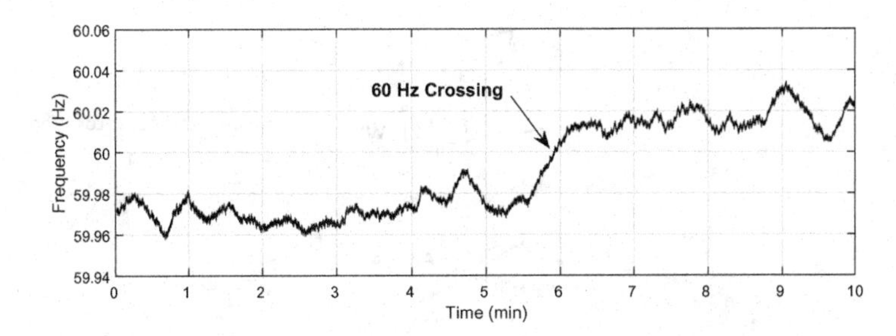

Figure 3.8 Frequency estimation based on the phase angle measurements in Figure 3.5.

The frequency estimations corresponding to the phase angle measurements in Figure 3.5 in Example 3.3 are shown in Figure 3.8 for $N = 120$. Frequency crosses 60 Hz at around minute six, i.e., around the time when there is a reverse in the pattern of changes in phase angle measurements, as we saw in Figure 3.5.

Increasing the window size N can result in obtaining more smooth frequency estimations. However, one should not use excessively long windows for frequency estimation to avoid degrading the accuracy of the estimated frequencies. During transient stability swings, the frequency of the power system may change rapidly. Thus, a long window may include significantly different frequencies over the window span, and this can result in error in frequency estimation.

3.3.3 Synchronized Frequency Measurements

Recall from Section 2.9 in Chapter 2 that major events can cause sudden changes in the frequency of the power system all across the interconnection, such as a sudden drop in frequency due to a major generation loss event in Example 2.23. Time synchronization across PMUs, together with the fact that PMUs can estimate frequency, can help us explain the system-wide impact of such frequency events.

Example 3.5 Consider the time-synchronized frequency measurements in Figure 3.9 that are made by FDRs in six different locations on the Western Interconnection in the United States as part of the FNET/GridEye project [133]. The frequency suddenly drops due to a major generator loss event. The lost generator is located in Southern California. As a result, frequency changes are first seen by the FDRs in Southern and Northern California. The relative timing of the frequency events at different locations can help estimate the location of the lost generator.

If synchronized frequency measurements are available, then we can also measure *frequency difference*, i.e., the difference in frequency of the power system at different locations across the same power system interconnection at the times when a major system-wide event happens; e.g., see [150].

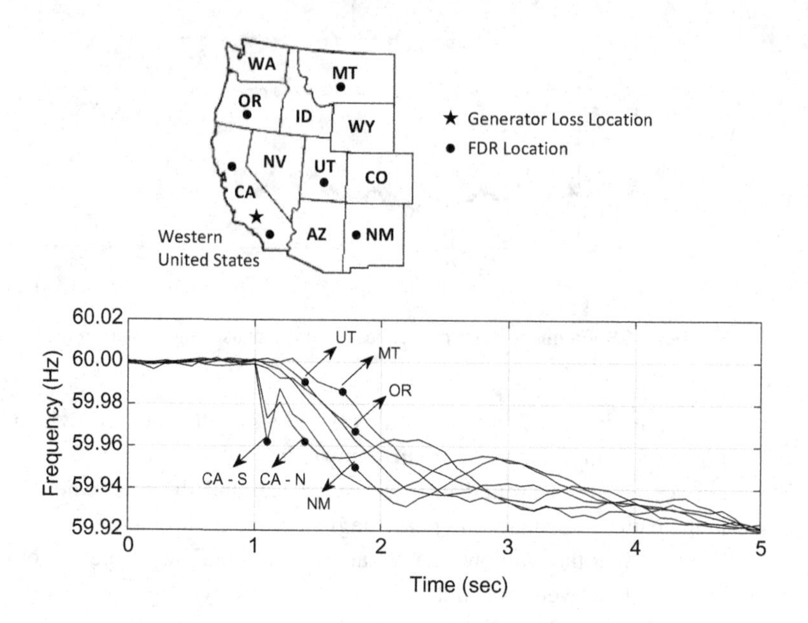

Figure 3.9 Time-synchronized frequency measurements during a major generator loss event on the Western Interconnection. The measurements are done by an FDR network.

3.4 Relative Phase Angle Difference

The Relative Phase Angle Difference (RPAD) is the difference between the phase angles in two synchrophasors. For example, given the voltage synchrophasors $V_1\angle\theta_1$ and $V_2\angle\theta_2$ at locations 1 and 2 on the same interconnection, such as at the two ends of a transmission line, the corresponding RPAD is obtained as

$$\text{RPAD} = \theta_1 - \theta_2. \tag{3.13}$$

Time synchronization is critical in order to calculate RPAD. If the two phasors are *not* synchronized, then RPAD calculation is practically useless.

In practice, RPAD is almost always calculated for voltage synchrophasors but not for current synchrophasors. Also, phase angle measurements must be unwrapped before they can be used to calculate RPAD; see Figure 3.7 in Section 3.3.

3.4.1 Approximate Relationship Between RPAD and Power Flow

Recall from Section 1.3.1 in Chapter 1 that the flow of real power between two buses on an AC power network depends on the voltage angle difference between the two buses. When the line impedance is dominantly inductive, as in the case of most transmission lines, the real power flow between bus 1 and bus 2 is approximated as

$$P_{12} = \frac{V_1 V_2}{X_{12}} \sin(\theta_1 - \theta_2), \tag{3.14}$$

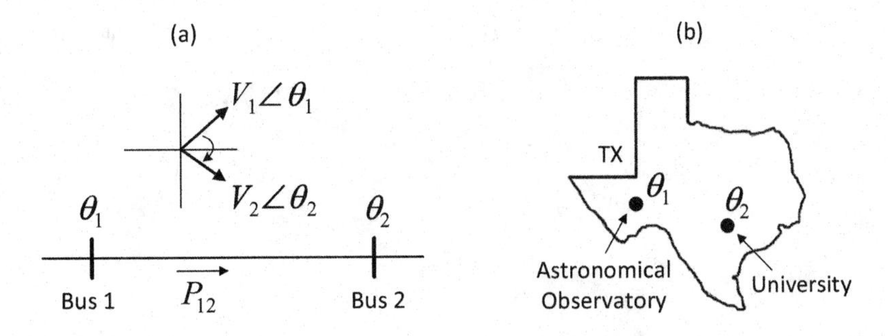

Figure 3.10 RPAD can be measured across (a) two ends of a transmission line; or (b) two regions on the same interconnection in the state of Texas.

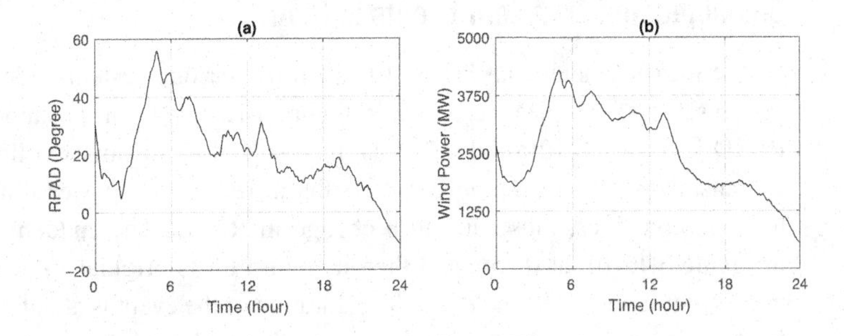

Figure 3.11 Comparing (a) the RPAD measurements between West Texas and the load centers in East Texas and (b) the amount of wind power generated in West Texas [113].

where X_{12} is the line inductance, $V_1\angle\theta_1$ is the voltage phasor at location 1, and $V_2\angle\theta_2$ is the voltage phasor at location 2; see Figure 3.10(a). From (3.14), active power flow in a power transmission line is nearly proportional to the sine of RPAD of voltages at the two terminals of the line. Hence, since many of the planning and operational considerations in a power network are concerned with the flow of active power, measuring RPAD across transmission lines is of great interest.

Note that (3.14) also sometimes provides an *approximation* of the active power flow between two *regions*, even if the measurements do not come exactly from two terminals of the same transmission line. An example is shown in Figure 3.10(b) for the case of the Texas synchrophasor network [113]. Here, location 1 is an astronomical observatory in West Texas, a region which has a relatively small population but is home to some of the largest wind farms in the United States. Location 2 is a university campus in East Texas, a region which is home to the state's largest population centers. RPAD measurements between these two locations, shown in Figure 3.11(a), are insightful when they are compared with the amount of wind power generation, shown in Figure 3.11(b). In fact, since the majority of the wind power that is generated in West Texas is consumed in East Texas, the overall shape of the wind power generation profile is roughly approximated by the overall shape of the active power flow (and RPAD) between the two regions.

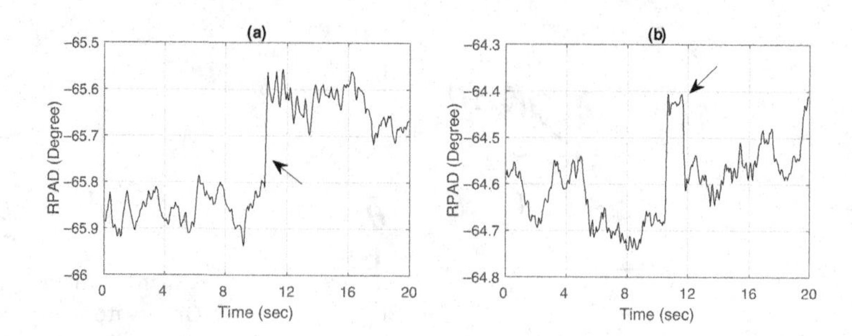

Figure 3.12 Two RPAD events: (a) sustained step change due to transmission line tripping; (b) temporary step change due to transmission line tripping and reclosing.

3.4.2 Sustained and Transient Events in RPAD

RPAD measurements can be used to identify various system-wide events in power systems. Two examples are shown in Figure 3.12 [113]. The sustained step change in RPAD in Figure 3.12(a), at 0.24°, is the result of a transmission line tripping event. A transmission line tripping event results in redirecting power flow to surrounding transmission lines, thus causing a change in RPAD. The sudden change in RPAD in Figure 3.12(b), at 0.25°, and then its return to its original value is the result of a transmission line reclosing event. The duration of the event is about 1.5 seconds. Here, the fault is rather temporary; therefore, the line reclosing operation is successful, and the power flow returns to its original value within a short period of time. RPAD also returns to its pre-fault value.

3.4.3 Impact of Inter-Area Oscillations on RPAD

Another important application of RPAD is to identify and characterize *inter-area oscillations*. Recall from Section 2.6.1 in Chapter 2 that an inter-area oscillation indicates an *unintentional periodic exchange of power* across different regions of a power grid. Based on the approximate relationship between RPAD and active power exchange that we saw in Section 3.4.1, oscillations in RPAD measurements across two regions can be considered as an indication of inter-area oscillations between the two regions. Inter-area oscillations often carry a frequency between 0.15 Hz and 1 Hz, or sometimes up to 2 Hz; depending on the interconnection.

Example 3.6 Two oscillation events in RPAD measurements are shown in Figure 3.13. For the damped transient event in Figure 3.13(a), the duration of the event is 4.4 seconds and the largest swing is 0.96°. The frequency of the oscillation, i.e., for its dominant mode, is 0.59 Hz. For the damped transient event in Figure 3.13(b), the duration is 4.8 seconds and the largest swing is 1.33°. The oscillation has a strong mode at 0.64 Hz. This oscillation was caused by a sudden loss of a 810 MW power generation unit [113]. The frequencies of both of the above oscillatory events fall within the typical range of inter-area oscillations.

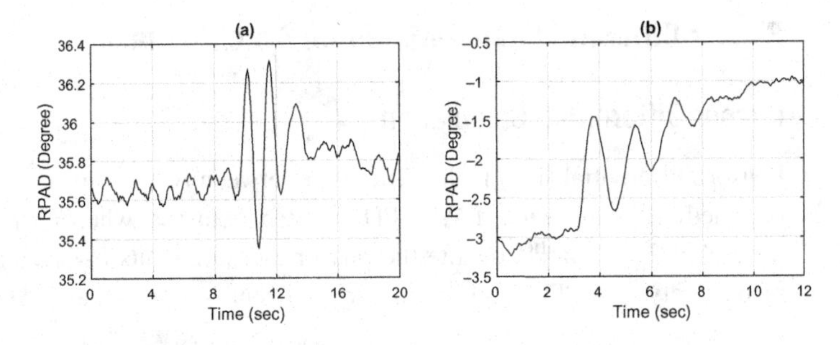

Figure 3.13 Two damping oscillation events identified by using RPAD measurements.

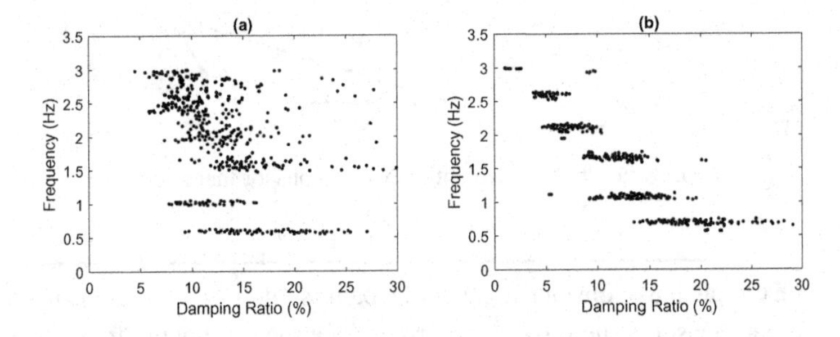

Figure 3.14 Oscillation modes that are captured in RPAD measurements across the two PMUs in Figure 3.10(b): (a) low-wind conditions; (b) high-wind conditions [113].

By examining all the oscillation events in RPAD measurements over several hours, we can obtain the *statistical characteristics* of the inter-area oscillations and other types of low-frequency oscillations in the power system.

Example 3.7 Figure 3.14 shows two scatter plots for the *frequency* versus *damping ratio* of the oscillation modes in the RPAD measurements across the two PMUs in Texas that we previously saw in Figure 3.10(b). The scatter plot in Figure 3.14(a) is obtained during *low-wind* conditions; where the wind power penetration is between 4.6% and 5%. The scatter plot in Figure 3.14(b) is obtained during *high-wind* conditions; where the wind power penetration is between 15% and 18%. We can see that the points in Figure 3.14(b) are more tightly clustered at certain frequencies, while the points in Figure 3.14(a) are more scattered, specially at frequencies above 1.5 Hz. Furthermore, some of the points in Figure 3.14(b) are shifted toward the left, i.e., they have smaller damping ratios, compared to the points in Figure 3.14(a). This means that some of the oscillation modes in the system are not damped as quickly during high-wind conditions as they do during low-wind conditions. This type of scatter plots can be used to characterize the oscillation modes in the system under different operating conditions; see the detailed study on this subject in [113].

3.5 Phasor Differential and Differential Synchrophasors

3.5.1 Phasor Differential Calculation

Phasor Differential (PD) is the difference between *two* phasor measurements that are obtained from the *same* PMU. PD is often obtained when an *event* occurs. Suppose X^{before} and X^{after} denote the phasor measurements *before* and *after* the event, respectively. The PD corresponding to the event is obtained as [151]:

$$\Delta X = X^{\text{after}} - X^{\text{before}}. \tag{3.15}$$

The geometry of the above relationship is shown in Figure 3.15.

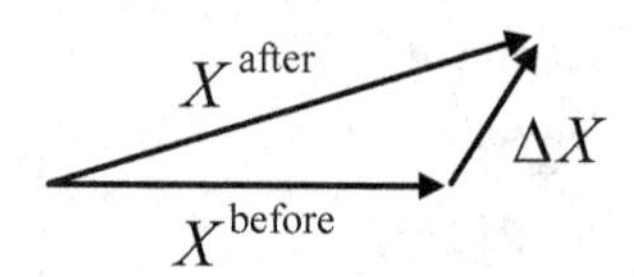

Figure 3.15 Phasor differential ΔX is a phasor that is obtained according to (3.15).

Example 3.8 Consider a capacitor bank switching event that is captured by a PMU that measures the phasor of current. The measurements for magnitude and the measurements for phase angle are shown in Figures 3.16(a) and (b), respectively. The current phasors before and after the event are measured as

$$I^{\text{before}} = 88.64\angle 335.73^\circ \tag{3.16}$$

and

$$I^{\text{after}} = 100.01\angle 307.81^\circ. \tag{3.17}$$

The PD corresponding to the above current phasors is obtained as

$$\Delta I = 100.01\angle 307.81^\circ - 88.64\angle 335.73^\circ = 46.83\angle -114.6^\circ. \tag{3.18}$$

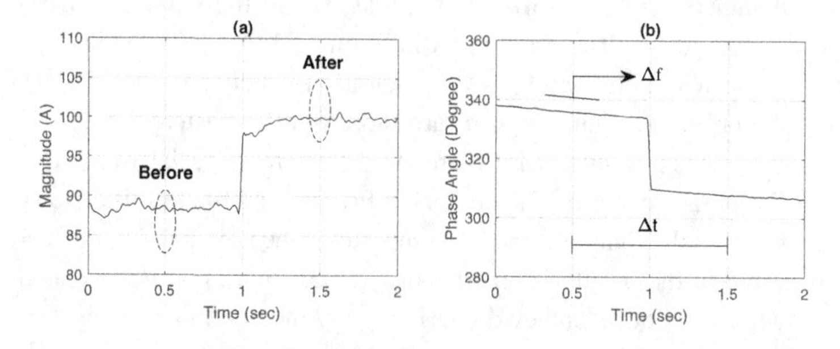

Figure 3.16 Changes in current phasor during a capacitor bank switching event [152]: (a) changes in magnitude; (b) changes in phase angle. See Examples 3.8 and 3.9 on how to model this event by using phasor differentials.

As we can see in Example 3.8, there is no strict definition for the timing of what we refer to as "before" the event. Similarly, there is no strict definition for the timing of what we refer to as "after" the event. Generally speaking, both X^{before} and X^{after} should be measured at steady-state conditions, but as close as possible to the moment when the event or the phenomenon of interest occurs.

Phasor differential can also be defined based on voltage phasor measurements. For example, for the capacitor bank switching event in Example 3.8, the voltage phasors before and after the event are measured as

$$V^{\text{before}} = 7266.7\angle 339.31^\circ \tag{3.19}$$

and

$$V^{\text{after}} = 7218.6\angle 335.55^\circ, \tag{3.20}$$

respectively. Accordingly, we can obtain:

$$\Delta V = 7218.6\angle 335.55^\circ - 7266.7\angle 339.31^\circ = 477.6\angle -118.35^\circ. \tag{3.21}$$

The phasor differentials in (3.18) and (3.21) model the impact of the event at the current and voltage phasor measurements at the location of the PMU.

Comparison with RPAD

Whereas RPAD is a scalar, PD is a *phasor*. RPAD is measured between two PMUs, but PD is measured for one PMU. RPAD uses only phase angle measurements, while PD uses both phase angle and magnitude measurements. RPAD is measured for voltage, but PD is measured for both voltage and current.

3.5.2 Impact of Off-Nominal Frequency on PD Calculation

Obtaining the PD is sometimes challenging due to the operation of the power grid at off-nominal frequencies. Recall from Section 3.3 that if Δf is *not* zero, then there is an offset in the measured phase angles across any two subsequent phasor measurements. From (3.6), and after reordering the terms:

$$\Delta\theta = 2\pi\ \Delta t\ \Delta f. \tag{3.22}$$

Here, Δt denotes the time between the two measurements, such as the time between measuring X^{before} and measuring X^{after} in the context of obtaining the PD. This offset can potentially cause error in measuring the PD corresponding to an event. Therefore, we may want to remove it before the PD is calculated.

Example 3.9 Consider the current phase angle measurements in Figure 3.16(b) in Example 3.8. There is a major sudden change in the phase angle due to the event. There is also a relatively small but continuous change in the phase angle that is *not* related to the event. It exists both before and after the event. It is rather due to the operation of the grid at an off-nominal frequency. In fact, by following the analysis in

Section 3.3.1, we can check the changes in the phase angle during the first 0.5 seconds in Figure 3.16(b), i.e., before the event occurs, and accordingly obtain the deviation in the off-nominal frequency as follows:

$$\Delta f = \frac{\Delta\theta}{360 \times \Delta t} = -0.0133 \text{ Hz}, \tag{3.23}$$

where $\Delta t = 0.5$ seconds and $\Delta\theta = 335.7^\circ - 338.1^\circ = -2.4^\circ$. Next, we can use (3.22) to make the following correction in the phase angle measurement for I^{after}:

$$307.81^\circ - 360 \times 1 \times -0.0133 = 312.61^\circ, \tag{3.24}$$

where $\Delta t = 1$ second is the time difference between I^{after} and I^{before}; as marked in Figure 3.16(b). Thus, the actual change in the phase angle that is caused by the event is less than what the initial measurement had suggested:

$$\Delta I = 100.01\angle 312.61^\circ - 88.6434\angle 335.73^\circ = 39.4\angle -109.41^\circ. \tag{3.25}$$

A similar correction can also be done in the voltage phase angle measurements.

3.5.3 Differential Synchrophasors

Recall that a PD is a phasor. It can be obtained at every PMU. Since PMU measurements are time-stamped, the PDs that are obtained at multiple PMUs can be time-synchronized. In particular, if X^{before} is measured at the same time at all PMUs, and X^{after} is measured at the same time at all PMUs, then the set of ΔX calculations that are provided by the PMUs form *differential synchrophasors*. Differential synchropasors capture how the voltage and/or current phasors are *simultanously* affected across the interconnected power system because of the *same* event.

Example 3.10 Consider the power system in Figure 3.17 which has five buses and four lines. Suppose two PMUs are installed at bus 1 and bus 5 to measure voltage phasors and current phasors at these two buses. Suppose an impedance switching event, such as a capacitor bank switching event or a motor load switching event, happens at Bus 3. The following PDs are obtained at the PMU at bus 1:

$$\begin{aligned} \Delta V_1 &= V_1^{\text{after}}\angle\theta_1^{\text{after}} - V_1^{\text{before}}\angle\theta_1^{\text{before}}, \\ \Delta I_{12} &= I_{12}^{\text{after}}\angle\phi_{12}^{\text{after}} - I_{12}^{\text{before}}\angle\phi_{12}^{\text{before}}. \end{aligned} \tag{3.26}$$

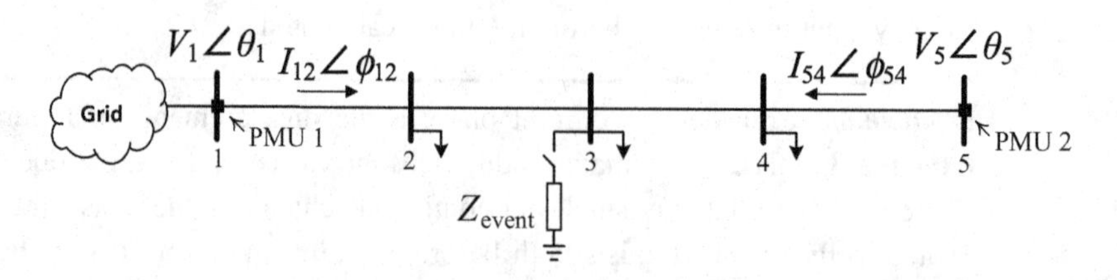

Figure 3.17 The PMUs at buses 1 and 5 can provide the differential synchrophasors in (3.28) to *simultaneously* capture the impact of the switching event at bus 3.

Similarly, the following PDs are obtained at the PMU at bus 5:

$$\begin{aligned} \Delta V_5 &= V_5^{\text{after}} \angle \theta_5^{\text{after}} - V_5^{\text{before}} \angle \theta_5^{\text{before}}, \\ \Delta I_{54} &= I_{54}^{\text{after}} \angle \phi_{54}^{\text{after}} - I_{54}^{\text{before}} \angle \phi_{54}^{\text{before}}. \end{aligned} \tag{3.27}$$

If the "before" and "after" instances are time-synchronized, then the above measurements result in obtaining the following vector of differential synchrophasors:

$$\begin{bmatrix} \Delta V_1 & \Delta V_5 & \Delta I_{12} & \Delta I_{54} \end{bmatrix}^T. \tag{3.28}$$

Some of the applications of differential synchropasors include event location identification, see Section 3.5.4, and tracking state estimation [153, 154].

3.5.4 Application in Event Location Identification

Again, consider the switching event in Example 3.10 in Section 3.5.3. Suppose the location of the event is unknown, i.e., we do *not* know the fact that the change in impedance occured at bus 3. The question is: Can we use the synchronized phasor measurements at PMU 1 and PMU 2 to identify the location of the event?

Recall that the impedance switching event in Example 3.10 was represented by the differential synchrophasors in (3.26)–(3.28). By using these differential synchrophasors and by taking the following four steps, we can identify the location of the event. The methodology in these four steps is developed in [151, 155].

Step 1: Forming the Equivalent Circuit Model

According to the *compensation theorem* [156], once an element changes in a circuit, such as due to a switch operation, the amount of changes in nodal voltages and branch currents on the circuit can be obtained by forming an *equivalent circuit*. In the equivalent circuit, the element that has changed is replaced by a *current source* that injects a current phasor that is equal to the *difference* in the current phasor that goes through the element *before* and *after* the event. Any other voltage source or current source in the circuit is replaced by its internal impedance.

The construction of the equivalent circuit is illustrated in Figure 3.18. Here, the event is a change in impedance Z_{event}. The voltage and current phasors *before* and *after* the event are shown in Figures 3.18(a) and (b), respectively. The equivalent circuit based on the compensation theorem is shown in Figure 3.18(c).

We can similarly construct the equivalent circuit corresponding to the event in Example 3.10. The equivalent circuit is shown in Figure 3.19. The only source in this equivalent circuit is the current source at bus 3 that represents the event. All nodal voltages are represented in terms of their phasor differences before and after the event; they are denoted by ΔV_1, ΔV_2, ΔV_3, ΔV_4, and ΔV_5. Similarly, all branch currents are represented in terms of their phasor differences before and after the event; they are denoted by ΔI_{12}, ΔI_{23}, ΔI_{34}, and ΔI_{45}. The line impedance at line segments

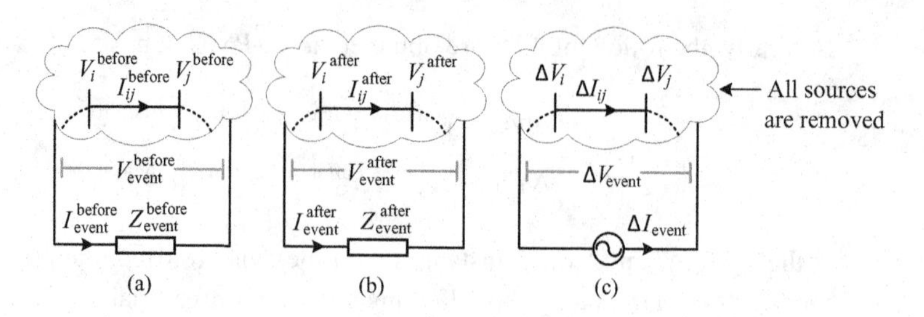

Figure 3.18 Event analysis based on the compensation theorem: (a) before the event occurs; (b) after the event occurs; (c) equivalent circuit [155].

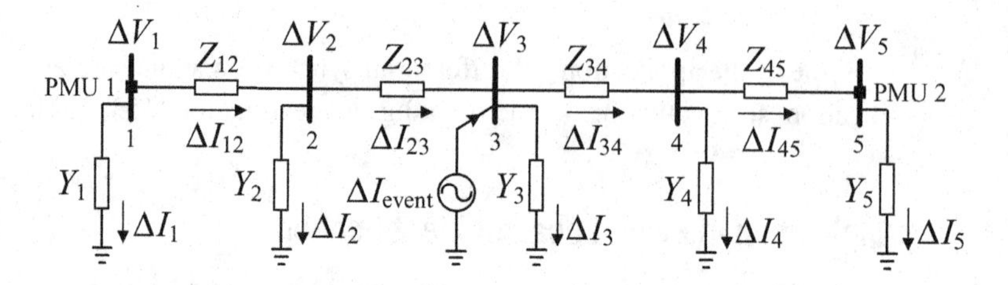

Figure 3.19 The equivalent circuit corresponding to the switching event in Example 3.10.

(1,2), (2,3), (3,4), and (4,5) are denoted by Z_{12}, Z_{23}, Z_{34}, and Z_{45}, respectively. The admittance at buses 1, 2, 3, 4, and 5, are denoted by Y_1, Y_2, Y_3, Y_4, and Y_5, respectively. The admittance at each bus is the combined admittance of any load and the internal admittance of any generator. The current that is drawn by the admittance at buses 1, 2, 3, 4, and 5 is denoted by ΔI_1, ΔI_2, ΔI_3, ΔI_4, and ΔI_5, respectively.

The differential synchrophasors that we obtained in (3.28) in Example 3.10 provide us with four measurements on the equivalent circuit in Figure 3.19. They include ΔV_1, ΔV_5, ΔI_{12}, and $\Delta I_{45} = -\Delta I_{54}$. The differential synchrophasors for ΔV_2, ΔV_3, ΔV_4, ΔI_{23}, ΔI_{34}, ΔI_1, ΔI_2, ΔI_3, ΔI_4, ΔI_5, and ΔI_{event} are still *unknown*. Furthermore, the *location* of current source ΔI_{event} is also *unknown*. Finally, the known parameters in this analysis include Z_{12}, Z_{23}, Z_{34}, Z_{45}, Y_1, Y_2, Y_3, Y_4, and Y_5. They are known because they can be obtained from the utility's basic circuit model [155].

Step 2: Nodal Voltage Calculation Starting from PMU 1

Let k denote the bus number where the switching event occurs. In Example 3.10, we have $k = 3$. As a hypothetical assumption, suppose we know parameter k; i.e., we know the location of the event. Suppose we even know phasor ΔI_{event}. Under these hypothetical assumptions, we can use the measurements in PMU 1, together with the known parameters, and successively apply the KVL and KCL to the equivalent circuit in Figure 3.19 to obtain:

$$\begin{aligned}
\Delta V_1 &= \Delta V_1 \quad \leftarrow \text{ Direct Measurement in (3.26)} \\
\Delta V_2 &= \Delta V_1 - \Delta I_{12}\, Z_{12} \\
\Delta V_3 &= \Delta V_2 - (\Delta I_{12} - Y_2\, \Delta V_2)\, Z_{23} \\
\Delta V_4 &= \Delta V_3 - (\Delta I_{12} - Y_2\, \Delta V_2 - Y_3\, \Delta V_3 + \Delta I_{\text{event}})\, Z_{34} \\
\Delta V_5 &= \Delta V_4 - (\Delta I_{12} - Y_2\, \Delta V_2 - Y_3\, \Delta V_3 + \Delta I_{\text{event}} - Y_4\, \Delta V_4)\, Z_{45}.
\end{aligned} \tag{3.29}$$

Notice that ΔI_{event} is used in the calculation of ΔV_4 and ΔV_5. Another note is that, in obtaining ΔV_3, we used the fact that $Y_2 \Delta V_2 = \Delta I_2$ and the fact that $\Delta I_{12} - \Delta I_2 = \Delta I_{23}$. Similar facts are used in obtaining ΔV_4 and ΔV_5.

However, since we do *not* actually know the location of the event, i.e., parameter k, and we do *not* know ΔI_{event}, we cannot write the equations in (3.29) in their current form. If we drop ΔI_{event} from the equations in (3.29), they become:

$$\begin{aligned}
\Delta V_1 &= \Delta V_1 \quad \leftarrow \text{ Direct Measurement in (3.26)} \\
\Delta V_2 &= \Delta V_1 - \Delta I_{12}\, Z_{12} \\
\Delta V_3 &= \Delta V_2 - (\Delta I_{12} - Y_2\, \Delta V_2)\, Z_{23} \\
\Delta V_4 &\neq \Delta V_3 - (\Delta I_{12} - Y_2\, \Delta V_2 - Y_3\, \Delta V_3)\, Z_{34} \\
\Delta V_5 &\neq \Delta V_4 - (\Delta I_{12} - Y_2\, \Delta V_2 - Y_3\, \Delta V_3 - Y_4\, \Delta V_4)\, Z_{45}.
\end{aligned} \tag{3.30}$$

Notice that the *equality* signs in the last two lines are now replaced with the *inequality* signs due to inevitably not being able to include ΔI_{event}. Of course, we cannot include ΔI_{event} in (3.30) because we do not know the location of the event.

Step 3: Nodal Voltage Calculation Starting from PMU 2

Similar to the analysis in Step 2, we can use the measurements in PMU 2 and successively apply the KVL to obtain the following:

$$\begin{aligned}
\Delta V_5 &= \Delta V_5 \quad \leftarrow \text{ Direct Measurement in (3.27)} \\
\Delta V_4 &= \Delta V_5 + \Delta I_{45}\, Z_{45} \\
\Delta V_3 &= \Delta V_4 + (\Delta I_{45} + Y_4\, \Delta V_4)\, Z_{34} \\
\Delta V_2 &\neq \Delta V_3 + (\Delta I_{45} + Y_4\, \Delta V_4 + Y_3\, \Delta V_3)\, Z_{23} \\
\Delta V_1 &\neq \Delta V_2 + (\Delta I_{45} + Y_4\, \Delta V_4 + Y_3\, \Delta V_3 + Y_2\, \Delta V_2)\, Z_{12}
\end{aligned} \tag{3.31}$$

Again, notice the *inequality* sign in the last two lines in (3.31).

Step 4: Minimum Discrepancy

Let us denote the expressions on the right hand side in (3.30) as

$$\widehat{\Delta V_1}, \quad \widehat{\Delta V_2}, \quad \widehat{\Delta V_3}, \quad \widehat{\Delta V_4}, \quad \widehat{\Delta V_5}. \tag{3.32}$$

Similarly, let us denote the expressions on the right hand side in (3.31) as

$$\widecheck{\Delta V_1},\ \ \widecheck{\Delta V_2},\ \ \widecheck{\Delta V_3},\ \ \widecheck{\Delta V_4},\ \ \widecheck{\Delta V_5}. \tag{3.33}$$

We define the *discrepancy* between the nodal voltage calculations in Step 2 at each bus and the nodal voltage calculations in Step 3 at the same bus as

$$\Phi_i = \widehat{\Delta V_i} - \widecheck{\Delta V_i},\ \ i = 1, \ldots, 5. \tag{3.34}$$

When it comes to calculating ΔV_3, the corresponding expression in (3.30) and the corresponding expression in (3.31) are both equal to ΔV_3; therefore, $\Phi_3 = 0$. However, that is not the case at *any other* bus. At any bus other than bus 3, either we face an inequality in (3.30); or we face an inequality in (3.31). Therefore, the discrepancy index Φ_i is not zero at any bus other than bus 3:

$$\begin{aligned} &\Phi_1 \neq 0 \\ &\Phi_2 \neq 0 \\ &\Phi_3 = 0 \\ &\Phi_4 \neq 0 \\ &\Phi_5 \neq 0. \end{aligned} \tag{3.35}$$

In other words, the discrepancy is nonzero at all buses, *except* at the event bus, which in this example is bus $k = 3$. This is always true for any event that changes an impedance, regardless of the number of buses or the values of the parameters [155].

In practice, the discrepancy may not be precisely zero at the event bus because of the error in phasor measurements and the inaccuracy in the known parameters. Therefore, instead of looking for a zero discrepancy at a bus, we look for the *minimum* absolute value of the discrepancy across all the buses in order to identify the location of the event. In this regard, parameter k is obtained as

$$k = \underset{i=1,\ldots,5}{\arg\min} \left|\Phi_i\right|. \tag{3.36}$$

The magnitude is used in the minimization, because discrepancy is a phasor.

Example 3.11 Consider the impedance switching event in Example 3.10 in Section 3.5.3. Suppose the voltage and current phasors before and after the event are measured as in Table 3.2. The line impedances are $Z_{12} = Z_{23} = Z_{34} = Z_{45} = 0.0025 + j0.0050$ per unit. The admittances at buses 2, 3, and 4 in the equivalent circuit are $Y_2 = 0.7 - j0.4$, $Y_3 = 0.5 - j0.3$, and $Y_4 = 1.0 - j0.5$ per unit. From (3.26) and (3.27), the differential synchrophasors for this event are obtained as

$$\begin{aligned}
\Delta V_1 &= 0.004708\angle 25.803^\circ, \\
\Delta I_{12} &= 0.950233\angle 115.902^\circ, \\
\Delta V_5 &= 0.014834\angle 6.937^\circ, \\
\Delta I_{45} &= 0.011738\angle -11.480^\circ.
\end{aligned} \tag{3.37}$$

From (3.30)–(3.34), we obtain:

$$\begin{aligned}
\widehat{\Delta V_1} &= 0.0047\angle 25.8029^\circ, & \widecheck{\Delta V_1} &= 0.0154\angle 8.598^\circ, \\
\widehat{\Delta V_2} &= 0.0098\angle 11.7582^\circ, & \widecheck{\Delta V_2} &= 0.0152\angle 7.987^\circ, \\
\widehat{\Delta V_3} &= 0.0150\angle 7.4890^\circ, & \widecheck{\Delta V_3} &= 0.0150\angle 7.505^\circ, \\
\widehat{\Delta V_4} &= 0.0204\angle 5.5197^\circ, & \widecheck{\Delta V_4} &= 0.0149\angle 7.116^\circ, \\
\widehat{\Delta V_5} &= 0.0258\angle 4.5440^\circ, & \widecheck{\Delta V_5} &= 0.0148\angle 6.937^\circ,
\end{aligned} \tag{3.38}$$

and

$$\begin{aligned}
&|\Phi_1| = 0.0110, \ |\Phi_2| = 0.0055, \ |\Phi_3| = 0.0000, \\
&|\Phi_4| = 0.0055, \ |\Phi_5| = 0.0110.
\end{aligned} \tag{3.39}$$

Thus, from (3.36), the location of the event is obtained as bus 3, which is correct.

Other Methods

The above method is important because it shows one of the key applications of differential synchrophasor measurements. However, there are also many other methods that one can use to identify the location of events. In particular, there is a rich literature on identifying the location of *faults* by using phasor measurements. A common approach is to try to estimate the *distance* between the location of the sensor and the location of the fault; e.g., see the methods in [157–159]. Other examples for the methods to identify the location of a fault include those outlined in [160–164]. There are also some, but significantly fewer, methods to identify the location of events that are *benign*, yet they can reveal how different components operate in the power system. For example, there are multiple studies on how to identify the location of a capacitor bank in power distribution systems, such as in [165, 166]. Note that the method that

Table 3.2 Synchrophasor measurements in per unit in Example 3.11.

Phasor	Before	After
$V_1\angle\theta_1$	$0.9926\angle 29.194^\circ$	$0.9973\angle 29.178^\circ$
$I_{12}\angle\phi_{12}$	$3.1738\angle 1.753^\circ$	$2.9169\angle 19.046^\circ$
$V_5\angle\theta_5$	$0.9570\angle 27.536^\circ$	$0.9709\angle 27.228^\circ$
$I_{45}\angle\phi_{45}$	$0.7566\angle 9.101^\circ$	$0.7676\angle 8.793^\circ$

we explained above based on the use of differential synchrophasor measurements can be used to identify the location of both faults and benign events; see [155].

3.6 Three-Phase and Unbalanced Phasor Measurements

Traditionally, most PMUs are installed on power transmission networks. Since the power system is mostly balanced at the transmission level, it is often not necessary for the PMUs to report phasor measurements on all three phases. They may report the phasor measurements on one phase only; or they may report only the *positive sequence* of the three-phase voltage and current phasors [139, 147].

However, when it comes to power distribution systems; the voltage and current phasors are commonly unbalanced; therefore, it is beneficial to measure and report the voltage and current phasors on *all three phases*. Accordingly, D-PMUs often do report phasors measurements on all phases; e.g., see [167].

An example for voltage phasor measurements at a three-phase load is shown in Figure 3.20. The unbalance in magnitude is evident in Figure 3.20(a). There is also a slight unbalance in phase angle in Figure 3.20(a); at a fraction of a degree. Note that to check the unbalance in phase angle, we need to use the measurements in Figure 3.20(b) to calculate the instantaneous phase angle difference between any two phases to see how much they deviate from 120°.

3.6.1 Symmetrical Components

A set of *unbalanced* three-phase voltage or current phasors can be expressed as the *sum* of three sets of *balanced* phasors, which are called *symmetrical components* [13]. Symmetrical components include the phasors in the *zero sequence*, the phasors in the *positive sequence*, and the phasors in the *negative sequence*.

Suppose V_A, V_B, and V_C denote the unbalanced three-phase voltage phasor measurements. Let us first define the following three new phasors:

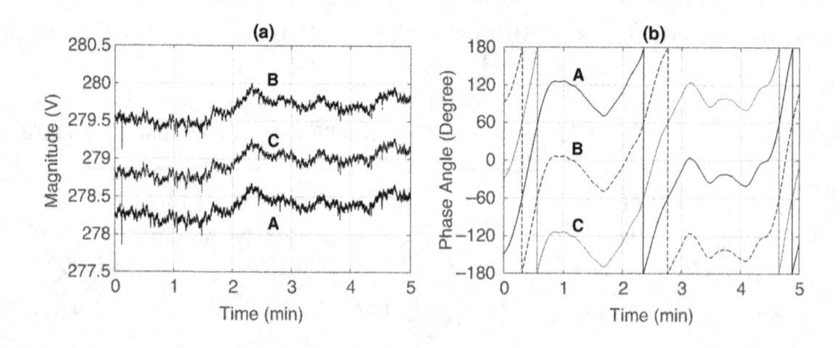

Figure 3.20 Unbalanced voltage phasor measurements: (a) magnitude; (b) phase angle.

$$
\begin{aligned}
V^0 &= (V_A + V_B + V_C)/3, \\
V^+ &= \left(V_A + \alpha V_B + \alpha^2 V_C\right)/3, \\
V^- &= \left(V_A + \alpha^2 V_B + \alpha V_C\right)/3,
\end{aligned} \tag{3.40}
$$

where

$$\alpha = e^{j2\pi/3}. \tag{3.41}$$

Vector α is a *phase shift* operator. By multiplying α to a phasor, we shift (increase) the phase angle by 120°. By multiplying α^2 to a phasor, we shift (increase) the phase angle by 240°. Note that, $\alpha^3 = 1$. Furthermore, we have:

$$1 + \alpha + \alpha^2 = 0. \tag{3.42}$$

The symmetrical components are defined as follows:

$$
\begin{aligned}
V_A^0 &= V^0, \\
V_B^0 &= V^0, \\
V_C^0 &= V^0,
\end{aligned} \tag{3.43}
$$

and

$$
\begin{aligned}
V_A^+ &= V^+, \\
V_B^+ &= \alpha^2\, V^+, \\
V_C^+ &= \alpha\, V^+,
\end{aligned} \tag{3.44}
$$

and

$$
\begin{aligned}
V_A^- &= V^-, \\
V_B^- &= \alpha\, V^-, \\
V_C^- &= \alpha^2\, V^-.
\end{aligned} \tag{3.45}
$$

Vectors V_A^0, V_B^0, and V_C^0 form the zero sequence; vectors V_A^+, V_B^+, and V_C^+ form the positive sequence; and vectors V_A^-, V_B^-, and V_C^- form the negative sequence.

From (3.40)–(3.45), we can reconstruct the original measurements as

$$
\begin{aligned}
V_A &= V_A^0 + V_A^+ + V_A^-, \\
V_B &= V_B^0 + V_B^+ + V_B^-, \\
V_C &= V_C^0 + V_C^+ + V_C^-.
\end{aligned} \tag{3.46}
$$

Example 3.12 Consider the following voltage phasor measurements:

$$
\begin{aligned}
V_A &= 278.0574\angle{-73.1170^\circ}, \\
V_B &= 276.8067\angle 167.2848^\circ, \\
V_C &= 278.5330\angle 47.1288^\circ.
\end{aligned} \tag{3.47}
$$

From (3.40), we can obtain:

$$\begin{aligned} V^0 &= 0.3469\angle{-76.9327^\circ}, \\ V^+ &= 277.7978\angle{-72.9014^\circ}, \\ V^- &= 1.0257\angle{-167.8488^\circ}. \end{aligned} \tag{3.48}$$

Accordingly, the symmetrical components are obtained as

$$V_A^0 = V_B^0 = V_C^0 = 0.3469\angle{-76.9327^\circ}, \tag{3.49}$$

and

$$\begin{aligned} V_A^+ &= 277.7978\angle{-72.9014^\circ}, \\ V_B^+ &= 277.7978\angle{167.0986^\circ}, \\ V_C^+ &= 277.7978\angle{47.0986^\circ}, \end{aligned} \tag{3.50}$$

and

$$\begin{aligned} V_A^- &= 1.0257\angle{-167.8488^\circ}, \\ V_B^- &= 1.0257\angle{-47.8488^\circ}, \\ V_C^- &= 1.0257\angle{72.1512^\circ}. \end{aligned} \tag{3.51}$$

We can confirm that (3.46) holds; i.e., the summation of the obtained symmetrical components on each phase results in the original phasors in (3.47).

If a set of three-phase measurements are *balanced*, then it has only the positive sequence, which comprises the exact same original three-phase measurements. Both the zero sequence and the negative sequence are zero for balanced three-phase phasor measurements; see Exercise 3.17.

Analysis of Faults

Symmetrical components can be used to analyze severe unbalanced system conditions, such as those that are caused by common faults. For example, in an *open phase* system, see Figure 3.21(a), or in a *phase-to-ground fault*, see Figure 3.21(b), there are

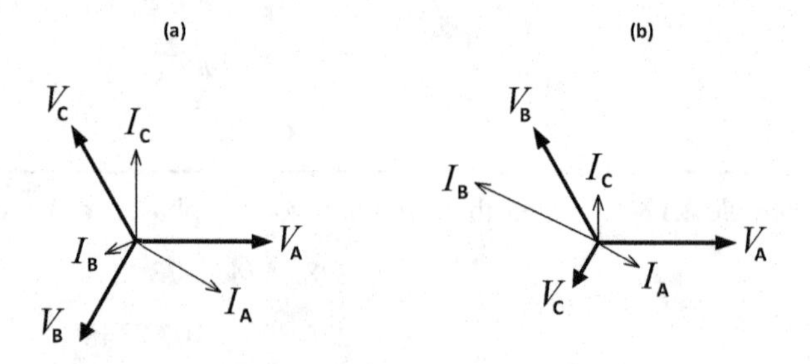

Figure 3.21 Severe unbalance can occur in voltage and/or current phasors during faults: (a) open phase; (b): phase-to-ground fault [171].

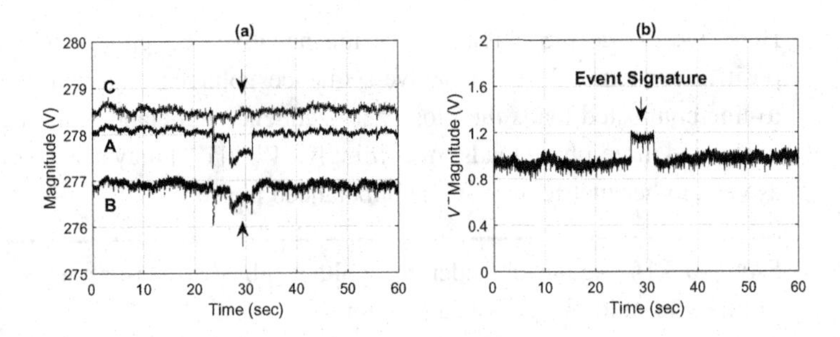

Figure 3.22 An unbalanced event: (a) the impact of the event on two phases of the three-phase voltage measurements; (b) the event signature on the negative sequence.

often major negative sequence currents and possibly major zero sequence currents; see [168–170].

3.6.2 Unbalanced Events

Apart from the faults that are often unbalanced, some *benign* events may also be unbalanced and affect only one or two phases. Unbalanced events can be detected by applying various event detection methods that we learned in Section 2.7.2 in Chapter 2 to the *negative-sequence* or *zero-sequence* measurements.

Example 3.13 Consider the three-phase voltage measurements that are shown in Figure 3.22(a). Only the magnitudes of the measured voltage phasors are shown here. An *unbalanced event* is marked on this figure with two arrows. This event affects only two phases, A and B. It does *not* affect Phase C. The magnitude of the corresponding negative sequence, i.e., the magnitude of V^-, is shown in Figure 3.22(b). We can see that the event has a very clear signature in this figure on the negative sequence. Any event detection method, such as the MAD method that we learned in Section 2.7.2 in Chapter 2, can be applied to the time series in Figure 3.22(b) in order to detect this unbalanced event.

3.6.3 Voltage Unbalance Factor

Recall from Section 2.8.2 in Chapter 2 that phase unbalance can negatively affect the operation of induction motors. We previously discussed PU in (2.30) as a metric to quantify the extent of phase unbalance. PU can also be calculated based on phasor measurements; see Exercise 3.18.

Since motors are influenced mostly by the negative-sequence voltage, an alternative metric to PU is the Unbalance Factor (UF), which is defined as [128]:

$$\mathrm{UF} = \frac{|V^-|}{|V^+|} \times 100\%. \tag{3.52}$$

Here, we divide the magnitude of the negative sequence to the magnitude of the positive sequence. While negative-sequence voltages impact motors and other line-to-line connected loads the most, zero-sequence unbalance can affect line-to-ground connected three-phase loads. Accordingly, $|V^0|/|V^+|$ may also be of potential interest as yet another metric to examine unbalanced phasors.

Example 3.14 Again, consider the voltage phasor measurements in Example 3.12. From (3.48) and (3.52), we can obtain

$$\text{UF} = \frac{1.0257}{277.7978} \times 100\% = 0.37\%. \tag{3.53}$$

3.6.4 Phase Identification

Recall from Section 2.8.3 in Chapter 2 that phase identification is an important and challenging problem in power distribution systems. However, if phasor measurements are available at the distribution level, i.e., when we use D-PMUs, then phase identification becomes a rather trivial problem.

Figure 3.23 shows the histogram of one day of RPAD measurements between the voltage phasor at a single-phase load and the voltage phasors at each of the three phases at the substation. RPAD measurements concentrate around $-120°$ on Phase A, $120°$ on Phase B, and $0°$ on Phase C. It is clear that the load is connected to phase C, because there is only a very small phase angle difference between the unknown phase at the load location and Phase C at the substation.

It is worth clarifying that the reason to have two separate clusters of bars in Figures 3.23(a) and (b) is the operation of a switched capacitor bank on this distribution feeder, which is energized for only a portion of the day [139]. Depending on whether or not the capacitor bank is energized, there can be some changes in the measured RPAD. However, these changes (about $2°$) are very small compared to the roughly $120°$ phase angle difference between any two phases that are *not* connected to each other. Thus, the operation of the capacitor bank in this example does *not* cause any ambiguity in phase identification based on RPAD.

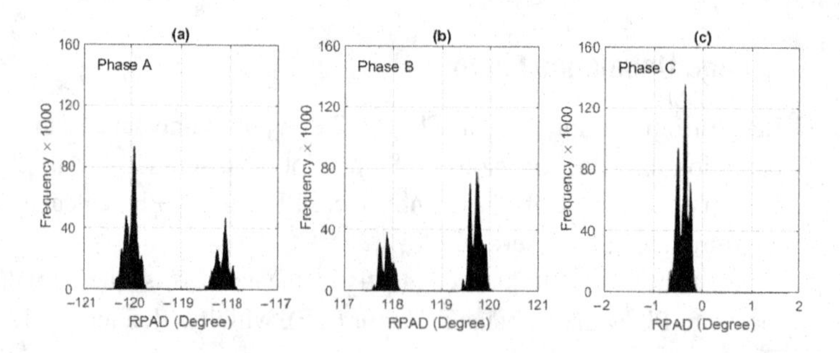

Figure 3.23 Phase identification by measuring RPAD between the unknown phase and each of the three known phases: (a) Phase A; (b) Phase B; (c) Phase C [139].

By using the above analysis, any change in phase connections in a power distribution system can be identified by D-PMUs almost instantly. Therefore, D-PMUs are ideal sensors to solve the phase identification problem. However, given the high cost of D-PMU, utilities may choose not to install D-PMUs at each load location across the distribution feeder. If only a few D-PMUs are available, then a combination of the above method together with the methods in Section 2.8.3 in Chapter 2 can be used to solve the phase identification problem.

We will also discuss phase identification in Section 5.9.2 in Chapter 5, and also in Sections 6.2 and 6.7 in Chapter 6.

3.7 Events in Phasor Measurements

We previously defined *events* in power systems as any change in any component that is worth studying; see Section 2.7 in Chapter 2. Events in power systems may affect both the magnitude and the phase angle in voltage or current phasors. When it comes to capturing the impact of an event on the *magnitude* of voltage or current, PMUs provide the same type of information that one can get from the conventional voltage and current sensors that we discussed in Chapter 2. However, what is truly unique about PMUs is their ability also to directly capture the impact of events on the *phase angle* of voltage or current. Therefore, in this section we focus on the analysis of events in phase angle measurements.

We will also discuss event classification. Together, the *event detection* methods that we learned in Section 2.7.2 in Chapter 2 and the *event classification* methods that we will learn in this section provide us with some basic tools that we can use to capture, analyze, and scrutinize different types of events.

3.7.1 Analysis of Events in Phase Angle Measurements

PMUs can reveal the impact of events on not only the magnitude but also the phase angle of voltage or current. An example is shown in Figure 3.24. Here, we examine the same event that we previously studied in Example 2.12 in Section 2.5.1 in Chapter 2. Recall that the event in this example is a *fault* on a power distribution feeder. The fault is caused by animal contact. We previously looked at the voltage measurements during this event at three different sensor locations: (1) at a load location on the same feeder where the fault occurs; (2) at the distribution substation that supplies the feeder where the fault occurs; and (3) at another substation that is located several miles away from the feeder where the fault occurs.

The magnitude and the phase angle measurements of the voltage phasors at the three aforementioned sensor locations are shown in Figures 3.24(a)–(c) and (d)–(f), respectively. We can see that the signature of the event is clearly visible on both magnitude and phase angle. Phase angle is affected not only at the first two sensor locations, which are close to the faulted location, but also at the third sensor location, which is relatively far from the faulted location.

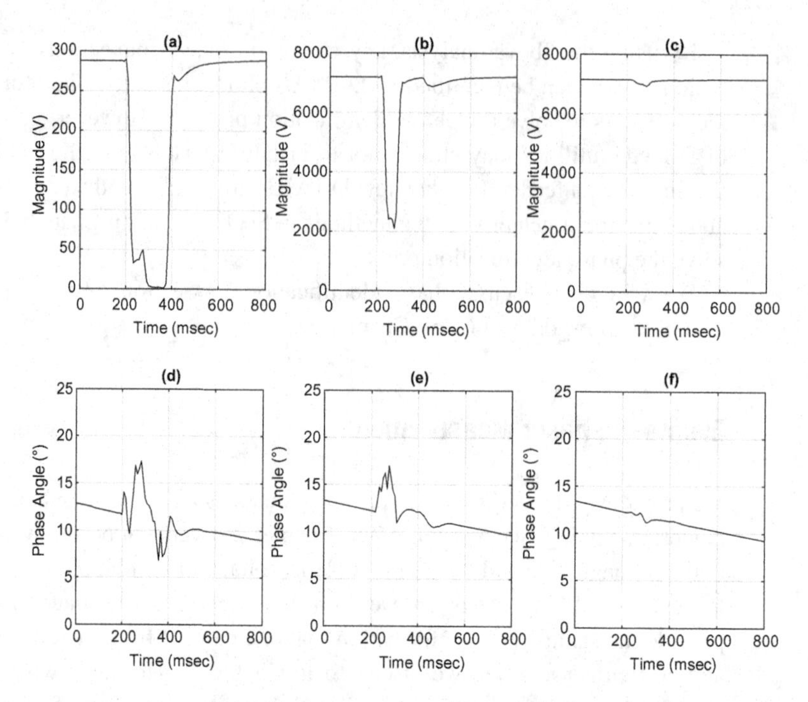

Figure 3.24 Phasor measurements during the animal-caused fault in Example 2.12 in Chapter 2: (a)(b)(c) magnitude of voltage phasors at the three sensor locations; (d)(e)(f) phase angle of the voltage phasors at the same three sensor locations [92].

Unwrapping Measurements

Consider the event that is captured in the phase angle measurements in Figure 3.25(a). At first glance, it may appear that the phase angle drops suddenly and drastically during this event. However, that is not really the case. In fact, the sudden drop in phase angle in this figure is simply due to the fact that the phase angle measurements in this example are wrapped around at 180° down to −180°.

The unwrapped version of the same event is shown in Figure 3.25(b). We can see that the event is still significant in its impact on phase angle, but it does not involve the type of sudden drop that the graph in Figure 3.25(a) may misleadingly suggest.

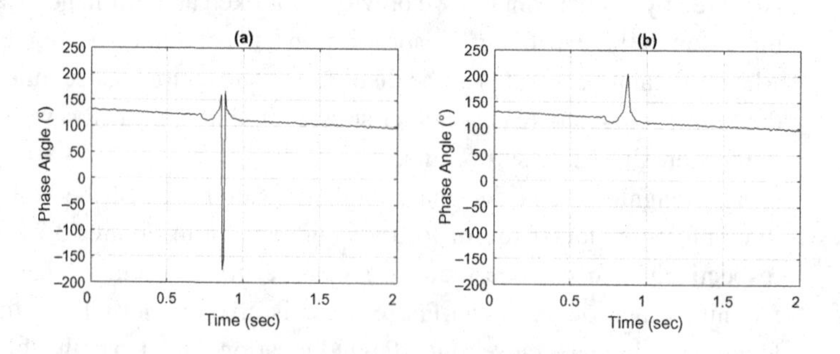

Figure 3.25 Events in phase angle measurements: (a) wrapped; (b) unwrapped.

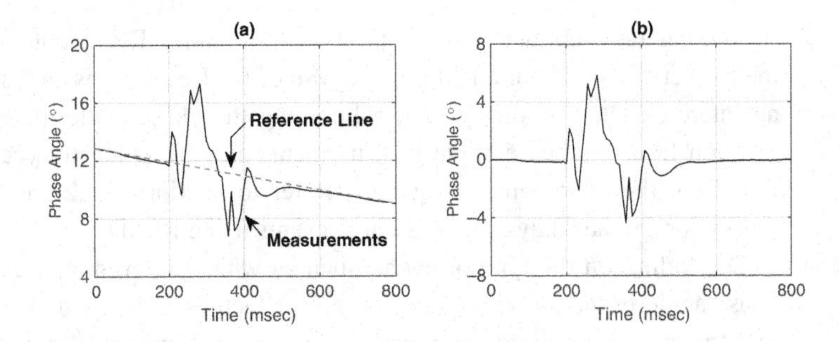

Figure 3.26 Removing the impact of off-nominal frequency from phase angle measurements during an event: (a) the reference line; (b) the result of subtraction.

Removing the Impact of Off-Nominal Frequency

Recall from Section 3.3 that the frequency of the power system directly affects how phase angle is measured. This creates some challenges when it comes to the analysis of events based on phase angle measurements. For example, consider the phase angle measurements in Figures 3.24(d),(e),(f). We can observe *two different types of changes* in the measurements in each of these three figures.

On one hand, some changes in the phase angle measurements *are* caused by the event; they appear between time $t = 200$ msec and time $t = 500$ msec. On the other hand, there are also some changes in the phase angle measurements that *are not* caused by the event; they take place *gradually* and for the entire period between time $t = 0$ msec and time $t = 800$ msec, which of course also includes the period between time $t = 200$ msec and time $t = 500$ msec.

We need a mechanism to distinguish the above two types of changes in phase angle measurements, because only the former changes represent the event. The latter changes represent only the off-nominal frequency of the power system.

One option to resolve this issue is to *remove* the impact of the off-nominal frequency. This can be done by forming an approximate *reference line*, an example of which is shown in Figure 3.26(a). The reference line, which is represented by a dashed line, is the line that goes through the point corresponding to the first measurement at time $t = 0$ on the top left of the figure and the point corresponding to the last measurement at time $t = 800$ msec on the bottom right of the figure. The purpose of this reference line is to approximate the impact of the off-nominal frequency on phase angle measurements. Accordingly, by *subtracting* the reference line from the measurements, i.e., by subtracting the dashed line from the solid line, we can obtain the curve shown in Figure 3.26(b). This curve is a good approximation of the changes in the phase angle measurements that are due *solely* to the *event* and not to the off-nominal frequency.

RPAD and Power Factor

An alternative to explicitly removing the impact of the off-nominal frequency is to use quantities that involve phasor angle measurements but are less sensitive to such impact. One example is RPAD. Another example is power factor.

The inherent subtraction in the process of obtaining RPAD automatically removes the impact of the off-nominal frequency. Since the frequency is almost the same across an interconnected power system, subtracting the phase angle measurements at one location from the phase angle measurements at another location can *cancel out* the impact of the off-nominal frequency. Refer to Sections 3.4.2 and 3.4.3 for further discussion on the analysis of wide-area events using RPAD.

Depending on the type of event, such as when the event causes a change in the phase angle in the *phasor of current*, power factor can be used as a measure to capture the change in phase angle with little impact from the off-nominal frequency of the system. Suppose $V\angle\theta$ and $I\angle\phi$ denote the phasor measurements for voltage and current, respectively. Recall that power factor is obtained as

$$\mathrm{PF} = \cos(\theta - \phi). \tag{3.54}$$

Again, the inherent subtraction in the process of obtaining the power factor automatically removes the impact of the off-nominal frequency. We will discuss the impact of events on the power factor in Section 5.2.1 in Chapter 5.

3.7.2 Event Classification

Recall from Example 2.11 in Chapter 2 that scatter plots can help us *visually* classify events into different groups that have *similar features*. In this section, we discuss how event classification can be done rather automatically.

Feature Selection

A critical task in classification is to choose adequate *quantitative features* that can help us distinguish among different classes. For instance, in Example 2.11 in Chapter 2, we created a scatter plot based on two features, ΔI_{inrush} and ΔI_{steady}.

When it comes to phasor measurements, the features can be defined as [124]:

$$V, \; I, \; \cos(\theta - \phi), \tag{3.55}$$

which include the magnitude of voltage, the magnitude of current, and the power factor, i.e., the cosine of the phase angle difference between voltage and current. An alternative to (3.55) for choosing the features is the following [118]:

$$V, \; I, \; VI\cos(\theta - \phi), \; VI\sin(\theta - \phi). \tag{3.56}$$

where the power factor is replaced with *active power* and *reactive power*.

Due to the impact of off-nominal frequency on measuring phase angles, phase angle measurements are usually *not* taken as features by themselves. Instead, we can use *phase angle differences* as features. One option is to use the phase angle difference between the voltage phasor and the current phasor, as in (3.55) and (3.56). Another option is to use the phase angle difference between two voltage phasors, i.e., as in RPAD. In fact, RPAD can be a useful feature for classification, for example, when we seek to classify inter-area oscillations; see Section 3.4.3.

The features in (3.55) and (3.56) can be defined *separately* on each phase (A, B, and C), or as the *average* across all phases. Features in *three-phase events* may also include symmetrical components. In fact, when it comes to the classification of *unbalanced events*, the magnitude of the *negative sequence*, i.e., V^-, and the magnitude of the *zero sequence*, i.e., V^0, can be useful features, because they reveal the unbalanced nature of the event; see Example 3.13 in Section 3.6.2.

Some arithmetic or statistical calculations based on the above various quantities can also be used as features. For example, one can use the *sign*, the *absolute value*, or the *differential* of the above features as alternative or additional features. One can also use the *minimum*, the *maximum*, the *mean*, the *median*, the *median absolute deviation*, or the *variance* of the above features over a given window of time as alternative or additional features.

Once all the features are selected, we can represent each event by the vector of its features. For instance, in Example 2.11 in Chapter 2, each event can be represented by the following vector, which contains two features:

$$\mathbf{x} = \begin{bmatrix} \Delta I_{\text{inrush}} \\ \Delta I_{\text{steady}} \end{bmatrix}. \tag{3.57}$$

Or, if we use the features in (3.55) for the events that are captured in phasor measurements $V\angle\theta$ and $I\angle\phi$, then each event can be represented by

$$\mathbf{x} = \begin{bmatrix} V \\ I \\ \cos(\theta - \phi). \end{bmatrix}. \tag{3.58}$$

The size of the vector of features is equal to the number of selected features.

Using SVM Classifier for Event Classification

Support vector machines (SVMs) are commonly used in machine learning as a tool for data classification [172–175]. In its basic form, SVM is a *linear* classifier. It works based on obtaining a *separating hyperplane* to classify the data.

As shown in Figure 3.27, the role of a separating hyperplane is to divide the data into *two* classes, which in this figure are denoted by Class I and Class II. All the points on one side of the hyperplane belong to the same class. If the events are represented by *two* features, then the separating hyperplane is a *line*, as in Figure 3.27(a). If the events are represented by *three* features, then the separating hyperplane is a *plane*, as in Figure 3.27(b). We can similarly define the separating hyperplanes in higher dimensions over *any number* of features.

The separating hyperplane has the following general formulation:

$$\mathbf{a}^T\mathbf{x} + b = 0, \tag{3.59}$$

where $\mathbf{a}$ is the vector of all the coefficients and b is a scalar. The size of vector $\mathbf{a}$ is the same as the size of vector $\mathbf{x}$, which is equal to the number of selected features.

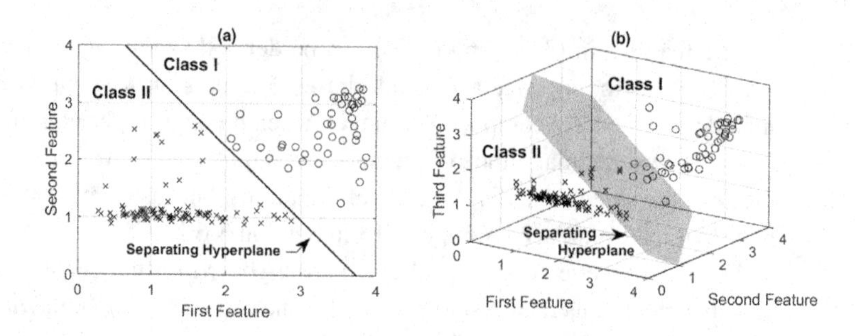

Figure 3.27 Separating hyperplane when the number of features is (a) two; (b) three.

Suppose we have captured an event, and its vector of features is $\mathbf{x}$. Suppose we do *not* know whether this event belongs to Class I or Class II. We can use the separating hyperplane in (3.59) to classify this event. If $\mathbf{a}^T\mathbf{x} + b < 0$, then the event belongs to Class I. If $\mathbf{a}^T\mathbf{x} + b > 0$, then the event belongs to Class II. If $\mathbf{a}^T\mathbf{x} + b = 0$, then the event can be placed in either class.

Example 3.15 Suppose we have captured four events. The number of features is two. These four events are represented by the following vectors of features:

$$\mathbf{x}_1 = \begin{bmatrix} 3.02 \\ 1.83 \end{bmatrix}, \quad \mathbf{x}_2 = \begin{bmatrix} 1.76 \\ 2.84 \end{bmatrix}, \quad \mathbf{x}_3 = \begin{bmatrix} 2.68 \\ 1.53 \end{bmatrix}, \quad \mathbf{x}_4 = \begin{bmatrix} 2.43 \\ 2.48 \end{bmatrix}, \tag{3.60}$$

respectively. Each event may belong to either Class I or Class II. Suppose we are provided with the following separating hyperplane for classification:

$$\mathbf{a} = \begin{bmatrix} -0.7902 \\ -0.6129 \end{bmatrix}, \quad b = 3.2635. \tag{3.61}$$

Accordingly, we can obtain

$$\begin{aligned}
\mathbf{a}^T\mathbf{x}_1 + b &= -0.2444 < 0 \;\rightarrow\; \text{Class I}, \\
\mathbf{a}^T\mathbf{x}_2 + b &= 0.1322 > 0 \;\rightarrow\; \text{Class II}, \\
\mathbf{a}^T\mathbf{x}_3 + b &= 0.2081 > 0 \;\rightarrow\; \text{Class II}, \\
\mathbf{a}^T\mathbf{x}_4 + b &= -0.1766 < 0 \;\rightarrow\; \text{Class I}.
\end{aligned} \tag{3.62}$$

Thus, we can conclude that the first and the fourth events belong to Class I, and the second and the third events belong to Class II.

Event classification can help put similar events into the same class so that we can focus on their characteristics, scrutinize them, and make conclusions.

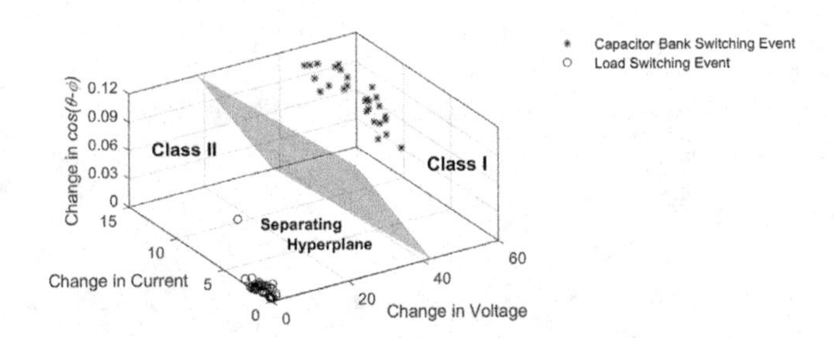

Figure 3.28 Classifying switching events in a power distribution system.

Example 3.16 Figure 3.28 shows two different types of switching events in a power distribution system: capacitor bank switching and load switching. They are characterized by three features based on the voltage and current phasor measurements at the substation: the change that is caused by the event in the magnitude of voltage; the change that is caused by the event in the magnitude of current; and the change that is caused by the event in the power factor. The events in Class I can be used to monitor the operation and the state of health of the capacitor bank; e.g., see the analysis in [152]. The events in Class II can be used to model the different types of loads in the system; e.g., see the analysis in [176].

The vectors of features for the two classes in Figure 3.28 are clearly separate from each other. However, that is not always the case, as we will see in the next example.

Example 3.17 Figure 3.29 shows two broad types of events in a three-phase power distribution system: balanced events and unbalanced events. Importantly, the labeling of these events is done manually by a power system expert who looked at the voltage phasor measurements and decided whether each event should be considered as a balanced event or an unbalanced event. The events are characterized by two features: the variance of the magnitude of the *negative sequence* of voltage over a window of two seconds; and the variance of the magnitude of the *zero sequence* of voltage over a window of two seconds. Here, the window of two seconds is basically the length of the window of the measurements that we previously used to detect the events; see Section 2.7.2 in Chapter 2 for the discussion on event detection. We can see in Figure 3.29(a) that the events that are labeled as unbalanced are scattered away from the origin, while the events that are labeled as balanced are concentrated close to the origin. However, unlike in Example 3.16, the two types of events are *not* separable based on their selected features. Nevertheless, we can still use a separating hyperplane to classify these events, as shown in Figures 3.29(b) and (c). Here, the classification is *not* precise, because some of the events show up on the *incorrect side* of the separating hyperplane. This is inevitable. These few events are considered to be *outliers*, as marked in Figure 3.29(c).

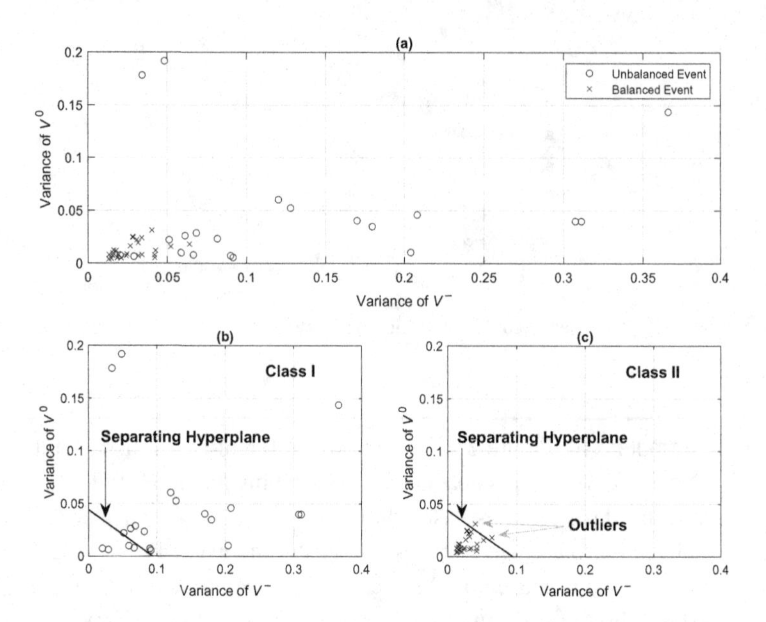

Figure 3.29 Classifying three-phase events as unbalanced and balanced: (a) all events; (b) the position of the separating hyperplane with respect to the unbalanced events; (c) the position of the separating hyperplane with respect to the balanced events.

One subtle point about Example 3.17 is that all the events in this example had *some* level of nonzero magnitude for the negative or zero sequences (or both). That means, *in theory*, they are all unbalanced events. However, *in practice*, many of these events are considered to be balanced events in the eyes of a power systems expert. The whole point of classification in this example is to figure out how to *mimic* the approach of the human intelligence that was taken by the power systems expert to label these events similarly, but rather automatically.

Training SVM Classifier for Event Classification

The above examples show how separating hyperplanes can be used for event classification. But how can we obtain the separating hyperplane?

In order to obtain $\mathbf{a}$ and b to model the separating hyperplane as in (3.59), we need to first manually label and classify a number of sample data.

Consider a set of m events that are already labeled as being in Class 1 or in Class 2. Each event i, where $i = 1, \ldots, m$, is represented by a vector of features $\mathbf{x}_i$. We define y_i as a *binary* variable to indicate the class of event i. If event i belongs to Class 1, then $y_i = -1$. If event i belongs to Class 2, then $y_i = 1$. We can use the following optimization problem to train the SVM classifier:

$$\begin{aligned} &\underset{\mathbf{a},b}{\text{minimize}} \quad \|\mathbf{a}\|_2 \\ &\text{subject to} \quad y_i\left(\mathbf{a}^T\mathbf{x}_i + b\right) \geq 1, \quad i = 1, \ldots, m. \end{aligned} \tag{3.63}$$

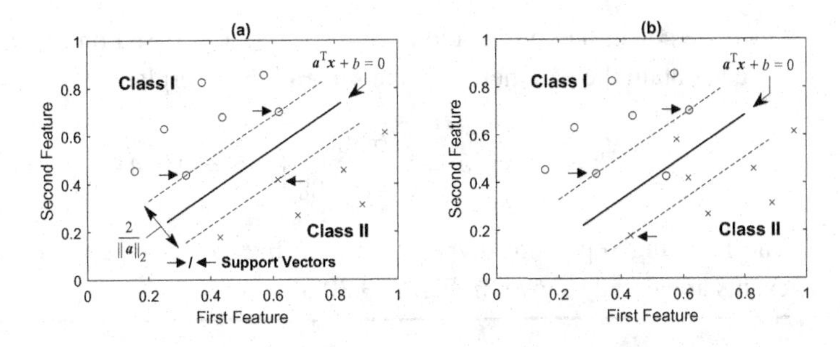

Figure 3.30 Training an SVM classifier with (a) hard margins based on (3.63), as in Example 3.18; (b) soft margins based on (3.67), as in Example 3.19.

The unknown optimization variables in (3.63) are $\mathbf{a}$ and b. The known parameters are $\mathbf{x}_i$ and y_i, for all $i = 1, \ldots, m$. For each event i, if $y_i = -1$, then the constraint in (3.63) requires that $\mathbf{a}^T \mathbf{x}_i + b \leq -1$; and if $y_i = 1$, then the constraint in (3.63) requires that $\mathbf{a}^T \mathbf{x}_i + b \geq 1$. Here, we enforce a *hard margin* of length 1 on each side of the separating hyperplane, to make a clear distinction between the events in Class I and the events in Class II. This is illustrated in Figure 3.30(a). Notice that there is no point inside the area between the two dashed lines.

As for the three points that are *on* one of the two dashed lines, i.e., the two circles and the one cross, they are referred to as the *support vectors* for the constructed separating hyperplane. That is why this method is referred to as SVM.

Equation (3.63) is a convex quadratic optimization problem [103]. It can be solved by using the convex programming solvers, such as the command `quadprog` in MATLAB [177] or by using the convex optimization toolbox CVX [178].

Example 3.18 Suppose we are provided with a set of 13 sample events to train an SVM classifier. The events are already labeled as belonging to either Class I (seven events) or Class II (six events). The number of features is two. The vectors of features for the seven events that belong to Class I are as follows:

$$\mathbf{x}_1 = \begin{bmatrix} 0.373 \\ 0.826 \end{bmatrix}, \quad \mathbf{x}_2 = \begin{bmatrix} 0.572 \\ 0.856 \end{bmatrix}, \quad \mathbf{x}_3 = \begin{bmatrix} 0.439 \\ 0.680 \end{bmatrix},$$
$$\mathbf{x}_4 = \begin{bmatrix} 0.620 \\ 0.702 \end{bmatrix}, \quad \mathbf{x}_5 = \begin{bmatrix} 0.251 \\ 0.631 \end{bmatrix}, \quad \mathbf{x}_6 = \begin{bmatrix} 0.321 \\ 0.437 \end{bmatrix}, \quad \mathbf{x}_7 = \begin{bmatrix} 0.156 \\ 0.454 \end{bmatrix}. \tag{3.64}$$

The vectors of features for the six events that belong to Class II are as follows:

$$\mathbf{x}_8 = \begin{bmatrix} 0.617 \\ 0.418 \end{bmatrix}, \quad \mathbf{x}_9 = \begin{bmatrix} 0.960 \\ 0.614 \end{bmatrix}, \quad \mathbf{x}_{10} = \begin{bmatrix} 0.829 \\ 0.457 \end{bmatrix},$$
$$\mathbf{x}_{11} = \begin{bmatrix} 0.681 \\ 0.268 \end{bmatrix}, \quad \mathbf{x}_{12} = \begin{bmatrix} 0.888 \\ 0.312 \end{bmatrix}, \quad \mathbf{x}_{13} = \begin{bmatrix} 0.431 \\ 0.178 \end{bmatrix}. \tag{3.65}$$

Once we solve the optimization problem in (3.63) based on the above training data, we can obtain the parameters of the separating hyperplane as

$$\mathbf{a} = \begin{bmatrix} 6.3004 \\ -7.1088 \end{bmatrix}, \quad b = 0.0841. \tag{3.66}$$

The resulting separating hyperplane and the vectors of features for all 13 training events are already shown in Figure 3.30(a).

The requirement to have a *hard margin* around the separating hyperplane based on the optimization problem in (3.63) may *not* always be achievable. For instance, recall from Example 3.17 that we *cannot* identify any line that can completely separate all the unbalanced events from all the balanced events in that example. In such cases, the optimization problem in (3.63) becomes infeasible, i.e., without solution.

The above issue can be resolved by replacing the hard margins in the SVM classifer with *soft margins*. This can be done by revising the formulation of the optimization problem in (3.63). The revised problem formulation is as follows:

$$\begin{aligned} \underset{\mathbf{a}, b, \zeta}{\text{minimize}} \quad & \|\mathbf{a}\|_2 + \lambda \sum_{i=1}^{m} \zeta_i \\ \text{subject to} \quad & y_i \left(\mathbf{a}^T \mathbf{x}_i + b \right) \geq 1 - \zeta_i, \quad i = 1, \ldots, m, \\ & \zeta_i \geq 0, \quad i = 1, \ldots, m. \end{aligned} \tag{3.67}$$

Here, for each event i, we define ζ_i as a *slack variable*, which measures the deviation from the initial hard margin of length 1 in order to achieve a feasible solution. Based on the second term in the objective function in (3.67), we seek to obtain the minimum for such a deviation. Parameter λ is a tuning parameter. If the selected features of a training event fall on the correct side of the separating hyperplane with respect to the label of the event, then $0 \leq \zeta_i < 1$; otherwise, $\zeta_i \geq 1$ [118]. In addition to resolving the issue with infeasible solutions that we had mentioned earlier, slack variables can help make the training process less sensitive to outliers, cf. [173, 175].

Equation (3.67) is also a convex quadratic optimization problem [103].

Example 3.19 Consider the same set of 13 training events in Example 3.18. Suppose we add the following two additional events to the training set:

$$\begin{aligned} \mathbf{x}_{14} &= \begin{bmatrix} 0.546 \\ 0.426 \end{bmatrix}, \quad y_{14} = -1 \\ \mathbf{x}_{15} &= \begin{bmatrix} 0.579 \\ 0.578 \end{bmatrix}, \quad y_{15} = 1. \end{aligned} \tag{3.68}$$

Event number 14 belongs to Class I, and event number 15 belongs to Class II. Equation (3.63) does *not* have a solution in this case. Instead, we should use Equation (3.67). We set $\lambda = 5$. The new separating hyperplane is obtained as

$$\mathbf{a} = \begin{bmatrix} 4.9723 \\ -5.6102 \end{bmatrix}, \quad b = -0.1444. \tag{3.69}$$

The slack variables are obtained as

$$\zeta_8 = 0.4216, \quad \zeta_{14} = 1.1805, \quad \zeta_{15} = 1.5082 \tag{3.70}$$

and

$$\zeta_i = 0, \quad i = 1, \ldots, 7, 9, \ldots, 13. \tag{3.71}$$

The resulting separating hyperplane is shown in Figure 3.30(b). Notice that the points corresponding to events 8, 14, and 15 are on the incorrect side of the separating hyperplane. These three events are considered as outliers.

Other Classifiers

The classification method that we discussed above was limited to having only two classes and also using only linear separation of the two classes. If the event classification involves more than two classes, then we can still use SVM classification, but we need to train *multiple separating hyperplanes*, where each separating hyperplane can separate one class from the rest of the classes. Furthermore, we can use nonlinear classification, where we replace the separating hyperplane with separating hyperbolic functions or polynomial functions; e.g., see [179–181].

SVM classification is considered a *supervised* learning method; because it requires us to first manually label some events to serve as training samples in order to obtain the classifier. There is a rich literature on supervised learning methods that can be considered for event classification; e.g., see [182–184].

There are also some powerful *unsupervised* classification methods that can be considered. In unsupervised classification, we do not need to first manually identify the samples of the data that belong to each class. We will see a popular method for *unsupervised* classification (clustering) in Section 5.4.3 in Chapter 5.

3.8 State and Parameter Estimation

PMU measurements can be used in a wide range of state and parameter estimation problems. In this section, we discuss the basics of state estimation using phasor measurements. We also see a few examples for parameter estimation, including the estimation of the topology of the power system. Some of the problems that we discuss in this section will also be discussed in future chapters for other types of measurements. For example, we will discuss the state estimation problem also in Section 4.5.1 in Chapter 4 and in Section 5.8 in Chapter 5.

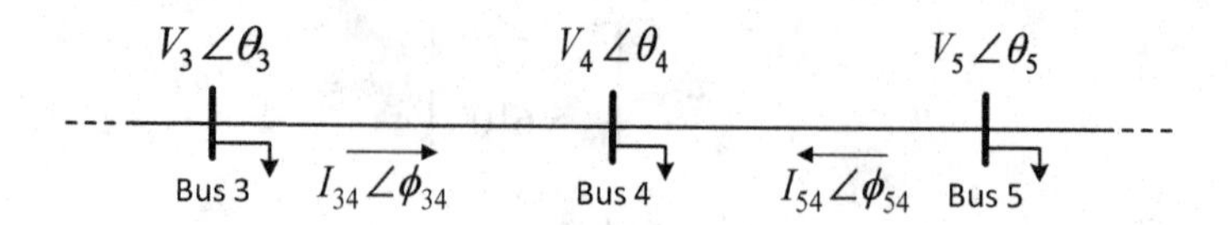

Figure 3.31 State estimation with voltage and current synchrophasors.

3.8.1 State Estimation Using PMU Measurements

State estimation in power systems is the problem of calculating the *states* of the system, i.e., voltage magnitudes and voltage phase angles at all buses, using proper measurements. Traditional state estimation, which we will discuss in Section 5.8 in Chapter 5, uses active and reactive power measurements. In contrast, in this section, the state estimation problem is solved by using PMU measurements.

If the PMU measurements that are available to us are limited to voltage synchrophasors (and no current synchrophasors), then there is practically no estimation, because the states are just *measured directly*. In such a case, the measurements do *not* provide *redundancy* for each other. If one PMU fails, then the measurements from other PMUs cannot be used to recover the troubled or missing data.

If *both* voltage and current synchrophasors are measured, then one can formulate a state estimation problem. For example, consider the network in Figure 3.31. The PMU at bus 3 measures V_3, θ_3, I_{34}, ϕ_{34}. The PMU at bus 5 measures V_5, θ_5, I_{54}, ϕ_{54}. No PMU is installed at bus 4. The following relationships hold:

$$I_{34}\angle\phi_{34} = (V_3\angle\theta_3 - V_4\angle\theta_4)\,(G_{34} + jB_{34}), \tag{3.72}$$

$$I_{54}\angle\phi_{54} = (V_5\angle\theta_5 - V_4\angle\theta_4)\,(G_{54} + jB_{54}). \tag{3.73}$$

Either (3.72) or (3.73) can be used individually to calculate the unknown states V_4 and θ_4. However, together, the two expressions in (3.72) or (3.73) can provide the redundancy that is needed to better address measurement noise. In fact, with the availability of current synchrophasor measurements, there is now redundancy to also estimate V_3, θ_3, V_5, and θ_5, even though they are measured directly.

Problem Formulation

The accuracy of a synchrophasor measurement is evaluated in terms of a *vector error*, which is a complex number indicating the difference between the measured synchrophasor and the true synchrophasor; see Section 3.9. Such a difference is easier to work with when it is presented in rectangular form $\epsilon_r + j\epsilon_i$. Let $\hat{X}\angle\hat{\theta}$ denote the measured synchrophasor of the true synchrophasor $X\angle\theta$. We define:

$$\hat{X}_r = \text{Re}\{\hat{X}\angle\hat{\theta}\} = \text{Re}\{X\angle\theta\} + \epsilon_r = X_r + \epsilon_r, \tag{3.74}$$

$$\hat{X}_i = \text{Im}\{\hat{X}\angle\hat{\theta}\} = \text{Im}\{X\angle\theta\} + \epsilon_i = X_i + \epsilon_i. \tag{3.75}$$

Subscripts r and i stand for real and imaginary, respectively. We can now relate the measurements to the state variables for the setup in Figure 3.31 as

$$\mathbf{z} = \mathbf{A}\mathbf{x} + \epsilon, \tag{3.76}$$

where

$$\mathbf{z} = \begin{bmatrix} \hat{V}_{r3} & \hat{V}_{i3} & \hat{I}_{r34} & \hat{I}_{i34} & \hat{V}_{r5} & \hat{V}_{i5} & \hat{I}_{r54} & \hat{I}_{i54} \end{bmatrix}^T, \tag{3.77}$$

$$\mathbf{x} = \begin{bmatrix} V_{r3} & V_{i3} & V_{r4} & V_{i4} & V_{r5} & V_{i5} \end{bmatrix}^T, \tag{3.78}$$

$$\epsilon = \begin{bmatrix} \epsilon^V_{r3} & \epsilon^V_{i3} & \epsilon^I_{r34} & \epsilon^I_{i34} & \epsilon^V_{r5} & \epsilon^V_{i5} & \epsilon^I_{r54} & \epsilon^I_{i54} \end{bmatrix}^T, \tag{3.79}$$

and

$$\mathbf{A} = \begin{bmatrix} 1 & 0 & 0 & 0 & 0 & 0 \\ 0 & 1 & 0 & 0 & 0 & 0 \\ G_{34} & -B_{34} & -G_{34} & B_{34} & 0 & 0 \\ B_{34} & G_{34} & -B_{34} & -G_{34} & 0 & 0 \\ 0 & 0 & 0 & 0 & 1 & 0 \\ 0 & 0 & 0 & 0 & 0 & 1 \\ 0 & 0 & -G_{54} & B_{54} & G_{54} & -B_{54} \\ 0 & 0 & -B_{54} & -G_{54} & B_{54} & G_{54} \end{bmatrix}. \tag{3.80}$$

Note that the expression in (3.76) is exact and *linear by construction*. That is, it is *not* a linearized approximation of any other expression.

The state estimation problem based on PMU measurements can be formulated as the following LS optimization problem:

$$\min_{\mathbf{x}} \|\mathbf{z} - \mathbf{A}\mathbf{x}\|_2 . \tag{3.81}$$

The solution of the above optimization problem is obtained as [103]:

$$\mathbf{x} = (\mathbf{A}^T\mathbf{A})^{-1}\mathbf{A}^T\mathbf{z}. \tag{3.82}$$

Any state estimation problem that is based solely on synchrophasor measurements can be formulated and solved similarly for any AC power network [140].

Example 3.20 Consider the 4-bus transmission network in Figure 3.32. Suppose we measure voltage synchrophasors at buses 1, 2, and 4, as well as current synchrophasors on transmission lines (1,3), (1,4), and (3,4). The true versus measured synchrophasors are listed in Table 3.3. The state estimation problem is formulated as in (3.81), where $\mathbf{x}$ is an 8×1 vector, $\mathbf{A}$ is a 12×8 matrix, and $\mathbf{z}$ is a 12×1 vector. The admittance for each transmission line is $0.5 - j10$ p.u. There is no shunt capacitor in the model of the transmission lines. By using the formulation in (3.82), the state estimation results are obtained as

$$V_1\angle\theta_1 = 0.8002 + j0.6505 = 1.0312\angle 39.1081^\circ, \tag{3.83}$$

$$V_2\angle\theta_2 = 0.8514 + j0.5181 = 0.9967\angle 31.3237^\circ, \tag{3.84}$$

$$V_3\angle\theta_3 = 0.8337 + j0.6340 = 1.0474\angle 37.2495^\circ, \tag{3.85}$$

$$V_4\angle\theta_4 = 0.8555 + j0.4641 = 0.9733\angle 28.4800^\circ. \tag{3.86}$$

The vector error for the estimated state at bus 3 is obtained as $1.0471\angle 50.5780^\circ - 1.0500\angle 50.3502 = -0.0051 + j0.0004$. Note that state estimation also estimates the states at buses 1, 2, and 4, which are also measured directly.

Table 3.3 The true and measured synchrophasors in Example 3.20.

Bus #	True Voltage (p.u.)	Measured Voltage (p.u.)
1	$1.0332\angle 38.8884^\circ$	$1.0297\angle 39.1346^\circ$
2	$0.9974\angle 31.5332^\circ$	$0.9967\angle 31.3237^\circ$
3	$1.0499\angle 37.0645^\circ$	–
4	$0.9755\angle 28.3416^\circ$	$0.9749\angle 28.4696^\circ$
Line #	**True Current (p.u.)**	**Measured Current (p.u.)**
1,2	$1.3522\angle 22.7486^\circ$	–
1,3	$0.3719\angle 67.6409^\circ$	$0.3707\angle 67.2757^\circ$
1,4	$1.9357\angle 19.2137^\circ$	$1.9519\angle 19.3943^\circ$
2,4	$0.5920\angle 11.1179^\circ$	–
3,4	$1.7117\angle 9.5889^\circ$	$1.7093\angle 10.1346^\circ$

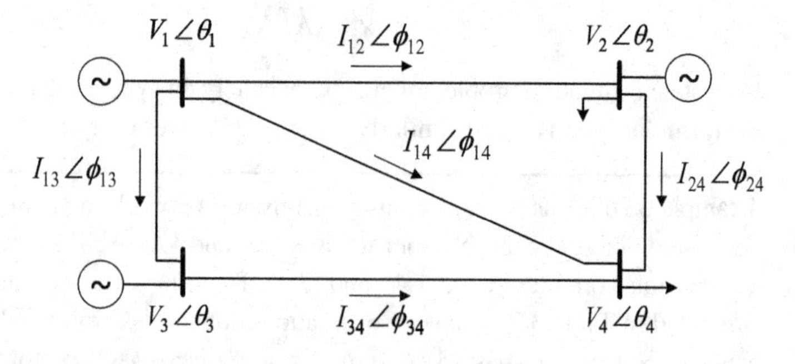

Figure 3.32 The 4-bus network in the state estimation problem in Example 3.20.

Incorporating the Distribution of Measurement Noise

The state estimation problem formulation in (3.81) is based on the assumption that all PMUs have the same level of measurement noise. However, if the PMUs have different probability distributions for their measurement noises that happen to be known in advance, then we can incorporate them into the state estimation problem formulation. For example, suppose the measurement noise follows a Gaussian distribution with zero mean and known variance. We can reformulate the state estimation problem in form of the following Weighted Least Square (WLS) problem:

$$\min_{\mathbf{x}} \|\mathbf{W}(\mathbf{z} - \mathbf{A}\mathbf{x})\|_2, \tag{3.87}$$

where $\mathbf{W}$ denotes the noise variance matrix, which is diagonal:

$$\mathbf{W} = \begin{bmatrix} 1/\sigma_1^2 & & 0 \\ & \ddots & \\ 0 & & 1/\sigma_m^2 \end{bmatrix}. \tag{3.88}$$

Here, m is the total number of PMUs and σ_i^2 is the variance for the measurement noise at PMU number i, where $i = 1, \ldots, m$. If a PMU has a large variance in measurement noise, then it is given a smaller weight in the WLS problem. The solution of the WLS problem in (3.87) is obtained as follows [103]:

$$\mathbf{x} = (\mathbf{A}^T \mathbf{W}\mathbf{A})^{-1}\mathbf{A}^T \mathbf{W}\mathbf{z}. \tag{3.89}$$

Observability

The ability to solve the state estimation problem for a given power system depends on the *number* and *location* of measurements, i.e., whether or not the unknown states or the unknown parameters are *observable* based on the available measurements. For instance, in Example 3.20, if we do not have access to the phasor measurements of current on transmission lines (1,3) and (3,4), then we cannot estimate the voltage phasor at bus 3. In that case, the state estimation problem in Example 3.20 would not be observable. When the states of the power system are not observable, it affects the *rank* of matrix $\mathbf{A}$; see Exercise 3.26.

Observability can be achieved by increasing the number of sensors and/or placing them at the right locations in the power system. In this regard, there is a rich literature on *sensor placement*; e.g., see [185–189].

Alternatively, or in addition to increasing the number of sensors, one can use techniques such as *probing* to enhance observability, as we will see in Chapter 6.

3.8.2 Parameter Estimation Using PMU Measurements

Another problem that is related to state estimation is *parameter estimation*. Recall that the system parameters, such as the admittance of conductors, are assumed to be known in state estimation. However, in practice, certain parameters may not be known,

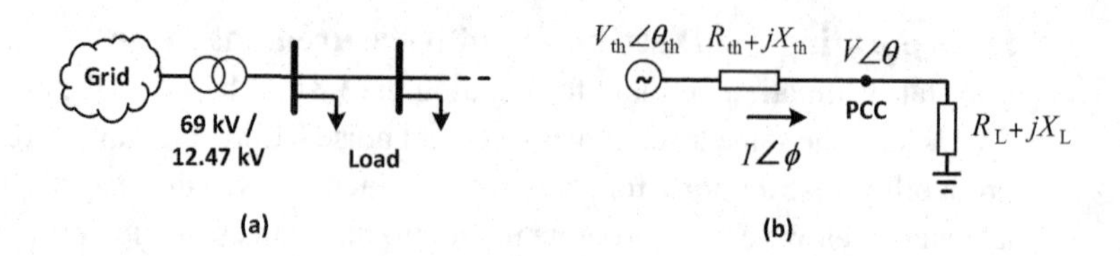

Figure 3.33 Parameter estimation using synchrophasor measurements at a PCC at a feeder head: (a) distribution network; (b) Thevenin equivalent circuit that is seen by the load.

or their values may change due to aging, atmospheric conditions, or grid conditions. Accordingly, parameter estimation might be necessary in order to calculate certain parameters of the power system by using measurements.

Consider the distribution network in Figure 3.33(a). Suppose the network is passive, i.e., it does not include any distributed generator. It only includes loads. The *load side* can be modeled as a load impedance $R_L + jX_L$. Also, the *utility side* can be modeled using its Thevenin equivalent circuit, as shown in Figure 3.33(b). Suppose a PMU is installed at the point of common coupling (PCC) to measure the voltage phasor $V\angle\theta$ and the current phasor $I\angle\phi$. On the load side, we can write:

$$V\angle\theta = (R_L + jX_L)\, I\angle\phi. \tag{3.90}$$

The above equation, together with the PMU measurements at PCC, are sufficient to estimate the load impedance. On the utility side, we can write:

$$V_{\text{th}}\angle\theta_{\text{th}} = V\angle\theta + (R_{\text{th}} + jX_{\text{th}})\, I\angle\phi. \tag{3.91}$$

The above equation, together with the PMU measurements at PCC, are *not* sufficient to estimate the utility impedance. This issue can be resolved if one can monitor the network under major load events, as explained in Example 3.21.

Example 3.21 Again, consider the network in Figure 3.33. Suppose there is an abrupt change in the load, e.g., due to a major load switching. Let $V^{\text{before}}\angle\theta^{\text{before}}$ and $I^{\text{before}}\angle\phi^{\text{before}}$ denote the PMU measurements *before* the load event; and $V^{\text{after}}\angle\theta^{\text{after}}$ and $I^{\text{after}}\angle\phi^{\text{after}}$ denote the PMU measurements *after* the load event. If the equivalent source $V_{\text{th}}\angle\theta_{\text{th}}$ and the equivalent impedance $R_{\text{th}} + jX_{\text{th}}$ do not change during the load event, then one can express (3.91) twice, i.e., based on the measurements *before* the load event and also based on the measurements *after* the load event. The two equations can then be solved to obtain [190]:

$$R_{\text{th}} + jX_{\text{th}} = \frac{V^{\text{after}}\angle\theta^{\text{after}} - V^{\text{before}}\angle\theta^{\text{before}}}{I^{\text{after}}\angle\phi^{\text{after}} - I^{\text{before}}\angle\phi^{\text{before}}}. \tag{3.92}$$

3.8.3 Topology Identification Using PMU Measurements

Topology identification (TI) is the problem of identifying the status of the switches in the network; thus, determining the correct network topology. The TI problem is a

Table 3.4 The synchrophasors that are studied in the topology identification problem.

Bus #	True Voltage (p.u.)	Measured Voltage (p.u.)
1	$1.0332\angle 38.8884°$	$1.0309\angle 39.0827°$
2	$0.9135\angle 16.9944°$	$0.9140\angle 16.8693°$
3	$1.0185\angle 34.5833°$	–
4	$0.9217\angle 22.2936°$	$0.9215\angle 22.4060°$
Line #	**True Current (p.u.)**	**Measured Current (p.u.)**
1,2	–	–
1,3	$0.7855\angle 28.7818°$	$0.7837\angle 28.4862°$
1,4	$3.0331\angle 12.0857°$	$3.0547\angle 12.2459°$
2,4	$0.8533\angle -163.0056°$	–
3,4	$2.2918\angle 6.4349°$	$2.2932\angle 6.9070°$

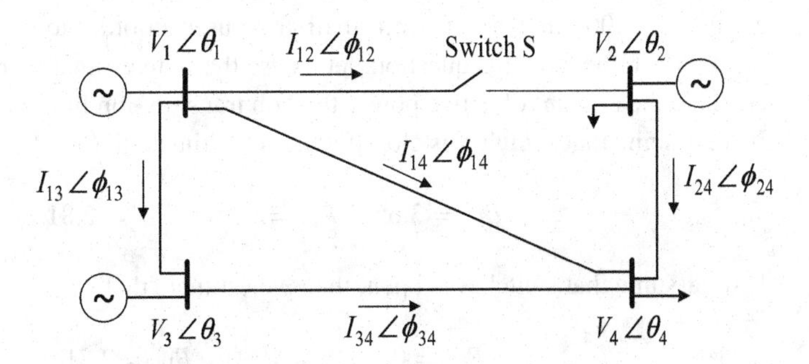

Figure 3.34 An example for the topology identification problem. There is a switch on line (1,2). If the switch is closed, then the network is the same as the one in Example 3.20.

special type of the parameter estimation problem, where the parameters to be estimated are the *closed* or *open* status for the switches in the network.

Again, consider the 4-bus power transmission network in Example 3.20 in Section 3.8.1. Suppose there is a switch, denoted by S, on transmission line (1,2), as shown in Figure 3.34. As in Example 3.20, suppose we measure voltage phasors at buses 1, 2, and 4 and current phasors on transmission lines (1,3), (1,4), and (3,4). Depending on the status of switch S, the phasor measurements can be very different. If switch S is *closed*, then the voltage and current phasors are as in Table 3.3. If switch S is *open*, then the voltage and current phasors are as in Table 3.4.

If there is a sensor to measure the current on transmission line (1,2), then we can directly check whether or not there is current flowing on this line. However, even without a direct measurement, we might be able identify the status of switch S and accordingly identify the topology of the network. This can be done by conducting state estimation results based on the measurements that are already available.

Suppose the results of the state estimation problem are as follows:

$$\begin{aligned} V_1\angle\theta_1 &= 1.0322\angle 39.0858^\circ, \\ V_2\angle\theta_2 &= 0.9140\angle 16.8693^\circ, \\ V_3\angle\theta_3 &= 1.0163\angle 34.7648^\circ, \\ V_4\angle\theta_4 &= 0.9203\angle 22.3802^\circ. \end{aligned} \tag{3.93}$$

If switch S is closed, then we expect the current phasor on line (1,2) to be:

$$I_{12}\angle\phi_{12} = (V_1\angle\theta_1 - V_2\angle\theta_2)\,(G_{12} + jB_{12})\,. \tag{3.94}$$

If switch S is open, then we expect that

$$I_{12}\angle\phi_{12} = 0\angle 0^\circ. \tag{3.95}$$

We need to decide whether (3.94) or (3.95) is correct.

Suppose we know that the capacity for active power generation of the power plant at bus 1 is 5.00 p.u. Can this information be used in order to decide between (3.94) or (3.95)? To answer this question, let us use the state estimation results in (3.93) and estimate the amount of active power flow on transmission lines (1,2), (1,3), and (1,4). If we assume that switch S is closed, then we estimate that

$$P_{12} = 3.66, \quad P_{13} = 0.80, \quad P_{14} = 2.81. \tag{3.96}$$

If we assume that switch S is open, then we estimate that

$$P_{12} = 0, \quad P_{13} = 0.80, \quad P_{14} = 2.81. \tag{3.97}$$

However, the estimations in (3.96) cannot be correct because they require the power plant at bus 1 to generate about 7.27 p.u. real power, which is way above its generation capacity at 5.00 p.u. Thus, we conclude that switch S is open.

The topology identification problem can sometimes be formulated as an optimization problem, whether as a stand-alone problem or as an extension of the state estimation problem. The idea is to introduce optimization variables that are *binary* such that they can capture the status of each switch; e.g., see [191, 192].

We will discuss the topology identification problem further in Section 4.5.2 in Chapter 4 and in Sections 6.2.1 and 6.7.1 in Chapter 6.

3.9 Accuracy in Synchrophasor Measurements

The assessment of the accuracy and performance of PMUs is an elaborate process. It requires taking into account the *performance class* of the PMU that is being used, as well as whether the phasor measurements are taken under *steady-state* operating conditions or *dynamic* operating conditions.

3.9.1 Performance Classes

According to the IEEE C37.118 Standard, the accuracy and performance of PMUs should be evaluated based on their class of performance. The above standard currently defines two classes of performance: P Class and M Class.

The P Class is intended for applications that require *faster response*. P Class PMUs are expected to have minimal internal filtering or other types of internal processing so as to speed up the reporting of the phasor measurements, even if it comes at the expense of some degraded measurement accuracy. The typical application of PMUs in this class is in power system *protection* applications [193, 194].

The M Class is intended for applications that require *higher accuracy*. M Class PMUs are expected to have enhanced internal filtering and other internal processing so as to improve measurement accuracy, even if it comes at the expense of some delay in reporting the measurements. The typical application of PMUs in this class is in *monitoring* systems, which require precise measurements [195, 196]. An example for the type of internal filters that might be different between the above two classes of PMUs is the details in their anti-aliasing filters; see Section 2 in Chapter 2.3.

Compared to the output of M Class PMUs, the output of P Class PMUs may contain additional aliasing components; see [141] for more details.

3.9.2 Steady-State Performance

Total Vector Error

The accuracy of a synchrophasor is most commonly evaluated in terms of its *total vector error* (TVE). Here, the *vector error* (VE) is the difference between the reported synchrophasor $\hat{X}\angle\hat{\theta}$ and the true synchrophasor $X\angle\theta$. VE is a complex number. TVE is then defined in form of a fraction as

$$\text{TVE} = \frac{|\hat{X}\angle\hat{\theta} - X\angle\theta|}{|X\angle\theta|}. \tag{3.98}$$

Example 3.22 A PMU reports a voltage phasor as $39657.214\angle 183.215°$. The true voltage phasor is $39832.582\angle 183.681°$. The TVE is obtained as

$$\begin{aligned}\text{TVE} &= \frac{|39657.214\angle 183.215782° - 39832.582\angle 183.680175°|}{|39832.582\angle 183.680175°|} \\ &= \frac{366.7791}{39832.582} = 0.92\%.\end{aligned} \tag{3.99}$$

Suppose $\hat{X} = X + \epsilon_m$ and $\hat{\theta} = \theta$, i.e., the error in phasor measurement is solely due to error in measuring magnitude. The TVE in this case is obtained as

$$\text{TVE} = \frac{|(X + \epsilon_m)\angle\theta - X\angle\theta|}{|X\angle\theta|} = \frac{|\epsilon_m|}{X}. \tag{3.100}$$

Next, suppose $\hat{X} = X$ and $\hat{\theta} = \theta + \epsilon_p$, i.e., the measurement error is solely due to error in measuring the phase angle. The TVE in this case is obtained as

$$\text{TVE} = \frac{|X\angle(\theta + \epsilon_p) - X\angle\theta|}{|X\angle\theta|} = 2|\sin(\epsilon_p/2)|. \tag{3.101}$$

The IEEE C37.118 Standard recommends limiting TVE at steady state to 1%. This requirement is the same for both P Class and M Class PMUs [141].

Frequency Error and Rate of Change of Frequency Error

Two other metrics that are often used to assess the accuracy of synchrophasor measurements are the *frequency error* (FE) and the *rate of change of frequency error* (RFE). FE is the difference between the theoretical frequency and the measured frequency for a given instant of time. RFE is the difference between the theoretical ROCOF and the measured ROCOF for a given instant of time. Both FE and RFE depend on the error in measuring the phase angle.

At steady-state conditions, it is recommended that FE be limited to 0.01 Hz for P class PMUs, and limited to 0.005 Hz for M class PMUs. It is also recommended that RFE be limited to 0.01 Hz/sec for both class P and class M PMUs [141].

Error in Time Synchronization

For accurate synchrophasor measurements, PMUs require reliable and accurate time synchronization, such as from the GPS satellites. Even a relatively small error in time synchronization may considerably increase TVE.

Example 3.23 At a 60 Hz power system, each cycle takes $1/60$ seconds $=$ 16.667 msec. Therefore, a 1 microsecond error in time synchronization results in the following error in measuring the phase angle:

$$\epsilon_p = 360^\circ \times 60 \times 10^{-6} = 0.022^\circ. \tag{3.102}$$

From (3.101), the corresponding TVE is about 0.04%.

3.9.3 Dynamic Performance

Certain metrics can be used to evaluate the accuracy and performance of PMUs during dynamic operating conditions.

Step Changes in Magnitude and Phase Angle

An important test to evaluate the dynamic performance of PMUs is to create a *step change* in either the magnitude or the phase angle of the phasor that is being measured; and then evaluate the corresponding phasor measurements that are reported by the PMU. Here, we discuss only a step change in the magnitude; however, a step change in the phase angle can be studied similarly.

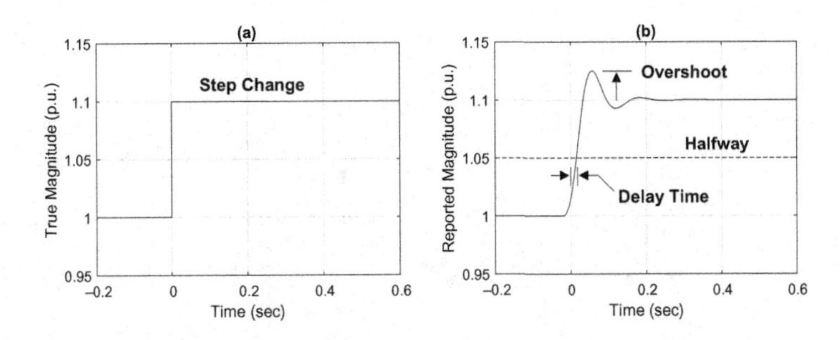

Figure 3.35 Dynamic accuracy assessment by doing a test with a step change: (a) the step change in the magnitude of the true phasor; (b) the reading of the PMU.

Consider a step change, from 1.0 p.u. to 1.1 p.u., that is made in the magnitude of a true phasor at time $t = 0$, as shown in Figure 3.35(a). Due to the internal dynamics of the PMU, what the PMU reports is *not* an identical instantaneous step change. It is rather a dynamic response, as shown in Figure 3.35(b).

The following three quantities are commonly examined in order to evaluate the dynamic response of a PMU under a step change [141]:

- Delay time;
- Overshoot (or undershoot);
- Response time.

The *delay time* is defined as the time interval between the instant that the step change occurs in the magnitude of the true phasor and the instant that the magnitude of the reported phasor measurement reaches a value that is *halfway* between the initial value and the final steady-state value. For the step change in Figure 3.35, the delay time is the time interval between $t = 0$ and $t = 0.016$ seconds, where the latter is the time instant at which the magnitude of the reported phasor measurement reaches 1.05 p.u., which is halfway between 1.0 p.u. and 1.1 p.u. Accordingly, the delay time in this example is 16 msec.

The *overshoot* is the maximum magnitude of the reported phasor measurements after the step change occurs until we reach the steady-state conditions. For the step change in Figure 3.35, the overshoot is 0.125 p.u. Note that the step change in Figure 3.35 is *positive*, i.e., the magnitude of the true phasor *increases* during the step change. If the step change is *negative*, i.e., the magnitude of the true phasor *decreases* during the step change, then we must check the *undershoot* instead of the overshoot.

The *response time* is the time it takes for the reported phasor measurements to reach the steady-state conditions after the step change. Here, the steady-state conditions are defined based on a specified accuracy limit. For instance, recall from Section 3.9.2 that TVE under steady-state conditions must be limited to 1%. Thus, the response time with respect to TVE is defined as the time interval between the instant that the measurements leave the required 1% limit in TVE due to the step change and the instant that they reenter and stay within the 1% limit in TVE. This is illustrated in Figure 3.36, which shows the TVE during the step change in Figure 3.35. The TVE is

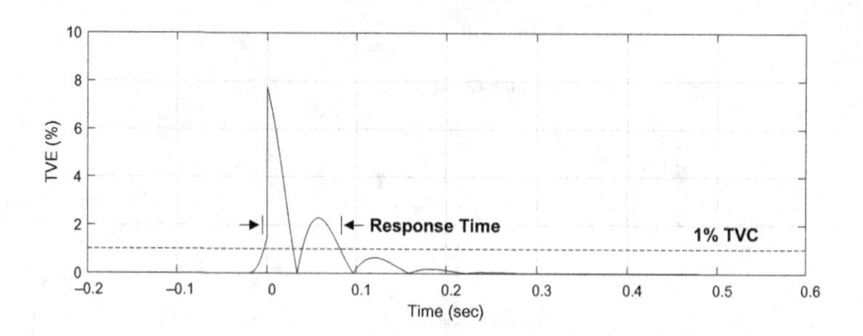

Figure 3.36 The TVE during the step change in Figure 3.35. The response time with respect to TVE is calculated based on the time interval when TVE exceeds 1% [141].

initially almost zero at steady state, well below the 1% limit. The step change causes the TVE to leave the 1% limit at $t = -0.003$ and spike at $t = 0$. It takes until $t = 0.082$ seconds for the TVE to go back to and stay below the 1% limit. The response time with respect to TVE is $0.082 - (-0.003)$ seconds = 85 msec.

It should be noted that, due to the common use of averaging filters in PMUs, even some of the readings of the phasors that are associated with a few instances *before* the step change are also affected by the step change. That is why the TVE starts increasing even for phasor measurements that are time stamped slightly before the step change occurs, i.e., at the negative time instances.

A P Class PMU must limit the overshoot or undershoot to 5% of the step change, the delay time to 1 divided by four times the reporting rate, and the response time to 1.7 divided by the nominal frequency of the system. For instance, consider the step change in Figure 3.35. Suppose the system frequency is 60 Hz and the reporting rate is 10 fps. A P Class PMU must limit the overshoot to $0.05 \times 0.1 = 0.005$ p.u., the delay time to $1/(4 \times 10) = 25$ msec, and the response time to $1.7/60 = 28.3$ msec. Here, the response time is the same with respect to the TVE requirements, the FE requirements, and the RFE requirements.

An M Class PMU has similar requirements as above, *except for the response time*. At 10 fps, the response time to meet the TVE requirement, the FE requirement, and the RFE requirement is 595 msec, 869 msec, and 1.038 seconds, respectively. We can see that these numbers are *much higher* than the 25 msec limit on the response time that we calculated in the previous paragraph at the same reporting rate for a P Class PMU. This is one of the main differences between the P Class and M Class PMUs. More details on the required maximum response time for M Class PMUs are available in Table 11 in [141].

Ramp Changes in System Frequency

The dynamic accuracy of phasor measurements also is usually tested during a *ramp change* in the true system frequency. The goal is to determine whether the TVE, FE, and RFE continue to stay within their required limits while the system frequency gradually changes. The details are available in [141].

3.9.4 Accuracy of Three-Phase Synchrophasors

The accuracy assessment methods that we discussed in Sections 3.9.2 and 3.9.3 are often used based on the assumption that either the true system is balanced or our focus is on measuring the symmetrical components. For example, both the step change and the ramp change that we discussed in Section 3.9.3 are explicitly required by the IEEE C37.118 Standard to be made by applying balanced three-phase step changes to balanced three-phase true phasors.

However, in principle, the same methods can also be used to assess the accuracy of the phasor measurements on each phase, A, B, and C.

Example 3.24 Suppose a PMU reports the three-phase voltage measurements as in (3.47) in Example 3.12 in Section 3.6. Suppose the *true* voltage phasors are

$$\begin{aligned} V_A &= 277.3742\angle{-73.1342^\circ}, \\ V_B &= 277.0118\angle 167.2712^\circ, \\ V_C &= 279.2083\angle 47.1329^\circ. \end{aligned} \tag{3.103}$$

The true positive sequence component is calculated as

$$V^+ = 277.8636\angle{-72.9100^\circ} \tag{3.104}$$

From (3.48) and (3.104), the TVE for the positive sequence is obtained as

$$\text{TVE}^+ = \frac{|277.7978\angle{-72.9014^\circ} - 277.8636\angle{-72.9100^\circ}|}{|277.8636\angle - 72.9100^\circ|} = 0.028\%. \tag{3.105}$$

From (3.47) and (3.103), the TVE for individual phases are obtained as

$$\begin{aligned} \text{TVE}_A &= \frac{|278.0574\angle{-73.1170^\circ} - 277.3742\angle{-73.1342^\circ}|}{|277.3742\angle{-73.1342^\circ}|} = 0.248\%, \\ \text{TVE}_B &= \frac{|276.8067\angle 167.2848^\circ - 277.0118\angle 167.2712^\circ|}{|277.0118\angle 167.2712^\circ|} = 0.078\%, \\ \text{TVE}_C &= \frac{|278.5330\angle 47.1288^\circ - 279.2083\angle 47.1329^\circ|}{|279.2083\angle 47.1329^\circ|} = 0.242\%. \end{aligned} \tag{3.106}$$

It is possible that the TVE for the positive sequence is below the limit that is required in the standard; but the TVE for some phases do exceed such limit.

3.9.5 Accuracy of RPAD and PD

The existing phasor measurement standards may not cover the accuracy of all the quantities that we discussed in this chapter. For example, TVE does not directly tell us what to expect for the accuracy of the RPAD measurements (see Section 3.4) or the accuracy of the PD measurements (see Section 3.5.1).

The accuracy of RPAD depends on the accuracy of measuring phase angles at the two locations where the voltage phasors are measured. If we use two PMUs, each of which is guaranteed to limit the error in phase angle measurements by 1°, then the error in RPAD is guaranteed to be limited to $2^\circ = 1^\circ + 1^\circ$.

As for the accuracy of PD, there is generally no direct relationship between the TVE for the two phasor measurements before and after the event and the TVE for the PD that is calculated based on the two phasor measurements [197].

Example 3.25 Suppose a PMU reports the voltage phasors before and after an event, as in (3.19) and (3.20), respectively. Suppose the *true* voltage phasors are

$$V^{\text{before}} = 7289.3\angle 339.79^\circ \tag{3.107}$$

and

$$V^{\text{after}} = 7237.9\angle 335.01^\circ. \tag{3.108}$$

Accordingly, the true phasor differential is obtained as

$$\Delta V = 7237.9\angle 335.01^\circ - 7289.3\angle 339.79^\circ = 607.98\angle -117.45^\circ. \tag{3.109}$$

The true phasor differential in (3.109) should be compared with the measured phasor differential in (3.21). While the TVE for the voltage phasor measurements, both before and after the event, is small and below 1%, the TVE for the phasor differential is very large:

$$\text{TVE}_{\Delta V} = \frac{|477.6\angle -118.35^\circ - 607.98\angle -117.45^\circ|}{|607.98\angle -117.45^\circ|} = 21.48\%. \tag{3.110}$$

Exercises

3.1 Verify that $\text{Re}\{X\angle\theta\}$ in (3.1) is equal to $X\cos(\theta)$ and $\text{Im}\{X\angle\theta\}$ in (3.2) is equal to $X\sin(\theta)$. Note that X is the phasor magnitude in RMS value.

3.2 Use (3.1) and (3.2) to obtain voltage phasor $V\angle\theta$ corresponding to the voltage wave in file `E3-2.csv`. Use a sampling rate of $N = 512$ samples per cycle; and a reporting rate of 60 fps, i.e., one phasor reading per cycle.

3.3 Repeat Exercise 3.2 for the voltage measurements in file `E3-3.csv`. Explain how the results here are different from the results in Exercise 3.2.

3.4 Repeat Exercise 3.2 for the voltage measurements in file `E3-4.csv`. Explain how the results here are different from the results in Exercise 3.2.

3.5 Consider the phase angle measurements in file `E3-5.csv`, which are given in the -180° to 180° range. The time stamps are in milliseconds in UTC.

(a) Present the measurements in the 0° to 360° range.

(b) Present the time stamps in year/month/day hour:minute:second.

3.6 Consider the measurements in Table 3.5 that are recorded by a PMU.

(a) What is the reporting rate of this PMU?

(b) Use (3.6) to estimate the frequency at each measurement.

Table 3.5 Voltage synchrophasor measurements in Exercise 3.6

UTC Time (millisecond)	Magnitude (V)	Phase Angle (°)
1286210900930	39830.183	359.7297
1286210900980	39831.177	359.8871
1286210901030	39832.088	0.0464
1286210901080	39833.174	0.1979

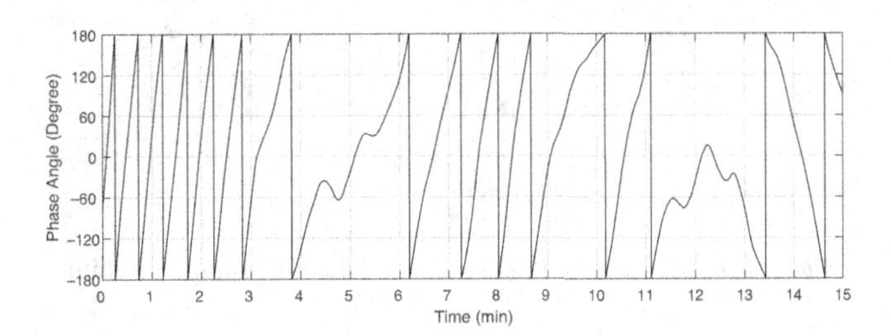

Figure 3.37 Voltage phase angle measurements in Exercise 3.7.

3.7 Consider the voltage phase angle measurements in Figure 3.37.
(a) How many times does frequency cross 60 Hz, the nominal frequency?
(b) Is the frequency mostly above 60 Hz, or below 60 Hz?

3.8 Consider the voltage phase angle measurements in file `E3-8.csv`.
(a) Estimate frequency using the linear model in (3.6).
(b) Apply the noise reduction command `smoothdata` in MATLAB [198] to the estimated frequency signal that is obtained in Part (a). Use the smoothing method `movmedian` and set the length of the `window` to be five.
(c) Plot the second-by-second average of the estimated frequency; before noise reduction, as in Part (a); and also after noise reduction, as in Part (b).

3.9 Again, consider the voltage phase angle measurements in Exercise 3.8. Estimate the frequency by applying the quadratic model in (3.11) to measurement windows of length one second, i.e., parameter $N = 20$.

3.10 Consider the phase angle measurements from two PMUs in file `E3-10.csv`.
(a) Plot the RPAD between the two PMUs versus time.
(b) How many events do you identify?
(c) How large in degrees is the size of each event?

3.11 Suppose two PMUs are installed at two ends of a 69 kV transmission line. The reactance of the transmission line is 0.541Ω. The voltage synchrophasor reading of the first PMU, on Phase A, is as in the first row in Table 3.1 in Section 3.2.1. The voltage synchrophasor reading of the second PMU, on Phase A, is as in Table 3.6. Estimate the amount and the direction of real power flow on this line. Assume that the transmission line is balanced.

3.12 A D-PMU is installed at the secondary side of a 69 kV–12.47 kV transformer at a power distribution substation. The voltage magnitude and the current

Table 3.6 Voltage synchrophasor reading at the second PMU in Exercise 3.11.

UTC Time (micro-second)	Magnitude (V)	Phase Angle (°)
1579101467916666	39821.487	183.318749

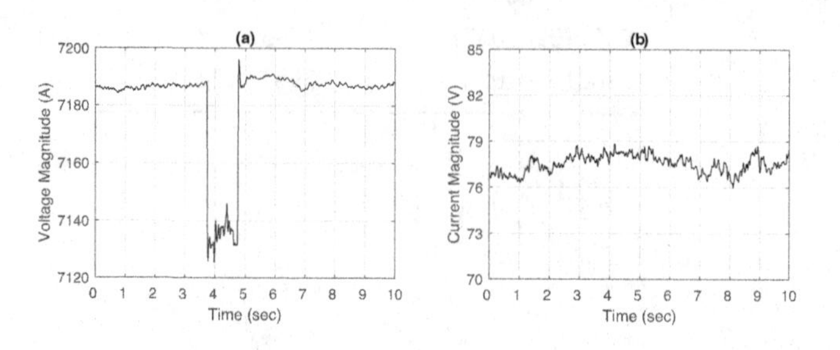

Figure 3.38 Measurements in Exercise 3.12: (a) voltage magnitude; (b) current magnitude.

magnitude that this D-PMU measures during a voltage event are shown in Figure 3.38. Explain the likely cause of the voltage event.

3.13 Obtain the frequency of *oscillations* in the RPAD measurements in file `E3-13.csv` using Fourier analysis. Indicate the likely cause of oscillations.

3.14 File `E3-14.csv` contains the current phasor measurements during an event. Obtain I^{before}, I^{after}, and ΔI; both *without* and *with* taking into account the impact of an off-nominal frequency that we discussed in Section 3.5.2.

3.15 File `E3-15.csv` contains the synchronized voltage phasor measurements that are obtained by two different PMUs during the same event.
(a) Obtain the differential synchrophasors corresponding to this event.
(b) Obtain the RPAD during this event.
(c) Compare your observations in Parts (a) and (b).

3.16 Consider the event location identification problem in Example 3.11. Suppose a different event occurs on the same power distribution feeder; and the synchrophasor measurements are obtained as in Table 3.7.
(a) Obtain $\Phi_1, \Phi_2, \ldots, \Phi_5$ as defined in (3.34).
(b) What bus is the location of this switching event?

3.17 Use (1.32) and (1.33) in Chapter 1 and also the definition of symmetrical components in (3.40) to show that the zero sequence and the negative sequence are both zero for balanced three-phase phasor measurements.

3.18 Consider the three-phase voltage synchrophasor measurements in `E3-18.csv`.
(a) Plot V^+, V^-, and V^0.
(b) Plot UF and obtain the maximum UF.
(c) Plot PU and obtain the maximum PU.

3.19 File `E3-19.csv` contains the current phasor measurements during a three-phase event. Is the event in these measurements a balanced or unbalanced event?

Table 3.7 Synchrophasor measurements in per unit in Example 3.16.

Phasor	Before	After
$V_1\angle\theta_1$	$0.9889\angle 28.796°$	$0.9958\angle 28.763°$
$I_{12}\angle\phi_{12}$	$3.1619\angle 1.354°$	$2.9292\angle 18.458°$
$V_5\angle\theta_5$	$0.9534\angle 27.138°$	$0.9741\angle 26.663°$
$I_{45}\angle\phi_{45}$	$0.7537\angle 8.703°$	$0.7701\angle 8.228°$

Elaborate your answer based on examining the magnitudes of the symmetrical components and also by examining the UF plot.

3.20 File `E3-20.csv` contains 10 minutes of synchrophasor voltage measurements at a three-phase reference location at a power distribution substation and also at a single-phase load location. All measurements are in per unit. The phase connection of the single-phase load is unknown. Identify the phase connection for the load by doing the following:

(a) Calculate the correlation coefficient between the voltage magnitudes, similar to the analysis in Section 2.8.3 in Chapter 2.

(b) Compare the voltage phase angles, as in Section 3.6.4.

3.21 File `E3-21.csv` contains phase angle measurements during an event.

(a) Plot the original phase angle measurements.

(b) Plot the unwrapped phase angle measurements.

(c) Plot the phase angle measurements after removing the impact of off-nominal frequency. Identify the reference line in your calculation.

3.22 File `E3-22.csv` contains the vector of quantitative features for a total of 50 events to train an SVM classifier. The events are already labeled to belong to Class I, Class II, or Class III. The number of features is two.

(a) Obtain three separating hyperplanes for this classification, where each separating hyperplane can separate one class from the rest of the classes.

(b) Suppose the two features are the magnitude of the voltage phasor differential ΔV and the magnitude of the current phasor differential ΔI corresponding to the event, respectively. Both features are presented in per unit. Use the separating hyperplanes in Part (a) to identify the class of an event that is characterized by the following PD measurements:

$$\begin{aligned} \Delta V &= 0.054 + j0.081 \\ \Delta I &= 0.52 + j0.65. \end{aligned} \tag{3.111}$$

3.23 Consider the vector of features for the same 50 events in Exercise 3.22. Suppose we add the following five additional events to the training set:

$$\mathbf{x}_{51} = \begin{bmatrix} 0.0254 \\ 0.1009 \end{bmatrix}, \quad \mathbf{x}_{52} = \begin{bmatrix} 0.0288 \\ 0.0797 \end{bmatrix}, \quad \mathbf{x}_{53} = \begin{bmatrix} 0.0386 \\ 0.0843 \end{bmatrix},$$
$$\mathbf{x}_{54} = \begin{bmatrix} 0.0353 \\ 0.1101 \end{bmatrix}, \quad \mathbf{x}_{55} = \begin{bmatrix} 0.0319 \\ 0.0919 \end{bmatrix}. \tag{3.112}$$

Events 51 and 55 belong to Class I. Event 52 belongs to Class II. Events 53 and 54 belong to Class III. Obtain the three separating hyperplanes for this classification by taking into account all 55 sample events. Are the resulting separating hyperplanes different from those in Exercise 3.22?

3.24 Obtain matrix **A** for the state estimation problem in Example 3.20. What is the rank of matrix **A**? Is it full column ranked?

3.25 Consider the 5-bus power transmission network in Figure 3.39. Suppose we measure voltage synchrophasors at buses 1, 2, 3, and 5, as well as current synchrophasors at lines (2,3), (2,4), (4,1), and (4,5). The measurements are given in Table 3.8. Estimate all states of the system.

3.26 Repeat Exercise 3.25 but this time assume that the voltage measurement at bus 3 and the current measurement on transmission line (2,3) are not available. Explain the results based on the rank of matrix **A**, as defined in (3.76).

Table 3.8 Synchrophasor measurements in Exercise 3.25.

Bus #	Measured Voltage (p.u.)	Line #	Measured Current (p.u.)
1	$1.1008\angle 20.6212°$	2,3	$0.5846\angle 14.6320°$
2	$1.0734\angle 16.8138°$	2,4	$0.6162\angle -16.9375°$
3	$1.0640\angle 14.1337°$	4,1	$1.3218\angle 171.1914°$
5	$1.0806\angle 19.2546°$	4,5	$1.0475\angle 170.0581°$

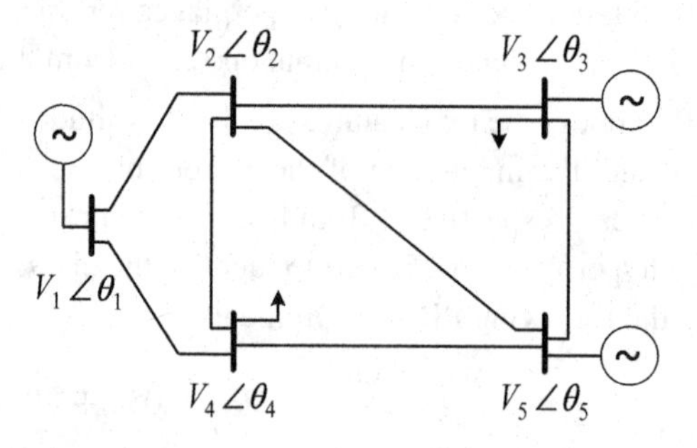

Figure 3.39 The 5-bus network in Exercise 3.25.

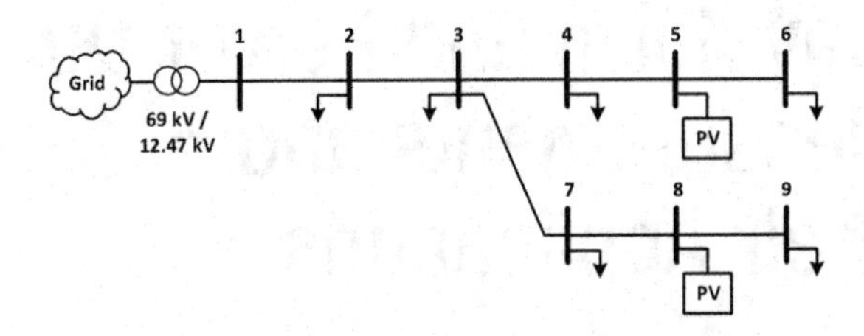

Figure 3.40 The balanced three-phase active distribution feeder in Exercise 3.27.

3.27 Consider a balanced three-phase power distribution system as shown in Figure 3.40. The per-phase impedance of each distribution line is $0.0412 + j0.0625\Omega$. Voltage phasors are measured (in volts) at selected buses:

$$\begin{aligned} V_1 &= 7200.0\angle 0^\circ, & V_2 &= 7173.4\angle{-0.0736^\circ} \\ V_4 &= 7141.7\angle{-0.1357^\circ}, & V_5 &= 7124.0\angle{-0.1263^\circ} \\ V_6 &= 7112.8\angle{-0.1541^\circ}, & V_7 &= 7133.9\angle{-0.1327^\circ} \\ V_9 &= 7121.6\angle{-0.1269^\circ} \end{aligned} \tag{3.113}$$

Current phasors are also measured (in amperes) at selected lines:

$$\begin{aligned} I_{21} &= 432.96\angle 139.9375^\circ, & I_{23} &= 346.30\angle{-43.4090^\circ} \\ I_{43} &= 160.31\angle 131.8792^\circ, & I_{45} &= 75.19\angle{-68.8901^\circ} \\ I_{65} &= 99.30\angle 150.5876^\circ, & I_{73} &= 137.31\angle 131.2470^\circ \\ I_{78} &= 55.64\angle{-100.9283^\circ}, & I_{98} &= 67.28\angle 148.7664^\circ. \end{aligned} \tag{3.114}$$

Estimate all the states of the system. Take bus 1 as the reference bus.

3.28 Verify the special case results for TVE in (3.100) and (3.101).

3.29 Two PMUs are installed at two ends of a transmission line. For both PMUs, TVE is limited to 1%, and it is solely due to potential error in measuring the phase angle. What is the maximum possible error, in degrees, in measuring RPAD between the voltage synchrophasors recorded by the two PMUs?

4 Waveform and Power Quality Measurements and Their Applications

So far, most of the discussions in this book have been based on the explicit or implicit assumption that the AC voltage and current signals are purely sinusoidal. However, this assumption may not hold in practice. Both voltage and particularly current may include *distortions* and take *non-sinusoidal* waveforms.

The instrument to measure voltage and current waveforms is the *waveform sensor*. Waveform sensors operate at very high sampling rates, such as at 256 samples per cycle, i.e., 15,360 samples per second [199]. This is much higher than the sampling rate of practically every other power system sensor, including PMUs in Chapter 3, e.g., compare it with 24 samples per cycle for a 10 fps PMU [133]. At such high sampling rate, a waveform sensor generates 3,981,312,000 samples per day from a single three-phase current signal. This is a huge amount of data to report. Therefore, in practice, most samples are *discarded* shortly after they are collected and as soon as they have gone through a light-weight analysis inside the sensor.

A waveform sensor may provide a *continuous* reporting of certain metrics for steady-state waveform distortion and power quality (see Section 4.1). In this regard, waveform sensors are sometimes referred to as power quality meters, or PQ meters. PQ meters may have other reporting features in addition to examining steady-state distortion in waveform measurements. For example, they also may analyze RMS voltage variations, frequency variations, voltage unbalance, and service interruptions.

A waveform sensor may also provide an *event-triggered* reporting of the voltage or current waveforms (see Sections 4.2–4.4). In this regard, different types of sensor technologies may act as waveform sensors that can capture events, such as most PQ meters and some other devices, such as Digital Fault Recorders (DFRs).

There are also some other classes or variations of waveform sensors that have been emerging recently, which we will discuss them in Sections 4.5 and 4.6.

4.1 Steady-State Waveform Distortion

Steady-state waveform distortions in power systems are often due to nonlinear loads. A load is nonlinear if its impedance changes with the applied voltage. The current that is drawn by a nonlinear load becomes non-sinusoidal, even if the applied voltage is sinusoidal. Static power converters, including AC to DC converters that are widely used by electronic loads, constitute the largest class of nonlinear loads. Other types of nonlinear loads include variable frequency drives (VFDs) and discharge lighting such as fluorescent lamps. Static power converters, such as the inverters that are used by distributed energy resources, such as PVs and batteries, may similarly contribute to current and voltage waveform distortions [200].

Even though *individual* inverters are often mandated to comply with harmonic emission limits set by industry standards, the *cumulative* effects of the harmonics that are injected by a large number of inverters may still cause considerable harmonic distortions in the power system that need to be accurately identified [201].

4.1.1 Measuring Harmonics

A non-sinusoidal signal $x(t)$ can be expressed in a Fourier series as follows:

$$x(t) = \sum_{h=1}^{\infty} \sqrt{2} X_h \cos(h\omega t + \phi_h). \tag{4.1}$$

The first term, corresponding to $h = 1$, is the *fundamental component*, which is a sinusoidal wave at fundamental frequency. The rest of the terms, corresponding to $h = 2, 3, 4$, etc., are the *harmonic components*, which are sinusoidal waves at harmonic angular frequencies $2\omega, 3\omega, 4\omega$, etc. Collectively, the harmonic waves create the distortions in the signal. A purely sinusoidal signal comprises only the fundamental component. Also, as a general note, a non-sinusoidal but symmetrically distorted signal comprises only the odd-numbered components; see Exercise 4.3.

In practice, waveform sensors use DFT to obtain the coefficients of the Fourier series in (4.1). This is because the DFT of a sampled periodic signal is proportional to the coefficients of the Fourier series of the continuous periodic signal.

The RMS value of the non-sinusoidal signal in (4.1) is obtained as

$$X_{\mathrm{rms}} = \sqrt{\sum_{h=1}^{\infty} X_h^2}. \tag{4.2}$$

If all harmonic components are zero, then $X_{\mathrm{rms}} = X_1$.

Example 4.1 Consider the voltage and current measurements for the motor load in Example 3.1 in Chapter 3. In an ideal scenario, the voltage and current signals are purely sinusoidal, as in Figure 3.1. However, in practice, the signals may look more like the ones in Figure 4.1. Here, the current signal is not sinusoidal:

$$\begin{aligned} i(t) = {} & 1.60\sqrt{2}\cos(\omega t - 0.7532) \\ & + 0.27\sqrt{2}\cos(3\omega t - 0.4323) \\ & + 0.15\sqrt{2}\cos(5\omega t + 3.3058). \end{aligned} \tag{4.3}$$

It includes the 3rd and the 5th harmonics. Its RMS value is obtained as

$$I_{\mathrm{rms}} = \sqrt{1.60^2 + 0.27^2 + 0.15^2} = 1.63 \text{ A}. \tag{4.4}$$

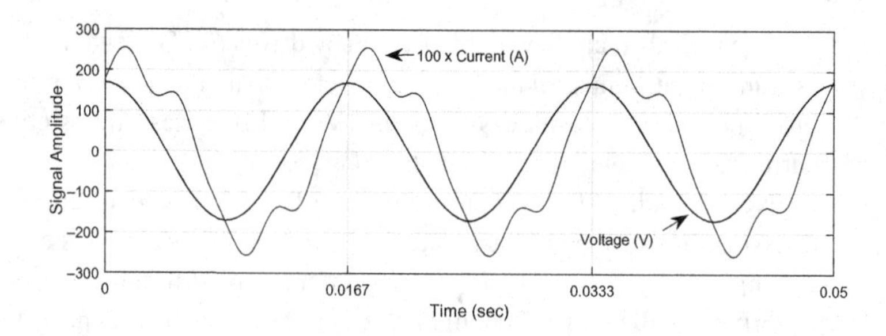

Figure 4.1 A sinusoidal voltage signal and a non-sinusoidal current signal.

The above RMS value is equal to the RMS value in Example 3.1. Therefore, a current sensor that measures only the RMS value is unable to distinguish the sinusoidal signal in Example 3.1 and the non-sinusoidal signal in this example.

PQ meters may provide weekly summery reports of the harmonic measurements based on standard power quality requirements, such as based on the EN 50160 standard on voltage characteristics [202]. These reports often distinguish *even* harmonics, *odd* harmonics that *are* multiples of three (called *triplen* harmonics), and *odd* harmonics that are *not* multiples of three (called *non-triplen* harmonics). The reason for such distinction is discussed in Exercise 4.8.

It is common to use a single quantity, the *total harmonic distortion* (THD), as a measure of the magnitude of harmonic distortion. THD is defined as

$$\mathrm{THD} = \sqrt{\sum_{h=2}^{\infty} \left(\frac{X_h}{X_1}\right)^2} \times 100\%. \tag{4.5}$$

For the voltage and current signals in Example 4.1, THD is 0% and 15.66%, respectively. THD is often calculated up to the 40th or 50th harmonic, i.e., for $h = 2, 3, \ldots, 40$ or $h = 2, 3, \ldots, 50$. From (4.2) and (4.5), if THD is presented as a fraction (not percentage), then the following relationship holds:

$$X_{\mathrm{rms}} = X_1\sqrt{1 + \mathrm{THD}^2}. \tag{4.6}$$

The harmonic distortion in current may be presented also in terms of another similar index, namely, the *total demand distortion* (TDD), which is defined as

$$\mathrm{TDD} = \sqrt{\sum_{h=2}^{\infty} \left(\frac{I_h}{I_{1,\,\max}}\right)^2} \times 100\%, \tag{4.7}$$

where $I_{1,\,\max}$ denotes the historical maximum RMS value of the non-harmonic current. It is often calculated as the average monthly peak RMS current over the past 12 months. Therefore, $I_{1,\,\max}$ is practically a constant. While THD can fluctuate with changes in the mix of harmonic versus non-harmonic loads, TDD is normalized such that it does *not* change due to changes in non-harmonic loads.

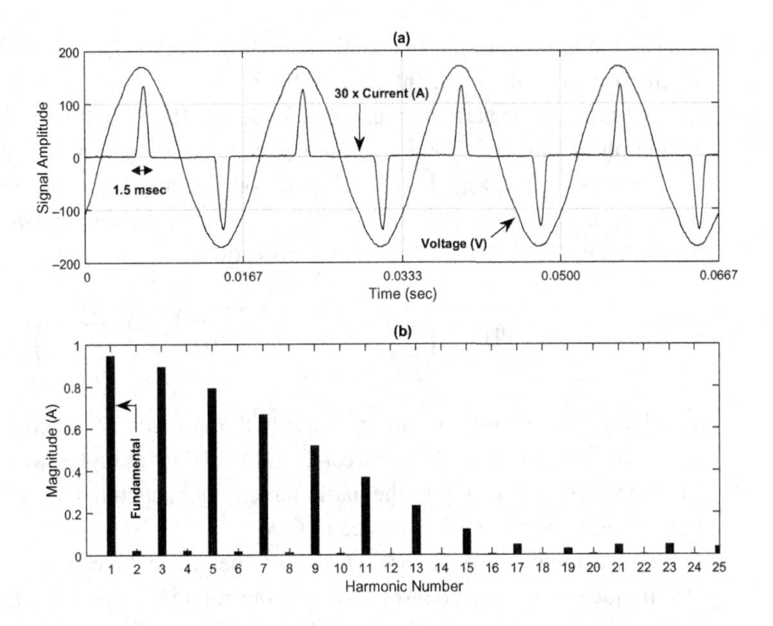

Figure 4.2 Analysis of steady-state waveform distortion: (a) voltage and current waveform measurements [203]; (b) harmonic spectrum for current up to the 25th harmonic.

Another quantity that is sometimes used to represent harmonic distortion in current waveform measurements is the *crest factor* (CF), which is defined as

$$\text{CF} = \frac{I_{\text{peak}}}{I_{\text{rms}}}. \tag{4.8}$$

It is the absolute peak of the instantaneous current signal. The peak value and the RMS value are taken from the same cycles. The absolute peak could occur in a positive half cycle or a negative half cycle. CF measures distortion at the moment that the maximum current is conducted. CF for a purely sinusoidal current waveform is $\sqrt{2} = 1.4142$. CF for a non-sinusoidal current waveform can be much higher, in particular for power electronics and VFD loads that rectify the AC voltage using diodes that conduct current only during the peaks of the AC voltage waveform.

Example 4.2 The voltage and current measurements for a single-phase power electronics load are shown in Figure 4.2(a). The current waveform is highly distorted, because current is drawn only during a small 1.5 msec time window in each half cycle. The harmonic spectrum for current is shown in Figure 4.2(b) up to the 25th harmonic. CF is 3.4635, which is much higher than 1.4142. THD is 161.8%.

In theory, a power electronics load with a high CF could draw its narrow current pulse anywhere across the voltage waveform. However, in practice, most power electronics loads use similar AC rectifier topology. Therefore, such pulses often coincide at the same point during each cycle, increasing the *aggregate* CF that is presented to distribution system equipment, such as transformers.

Most indices for harmonic analysis, such as THD, TDD, and CF, do *not* account for the phase angles of the harmonic components. However, the phase angle of the harmonic waveforms *do* have an impact on the total distorted current or voltage waveforms. Therefore, one may also use the *phasor harmonic index* (PHI), which is developed in [204], to assess harmonic distortion:

$$\text{PHI} = \left(\sum_{h=1}^{\infty} X_h |\cos(\phi_h - \phi_1)|\right) \Big/ \left(\sum_{h=1}^{\infty} X_h\right). \tag{4.9}$$

If all harmonic components are in-phase with the fundamental component, then PHI is 1; otherwise, it is a number between 0 and 1. PHI gives higher weight to phase angle differences corresponding to the harmonic components that have higher magnitudes. The PHI in Example 4.1 is obtained as 0.964.

If a measurement signal has a DC offset, then it can be expressed as a harmonic at zero frequency, i.e., as a constant outside the summation in (4.1). Electric noise may also be expressed as harmonics, at high frequencies, up to 200 kHz.

4.1.2 Measuring Inter-Harmonics

If all cycles of the distorted waveform are identical, as in Example 4.1, then the waveform can be expressed as in (4.1), where the frequencies of all harmonic components are multiples of the fundamental frequency. However, in practice, distortions are often *not* exactly identical across cycles. Therefore, the waveform may include components that are *not* integer multiples of the fundamental frequency. These components are referred to as *inter-harmonics*.

Let C denote the harmonic calculation window size, i.e., the number of cycles of the measured signal $x(t)$ that are considered in calculating the harmonic and inter-harmonic components. The signal can be expressed as

$$x(t) = \sum_{n=1}^{\infty} \sqrt{2} X_n \cos((n/C)\omega t + \phi_n). \tag{4.10}$$

The terms where n is a multiple of C correspond to harmonics. The terms where n is not a multiple of C correspond to inter-harmonics. Note that $n = C$, and not $n = 1$, corresponds to the fundamental component. Since inter-harmonics are not periodic at the fundamental frequency, they can be seen as a measure of the non-periodicity of the power system waveform. In (4.1), we measure one cycle and assume that it is repeated in all other cycles. In (4.10), we measure a window of C cycles and assume that the entire such window is repeated. That explains why $1/C$ is the smallest fraction of the fundamental frequency being considered.

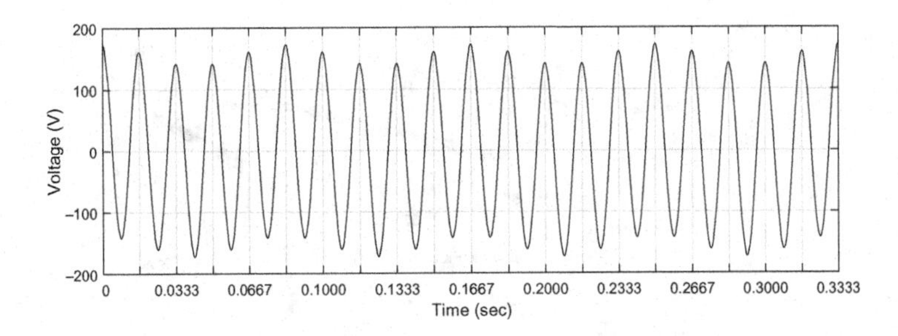

Figure 4.3 A voltage waveform with significant inter-harmonics; see Example 4.3.

Example 4.3 Consider the voltage waveform in Figure 4.3. A total of $C = 20$ cycles are captured in this figure. Therefore, one can obtain the inter-harmonics at $60/20 = 3$ Hz resolution. This waveform can be expressed as

$$v(t) = 110\sqrt{2}\cos(\omega t) + 12\sqrt{2}\cos((1/5)\omega t). \tag{4.11}$$

The inter-harmonic spectrum of $v(t)$ has two bins. One corresponds to the fundamental component, at 60 Hz, i.e., $n = 20$, with magnitude $110\sqrt{2}$, and one corresponds to the inter-harmonic component at 12 Hz, i.e., $n = 4$, with magnitude $12\sqrt{2}$.

Inter-harmonics cause variations in the per-cycle RMS value, calculated over each fundamental period. This can cause *light flicker*. PQ meters may provide inter-harmonic measurement summary reports in comparison with the limits on inter-harmonic magnitudes to prevent flicker, such as based on the IEEE 519-2014 or the IEC 61000-2-2 standards [205, 206].

According to the IEC 61000-4-7 Standard [207], the inter-harmonic components are calculated over a sampling window of $C = 12$ cycles. The spectrum is divided into $60/12 = 5$ Hz bands. Every 12th frequency bin is a multiple of 60 Hz; thus, it corresponds to a harmonic, i.e., 120 Hz, 180 Hz, 240 Hz, etc.

Inter-harmonics can be caused by various sources, including by powerline communications (PLC) carriers; see Section 6.5 in Chapter 6. For example, if a PLC-based smart meter communications system operates at 555 Hz and 585 Hz, PLC carriers coincide with two standard inter-harmonics between the $\lfloor 555/60 \rfloor = 9$th and the $\lceil 585/60 \rceil = 10$th harmonics. They are clearly visible in the inter-harmonic graph in Figure 4.4, and they are much lower in amplitude than the 9th or 11th harmonics but higher than any inter-harmonics between these two harmonics. The 10th harmonic is not shown in this figure.

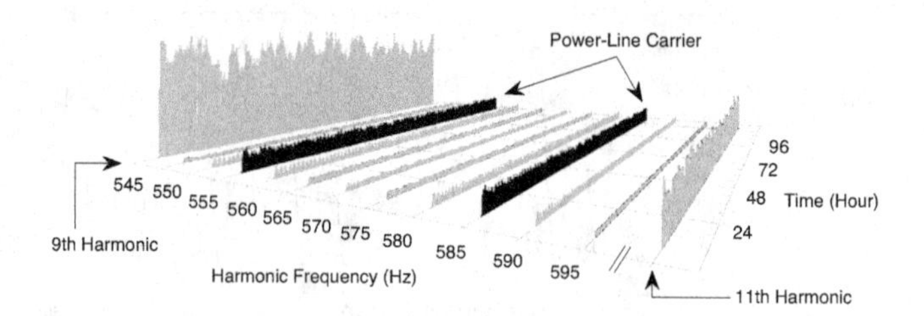

Figure 4.4 Inter-harmonic spectrum in presence of PLC carriers [208].

4.1.3 Measuring Notching

Voltage notching is a periodic waveform distortion. It often lasts less than half a cycle; see Figure 4.5. It occurs due to the normal operation of certain power electronics devices such as line-commutated thyristors and silicon-controlled rectifiers, when the current is commutated from one phase to another. During the notching period, there exists a momentary short-circuit between the two phases that results in reducing the line circuit, thus creating a notch shape in the voltage waveform. The amount of voltage reduction depends on the system impedance.

In principle, voltage notching can be analyzed via harmonics and inter-harmonics. If the notch is identical from cycle to cycle, which is often the case in practice, then it creates harmonics. Otherwise, it also creates inter-harmonics.

However, in practice, given the specific and simple shape of notching in time-domain, it is easier to quantify it from the raw waveform data, using the following four parameters: *notch depth*, which is the average depth of the notch from the theoretical sinusoidal waveform at fundamental frequency; *notch width*, which is the time duration of the notch; *notch area*, which is the product of the notch depth and the notch width, i.e., the area of the missing piece of the sinusoidal waveform; and *notch position*, which is where the notch occurs on the sinusoidal waveform.

Example 4.4 Again, consider the voltage measurements in Figure 4.5. A notch is positioned close to the peak at each positive half cycle. From the measurements on the figure, the notch depth is approximately calculated as $(51 + 64) / 2 = 57.5$ V, or 33.9% of the peak voltage. The notch width is 2.4 msec, or 2400 microseconds. Therefore, the notch area is $57.5 \times 2400 = 138{,}000$ volt-microsecond. According to the IEEE 519-2014 Standard, the notch depth and notch area should be limited to 20% and 22,800 volt-microsecond, respectively, assuming a 480 V system. For a 120 V signal, the limit on notch area must be adjusted to $(120/480) \times 22{,}800 = 5700$ volt-microsecond. This notch exceeds the limits for both notch depth and notch area.

Voltage notching may create parallel resonances or cause failure or damage to capacitor banks [209]. It is important to properly monitor voltage notching.

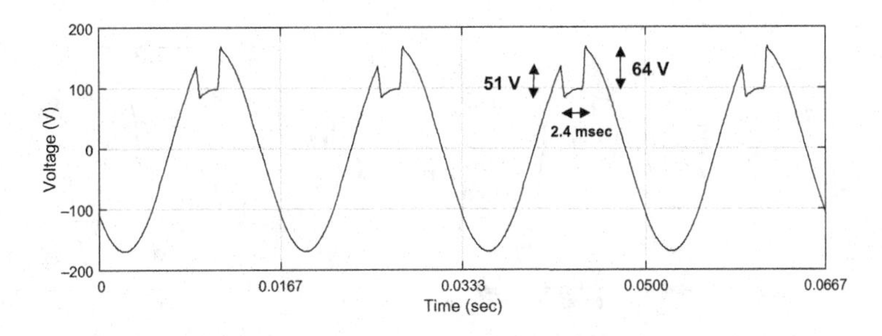

Figure 4.5 Notch in a non-sinusoidal voltage waveform [209].

4.2 Event-Triggered Waveform Capture

The previous section focused on steady-state waveform distortions. In this section, we examine momentary waveform distortions, i.e., waveform *events*.

Given their very high sampling rate, waveform sensors generate a huge amount of data in a relatively short amount of time. Therefore, it is practically impossible, and for the most part unnecessary, to store and report all collected data. Instead, the waveform data is *captured* only if the signal waveform is somewhat *unusual* and thus worthy of further examination. A captured event can be studied in real time, or it can be stored for further investigation in the future. An event-triggered waveform capture may also be referred to as a *waveform snapshot*.

Capturing Events in Waveform Measurements

Once an event is detected, i.e., once we notice something unusual in the waveform signal, the waveform sensor stores the waveform data over several cycles, starting from a certain number of cycles *before* the event, denoted by C_{before}, and ending by a certain number of cycles *after* the event, denoted by C_{after}.

Example 4.5 Consider the voltage measurement across 10 cycles in Figure 4.6(a). The waveform is sinusoidal in all but one cycle, i.e., cycle number 5. The waveform is highly distorted during this cycle, as shown in Figure 4.6(b). The distortion is due to a momentary *ringing* event that was caused by resonances formed between a capacitor bank and an inductive load during an upstream fault in a distribution system [210]. Waveform capture starts at $C_{\text{before}} = 4$ cycles before the distorted cycle and ends at $C_{\text{after}} = 5$ cycles after the distorted cycle. The damping oscillations in this event will be analyzed later in Example 4.18.

Event Detection in Waveform Measurements

There are different methods to *detect* an event in order to trigger capturing its waveform. The main idea in most waveform event detection methods is to *compare* the measured waveform with a reference waveform. The reference waveform is supposed

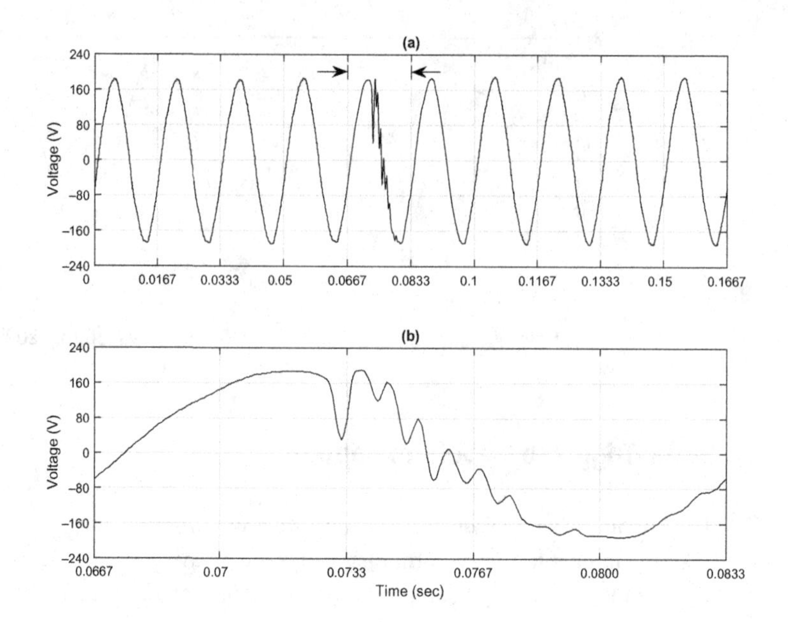

Figure 4.6 A ringing event in voltage waveform: (a) waveform capture starts a few cycles before and ends a few cycles after the event; (b) distorted cycle [210].

to represent the *normal* behavior of the system. If there is a considerable difference between the measured waveform and the reference waveform, then it can infer *abnormal* behavior, which in turn can trigger an event capture.

In practice, it is very common to simply compare *two consecutive cycles*, i.e., to take the waveform in the previous cycle as the reference to detect an event in the waveform in the present cycle. For instance, cycle number 5 in Figure 4.6(a) is deemed as an event because the waveform in this cycle is considerably different from the (reference) waveform in cycle number 4. Of course, one can also make the comparison with the waveform in two or three cycles back, instead of with the waveform in the immediate previous cycle; see Exercise 4.10.

There are different ways to compare two cycles of measured waveforms. Next, we go through several options and discuss their differences.

4.2.1 Comparing THD

Given that waveform events are often due to waveform distortion, a good option to detect an event is to compare waveforms based on their THD values. For instance, the ringing event in Example 4.5 was detected through measuring the THD for each cycle. If one applies DFT to each individual cycle, then the THD corresponding to cycle number 5 is obtained as 16%, while the THD corresponding to other cycles is less than 1%. Therefore, one can set a *threshold* to identify a waveform distortion event. Of course, what matters here is the *change* in THD from one cycle to the next, denoted by ΔTHD, not the THD itself. This is because a high THD in a cycle does not necessarily mean the presence of an event. It could also indicate a steady-state waveform distortion; see Section 4.1. Therefore, it is rather a change in THD that indicates a change in waveform. In this regard, we can check the following inequality to detect an event:

$$|\Delta\text{THD}| \geq \alpha_{\text{THD}}, \tag{4.12}$$

where α_{THD} is a predetermined threshold. Note that both positive and negative changes are of interest here because both indicate changes in waveshape. A negative change in THD indicates reduction in waveform distortion.

4.2.2 Comparing RMS

Another common approach in event detection is to compare the RMS value of the two waveforms. Let ΔRMS denote the *change* in the RMS value between the two cycles. An event is detected if the following inequality holds:

$$|\Delta\text{RMS}| \geq \alpha_{\text{RMS}}. \tag{4.13}$$

It is common to normalize $|\Delta\text{RMS}|$ with respect to the RMS value of the reference cycle, i.e., the first cycle in the case of comparing two consecutive cycles. In that case, the predetermined threshold α_{RMS} must be in percentage.

This approach is good for capturing events associated with voltage sags, swells, and some faults. For instance, consider the voltage sag in Figure 2.7 in Chapter 2. A regular voltage sensor would provide a curve such as those in Figure 2.7(c), depending on the size of its measurement window, as in Figure 2.7(b). However, a waveform sensor would provide the actual waveform, as in Figure 2.7(a). Of course, the level of details in the captured waveform depends on the sampling rate of the sensor.

Example 4.6 Consider the voltage measurement in Figure 4.7(a). The impulse in the second cycle is due to a lightning strike. The resolution of the waveform signal in this figure is 256 samples per cycle, or 65 microseconds per sample. The impulse appears to peak at 483 V. Next, consider the same waveform, but this time captured by a waveform sensor with a 1 MHz sampling rate, or 1 microsecond per sample. The impulse now appears to peak at 1104 V. The second sensor provides a more accurate representation of this extremely fast impulse event.

Comparing the RMS values of two waveforms may *not* always be effective for detecting events associated with waveform distortions. For instance, for the event in Example 4.5, the RMS value corresponding to cycle number 5 is 132.13 V and the RMS value corresponding to cycle number 4 is 129.11 V. Therefore, we have:

$$\Delta\text{RMS} = 3.02\text{ V} = 2.34\%. \tag{4.14}$$

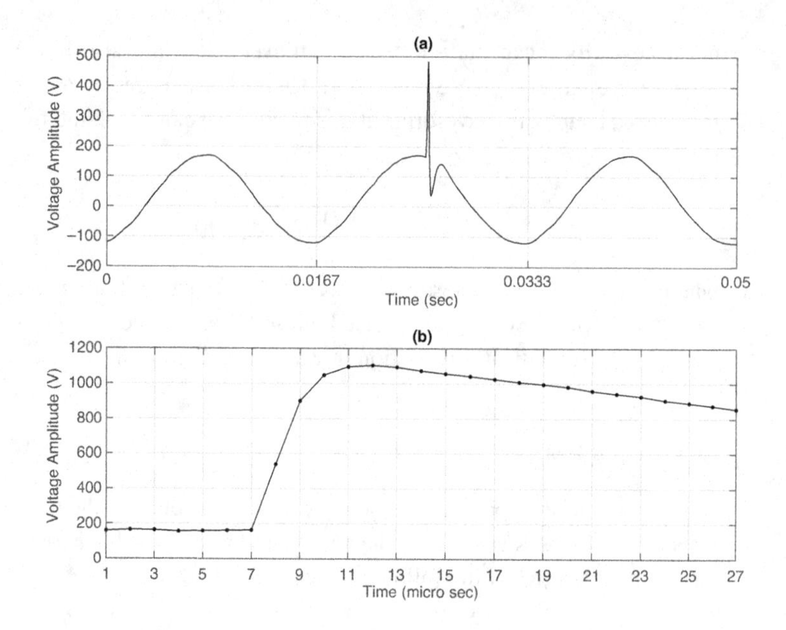

Figure 4.7 An impulse event in voltage waveform captured by two waveform sensors with different sampling rates: (a) 15.36 kHz; (b) 1 Mhz [211].

This is a very small change. Such small change in the RMS value could very well be due to a small change in load, as opposed to an event that is worth studying. Therefore, checking the RMS value does not help in detecting this event.

4.2.3 Point-to-Point Comparison

Two waveform cycles can also be compared point-to-point, i.e., sample-by-sample. An event is detected if the sample-by-sample difference between the two waveforms exceeds a *magnitude threshold* and it lasts longer than a minimum *duration threshold*. However, there are several challenges in using this approach.

One challenge in conducting point-to-point waveform comparison is the need to precisely identify the beginning and the end of each cycle. This can be done by *detecting the zero-crossing points*; see Appendix A1 in [212]. However, identifying the boundaries of each cycle is often prone to error. Another challenge is when the two waveform cycles do *not* have the same frequency, due to the changes in the frequency of the power system; see Section 2.9 in Chapter 2. The remedy is to do *frequency variation correction*; cf. [213]. However this type of correction itself is prone to error. In fact, when it comes to waveforms that carry high distortion, it is quite likely that the zero-crossing points are identified incorrectly and the frequency variation is not properly addressed, thus creating errors in conducting the point-to-point comparison of the two waveform cycles. Last but not least, in practice, it is very unlikely that two consecutive voltage waveform cycles, let alone two consecutive current waveform cycles, are exactly identical. At the very least, there is the impact of measurement noise. As a result, event detection based on point-to-point comparison is very sensitive to the choice of the threshold values.

All in all, point-to-point comparison may not be a good approach in practice. It may catch a lot of inconsequential events or miss important events [212].

4.2.4 Comparing Sub-Cycle RMS

This method is a trade-off between the point-to-point comparison and the comparison based on the RMS value. In this method, each of the two waveform cycles is divided into M segments, where M is between 4 and 16. The RMS value is calculated for each segment for each waveform cycle. The difference between the RMS values is then calculated across the two waveforms in order to obtain:

$$\Delta\text{RMS}[1], \ldots, \Delta\text{RMS}[M]. \tag{4.15}$$

An event is detected if the following condition holds for *any* $i = 1, \ldots, M$:

$$|\Delta\text{RMS}[i]| \geq \alpha_{\text{SCRMS}}. \tag{4.16}$$

As in (4.13), it is common to normalize $|\Delta\text{RMS}[i]|$ with respect to the RMS value of sub-cycle i in the reference cycle. In that case, the threshold α_{SCRMS} is set as a percentage. If $M = 1$, then the above method reduces to comparing the RMS values of the two waveform cycles. If M is equal to the number of samples in each waveform cycle, then the above method reduces to point-to-point comparison.

Example 4.7 Consider the ringing event in the voltage wave in Example 4.5. The waveform in cycle number 5 is shown in Figure 4.8(a). The waveform in cycle number 4 is shown in Figure 4.8(b). Suppose $M = 10$. The absolute differences between the RMS values of the corresponding sub-cycles across the two waveforms are shown in Figure 4.8(c). All values are normalized and presented in percentage. We can see that some of the sub-cycle RMS values have gone up to almost 20%. This is in sharp contrast with the comparison of the full-cycle RMS values in (4.14), where the absolute difference was only 2.34%. If the detection threshold is set to 15%, then the RMS value in sub-cycles 5 and 6 trigger event detection.

4.2.5 Differential Waveform

This method investigates the abnormal components that are *superimposed* to the normal voltage or normal current waveforms during an event. It works based on obtaining the following *differential waveform*:

$$\Delta x(t) = x(t) - x(t - NT), \tag{4.17}$$

where $x(t)$ is the measured current waveform or voltage waveform; T is the waveform interval, i.e., $T = 1/60$ second for a 60 Hz waveform; and N is a small integer number,

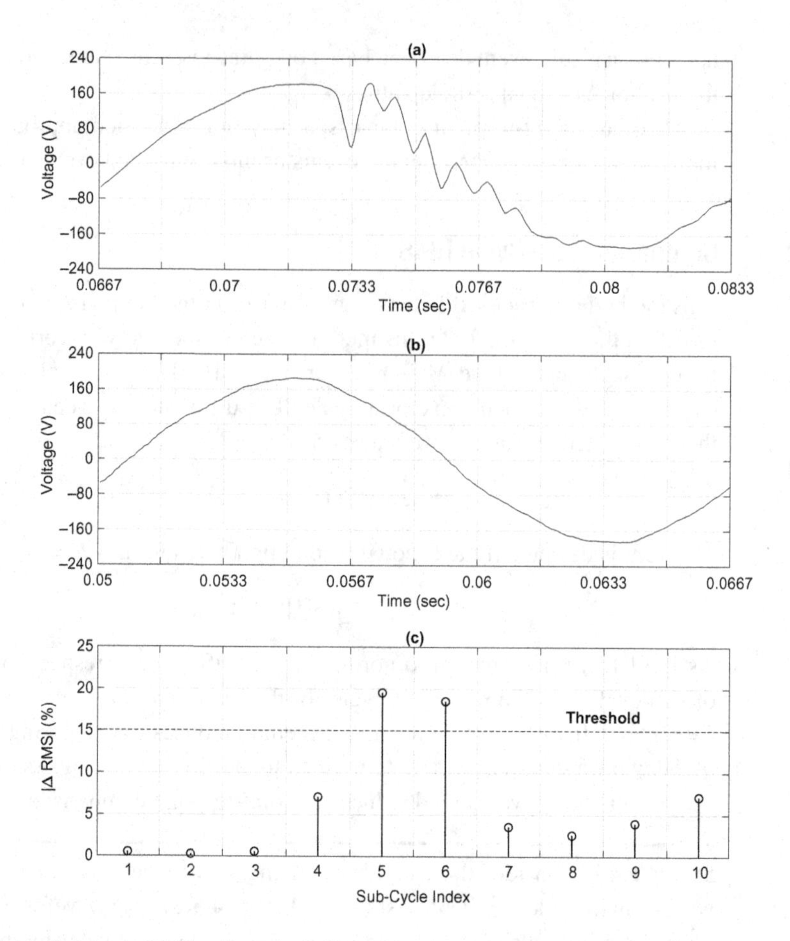

Figure 4.8 Sub-cycle RMS value calculation with $M = 10$ divisions: (a) the current cycle; (b) the previous cycle; (c) the difference of RMS values for the 10 sub-cycles.

such as 1, 2, 3, 4, or 5. Basically, the differential waveform is the difference between the original waveform measurement and a *delayed version* of the same waveform measurement, where the delay is equal to N cycles.

Example 4.8 Consider the current waveform measurement in Figure 4.9(a). There is an event in the second cycle, as marked on the figure. A delayed version of the current waveform measurement is shown in Figure 4.9(b). The delay is at $N = 1$ cycle. The corresponding differential waveform is shown in Figure 4.9(c). It is obtained by subtracting the delayed waveform in Figure 4.9(b) from the original waveform in Figure 4.9(a). We can see that the event has created *two* distinct blips in the differential waveform, which are denoted by ① and ②. Note that both of them are associated with the *same* event. The first blip is due to comparing the event cycle with the previous normal cycle, where the normal cycle serves as the reference. The second blip is due to comparing the next normal cycle with the event cycle, where the event cycle serves as the reference.

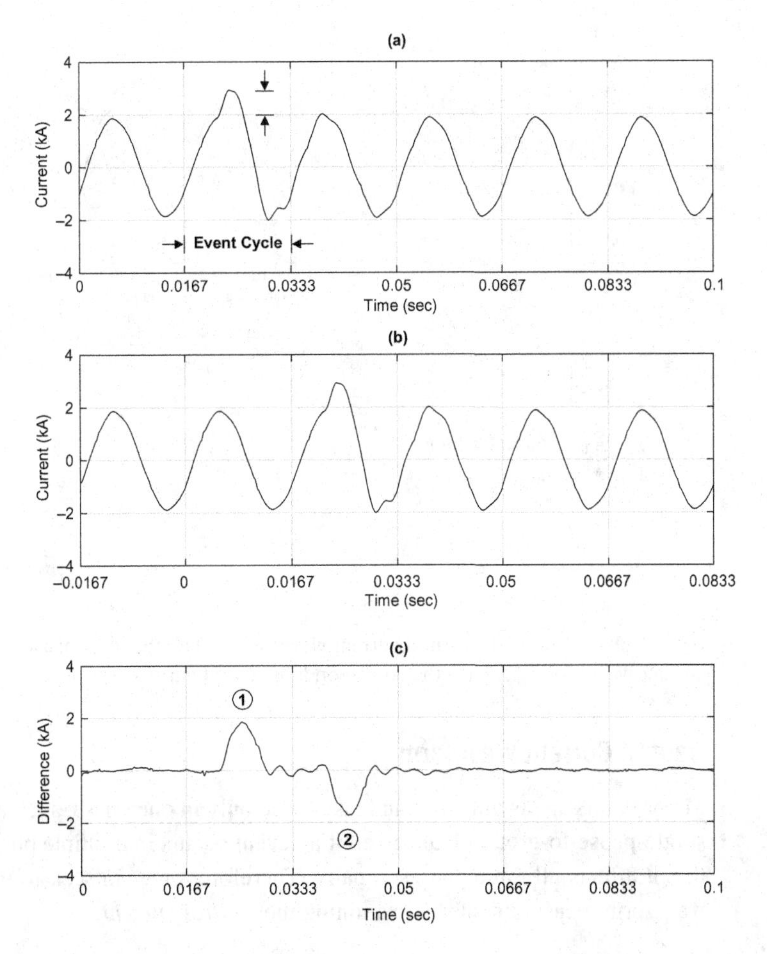

Figure 4.9 Obtaining differential waveform: (a) the original waveform; (b) the original waveform that is delayed by $N = 1$ cycle; (c) the corresponding differential waveform.

There are multiple options to use the differential waveform to detect an event. One is to check the absolute value of the instantaneous waveform. In this option, an event is detected if the absolute value of the differential waveform exceeds a *magnitude threshold* and it lasts longer than a minimum *duration threshold*. For instance, for each of the two blips in Figure 4.9(c), the absolute value exceeds 1.5 kA, and it lasts for over a quarter of a cycle. It is clear that this first option is somewhat similar to the point-to-point comparison that we saw earlier in Section 4.2.3.

Another option is to treat the differential waveform as a signal by itself. Accordingly, we can apply methods such as Fourier analysis or Wavelet analysis to the differential waveform in order to perform a frequency-domain analysis. A threshold can be applied to the magnitude of the fundamental frequency component, or just the magnitude of the dominant frequency component of the differential waveform based on each cycle or multiple consecutive cycles [212].

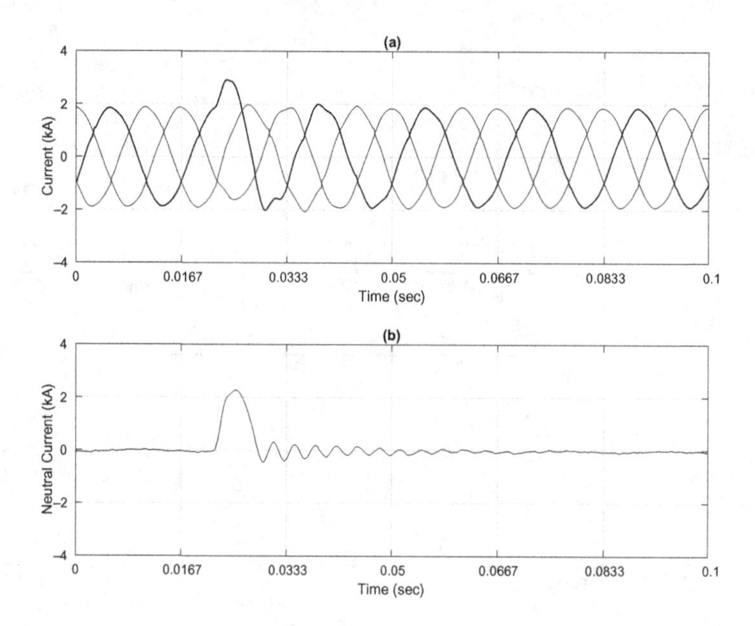

Figure 4.10 Obtaining the neutral current waveform: (a) all the three phases of the original current waveform [215]; (b) the corresponding neutral current waveform.

4.2.6 Neutral Current Waveform

Many events are asymmetric and take place only in one phase, such as in the case of a single-phase-to-ground fault. Even if an event occurs in multiple phases, it is unlikely that it affects all three phases equally. Therefore, one can seek to detect an event in waveform measurements by examining the *neutral current*:

$$i_{\mathrm{N}}(t) = i_{\mathrm{A}}(t) + i_{\mathrm{B}}(t) + i_{\mathrm{C}}(t), \tag{4.18}$$

where $i_{\mathrm{A}}(t)$, $i_{\mathrm{B}}(t)$, and $i_{\mathrm{C}}(t)$ denote the waveform measurements on Phase A, Phase B, and Phase C, respectively. Of course, the analysis of neutral current waveform is relevant only to three-phase waveform measurements.

Example 4.9 Again, consider the current waveform measurements in Example 4.8. All three phases of the original waveform are shown in Figure 4.10(a). From (4.18), the corresponding neutral current waveform is obtained as in Figure 4.10(b). The event creates a significant signature in the neutral current waveform during the event cycle. It also creates some transient oscillations that damp down within three to four cycles after the event occurs.

There are a few advantages in analyzing the neutral current waveform instead of (or in addition to) the differential current waveform. One advantage is that the neutral current waveform does *not* manifest the confusing second blip that we saw in Figure 4.9(c). Another advantage is in the cases of capturing fault waveforms in power distribution systems. Note that, in a power distribution system, the superimposed faulted phase current typically contains a component of the load current, which is unrelated to the fault. On the contrary, the neutral current during the fault consists almost of only the fault current; because the mostly balanced load current in the three phases cancel out each other in the neutral current [214].

As in the case of differential waveform, neutral current waveform can be analyzed not only in time-domain but also in frequency-domain.

4.2.7 Other Factors and Methods

An event in waveform measurements also can be defined based on a change in frequency. This will allow capturing the waveform during frequency events, such as the one in Example 2.23 in Chapter 2. An event also may be defined based on a *combination* of multiple parameters, such as a certain change in THD and a certain change in RMS value. If proper communications and precise time synchronization are facilitated across multiple waveform sensors, then one may also define an event based on *simultaneous* change of waveform parameters at different locations. Time synchronization can be facilitated by the GPS; see Section 3.2 in Chapter 3.

Selecting the right thresholds is critical in event-triggered waveform captures. If the thresholds are set too high, then important events can be missed. If they are set too low, then many false triggers can be generated for innocuous events.

All trigger mechanisms work in parallel. Therefore, one event may trigger multiple trigger mechanisms, such as THD and RMS value, at the same time.

4.3 Analysis of Events and Faults in Waveform Measurements

Once an event is detected and captured, the next step is to analyze the extracted waveform in order to obtain some sort of *useful information* from the event, such as to determine the cause of the abnormality. In most cases, the event in waveform measurements is due to minor power quality issues. However, there are also events that are due to faults or incipient faults. Analysis of such events is particularly important because some events show whether the power system and its various equipment are in a safe and healthy state of operation.

Accordingly, in this section, we will review several *waveform signatures* that are due to faults or incipient faults. Incipient faults are usually *self-clearing* and last for only a *short period of time*, such as only a fraction of a cycle or up to only two or three cycles. They are often extinguished for various physical reasons before the utility protective devices have time to operate. Nevertheless, capturing incipient faults is important because they may indicate a potential catastrophic fault in the future, i.e., a major failure that is still in its early stages, but it may get worse.

Note that, while most of the typical power quality disturbances manifest as changes in voltage waveforms, equipment failures and incipient faults rather demonstrate unique signatures in current waveforms [212]. Therefore, for the discussions in this section, attention should be paid not only to the voltage waveforms but also particularly to the current waveforms.

4.3.1 Faults in Underground Cables

Many utilities use underground cables in urban areas. Failures in underground cables are gradual and take place over a period of time. They are often caused by moisture penetration into the cable splice, which results in cable insulation breakdown. The water produces an arc; but then the arc quickly evaporates water, which in turn extinguishes the arc, making the fault *self-clearing* [216].

The self-healing nature of the above fault means that it does *not* trigger any over-current protection device; therefore, it can go unnoticed for a while. However, this type of incipient faults may ultimately turn into permanent faults after self-clearing many times and gradually damaging the cable. Once they turn permanent, they will cause the operation of the over-current protection devices; which often means losing service for several utility customers.

Example 4.10 Figure 4.11(a) shows the voltage and current waveforms during a fault in an underground cable. The fault is self-clearing, and it lasts for less than a cycle. Figure 4.11(b) shows the voltage and current waveforms during another fault in the same underground cable. This fault takes place only about 1.5 hours after the first fault. The second fault is also self-clearing and lasts for less than a cycle. Figure 4.11(c) shows the voltage and current waveforms during yet another fault in the same underground cable. This fault takes place about two days after the first two

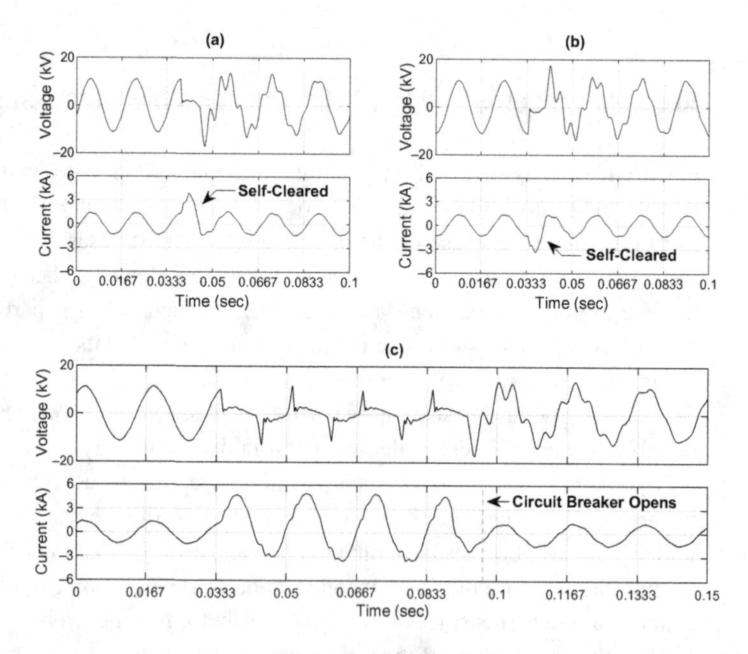

Figure 4.11 Voltage and current waveform measurements during several events in an underground cable: (a) a self-clearing incipient fault; (b) another self-clearing incipient fault that occurred 1.5 hours after the first one; (c) a permanent fault that occurred two days later and was cleared by a circuit breaker [214].

faults. This time the fault lasts for three cycles, and it is cleared by an over-current circuit breaker which isolated the faulted area. While the first two faults were incipient faults, they were followed by a permanent fault.

Faults in underground cables can happen in the cable insulation, cable splice, or cable termination. Cable splice is where two cables are connected. Cable termination is where the underground cable is brought above the ground. Depending on where the fault occurs, it may demonstrate different waveforms. For example, Figure 4.12 shows the waveform measurements during a fault at a cable termination location. The fault took place on one phase and lasted for about five cycles before it was cleared by a circuit breaker opening. Notice the *impulses* in the faulted voltage waveform during the first two cycles of the event. However, not all cable termination faults demonstrate impulses in the faulted phase voltage waveform. Nevertheless, there do exist some methods to apply to the captured waveform measurements of the fault in order to identify whether the fault took place at the cable joint, cable insulation, or cable termination; cf. [217].

4.3.2 Faults in Overhead Lines

Failures in overhead transmission and distribution lines are often due to short-circuit conditions that can be caused in different ways, such as due to *tree contact*, *animal contact*, *traffic accidents*, or *lightning*. Some of these causes are inherently sudden, with no precursor conditions, such as traffic accidents. However, some other causes

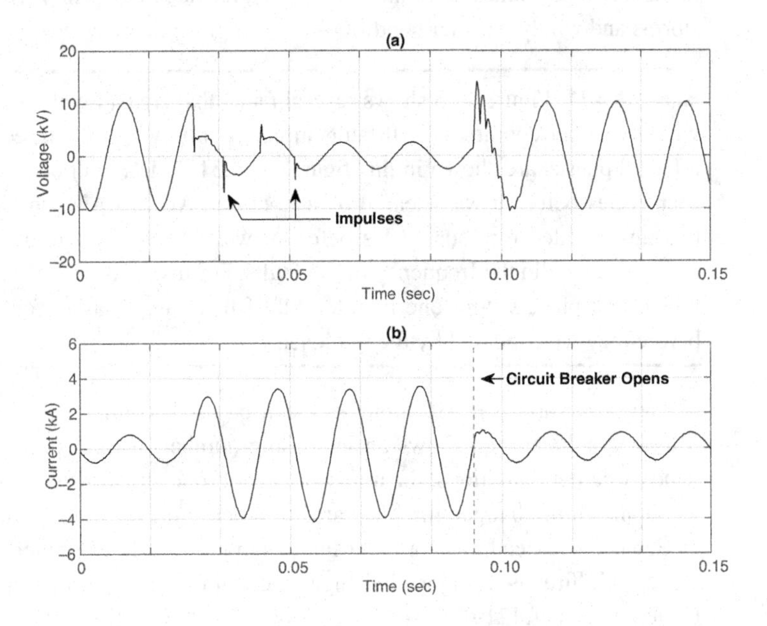

Figure 4.12 Waveform measurements during a fault in a cable terminator device [217]: (a) voltage; (b) current. Notice the impulses on the faulted voltage waveform.

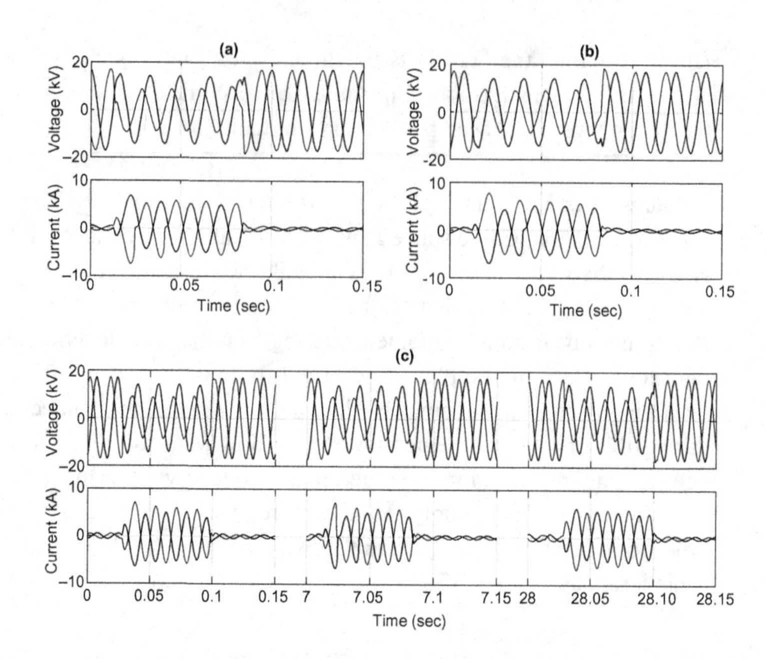

Figure 4.13 Voltage and current waveform measurements on two phases during several faults that occurred on an overhead power line and cleared by re-closers on a windy day in a span of 30 minutes: (a) one fault occurred at 4:31 AM; (b) one fault occurred at 4:53 AM; (c) three faults occurred at 5:01 AM [218]. Only two of the three phases are shown here.

may manifest as *incipient faults*. For instance, tree contacts can sometimes be incipient faults that repeat and evolve into a major outage due to growth in vegetation or during storms and windy weather conditions.

Example 4.11 Figure 4.13 shows several faults that are caused by *tree contacts* during windy weather conditions. All faults affected only two of the three phases. Only the affected phases are shown in this figure. At 4:31 AM, a fault occurred that lasted for four cycles before it was cleared by a re-closer. At 4:53 AM, another fault occurred that again lasted for four cycles before it was cleared by a re-closer. As the storm intensified, so did the frequency of the faults. For instance, at 5:01 AM, three separate faults took place within one minute. All of them again lasted for about four cycles before they were cleared by the re-closer.

The observations in Example 4.11 can be used to improve safety. For example, if a geographical area shows an increasing number of tree contacts during dry and windy weather conditions, the utility may choose to shut down power in the high-risk locations within that geographical area in order to prevent *wildfire* that can be caused by downed overhead lines or other failures in power grid equipment. Preventing grid-caused wildfires is a major challenge in certain places, such as in California in the United States; cf. [219].

Another application of the above analysis is to plan cutting vegetation around overhead lines when the number of faults caused by tree contacts exceeds a certain threshold; also see Section 7.4.2 in Chapter 7 for a related discussion.

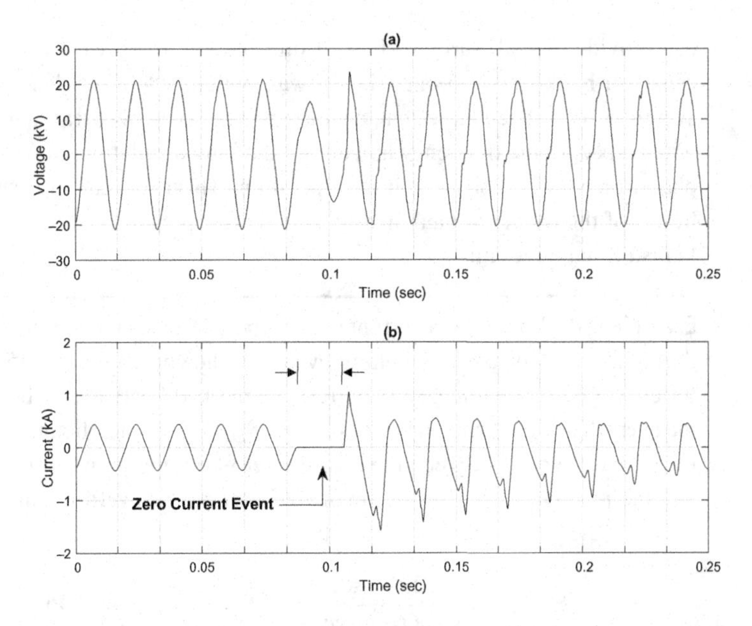

Figure 4.14 Waveform measurements during an incipient fault on a transformer tap changer: (a) voltage; (b) current. The fault caused a momentary zero current event [220].

4.3.3 Faults in Transformers

Failures in transformers can take place in various components, such as tap changer, bushing, winding, etc. Early detection of these failures can allow for corrective actions to prevent costly outages and reduce downtimes.

Example 4.12 The voltage and current waveform measurements during a transformer tap changer incipient failure are shown in Figure 4.14. The fault occurred during tap changing, and it caused zero current. Only the faulted phase is shown here. Initially, the fault occurred occasionally, and the duration of the zero current period was less than one cycle. However, over several days, the abnormality took place more frequently and repeated multiple times every day. The duration of the zero current period also gradually increased to over one cycle. When technicians inspected the tap changer, they discovered a pin that was shearing and causing arcing during the travel of the tap changer. Utility technicians reported that the transformer would have faced a catastrophic failure within two weeks of the inspection if the arcing had not been detected and fixed [220].

4.3.4 Faults in Capacitor Banks

Capacitors are widely used in both transmission and distribution networks in order to provide reactive power and voltage support. As a result, a capacitor switching transient is one of the most common equipment switching events in power systems. Analyzing the voltage and current waveform measurement during capacitor bank switching can provide us with various kinds of information. First, we can identify whether the capacitor bank was *energized*, i.e., it was switched "on," or *de-energized*, i.e., it was switched "off." Second, we can characterize the switching transient behavior, such as with respect to the frequency and duration of any oscillations, switching angle at each phase, or balanced or unbalanced operation. Third, we can evaluate the state of the health of the capacitor bank and its various components and identify the presence of faults or incipient faults.

Example 4.13 Voltage and current waveforms are measured at a distribution substation during a capacitor bank switching event, as shown in Figure 4.15. The phase angle difference between voltage and current are marked for Phase A, both *before* and *after* the event. They are denoted by $\vartheta_{\text{before}}$ and ϑ_{after}. They are obtained by first calculating the time difference between the positive zero-crossing point of the voltage waveform and the positive zero-crossing point of the current waveform, and then dividing the result by the waveform time interval:

$$\vartheta_{\text{before}} = \frac{0.9231\ \text{msec}}{16.6667\ \text{msec}} \times 360 = 19.94^\circ \quad \Rightarrow \quad \text{PF} = 0.940 \tag{4.19}$$

and

$$\vartheta_{\text{after}} = \frac{0.4103\ \text{msec}}{16.6667\ \text{msec}} \times 360 = 8.86^\circ \quad \Rightarrow \quad \text{PF} = 0.988. \tag{4.20}$$

The power factor has considerably increased after the event. Therefore, we can determine that the capacitor bank is *energized.* Similar analysis can be done on Phase B, where PF increases from 0.950 to 0.992, and on Phase C, where PF increases from 0.955 to 0.995. The capacitor bank has an almost balanced operation.

The transient behavior that happened during the capacitor bank switching event in Example 4.13 is generally considered normal. Note that, during the energization of a capacitor bank, its capacitance interacts with the inductance of the rest of the power system; which gives rise to transient oscillations. The oscillations in voltage may peak between 1.1 and 1.4 per unit. The frequency of the transient oscillations is typically between 300 Hz and 1000 Hz. For the transient oscillations in Example 4.13, the frequency is about 300 Hz [221].

Capacitor switching transients can be minimized by energizing the capacitor at or near voltage zero, i.e., at or near a zero-crossing point of the voltage waveform. Performing switching at such precise times can be achieved by using a mechanism called a *synchronous switching control.* In practice, if the phase angle of voltage at the time of energizing each phase is between -5° and 5° of voltage zero on that phase, then switching is considered synchronous [221].

The switching phase angles across the three phases during the capacitor bank energizing event in Example 4.13 are shown in Figure 4.16. The closing at Phase A is premature by 89.6°; the closing at Phase B is premature by 23.9°; and the closing at Phase C is delayed by 37.2°. It is clear that synchronous closing control was not

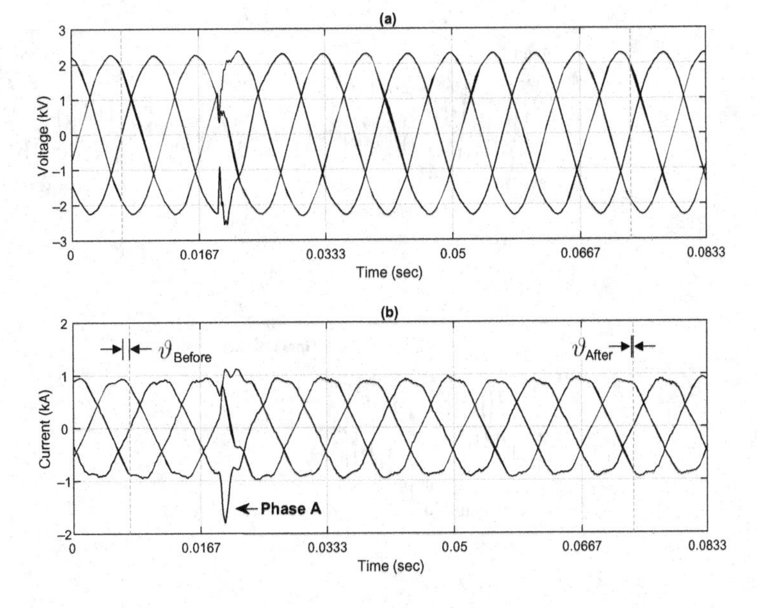

Figure 4.15 Waveform measurements during a capacitor bank switching event [221]: (a) voltage; (b) current. By examining the phase angle differences before and after the event, i.e., $\vartheta_{\text{before}}$ and ϑ_{after}, we can determine that the capacitor bank is energized.

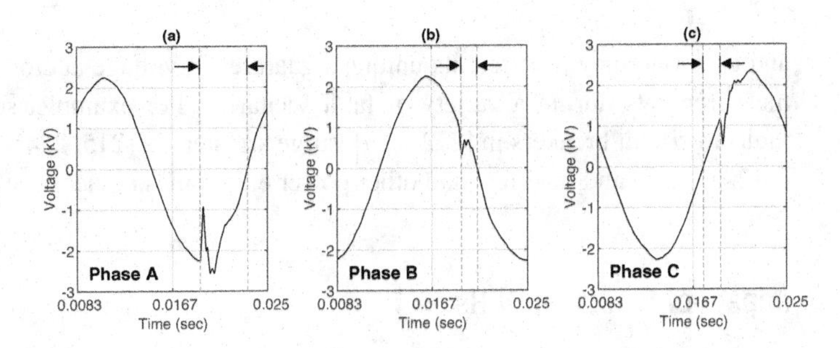

Figure 4.16 Obtaining the switching phase angles for the event in Figure 4.15.

successful, particularly at Phase A. This created a relatively major transient behavior on Phase A, as we previously saw in Figure 4.15.

We can similarly diagnose potential issues with capacitor switching when it is de-energized. Generally speaking, a normal capacitor de-energizing event should *not* produce any major switching transient. However, if the switch contacts are contaminated or faulty, then the switch may *re-strike* while the capacitor is being de-energized [222]. One such example is discussed in Exercise 4.12.

4.3.5 Faults in Other Devices and Equipment

The above are only a few examples of the fault signatures that can be captured and analyzed using waveform sensors. The power system has many devices and equipment,

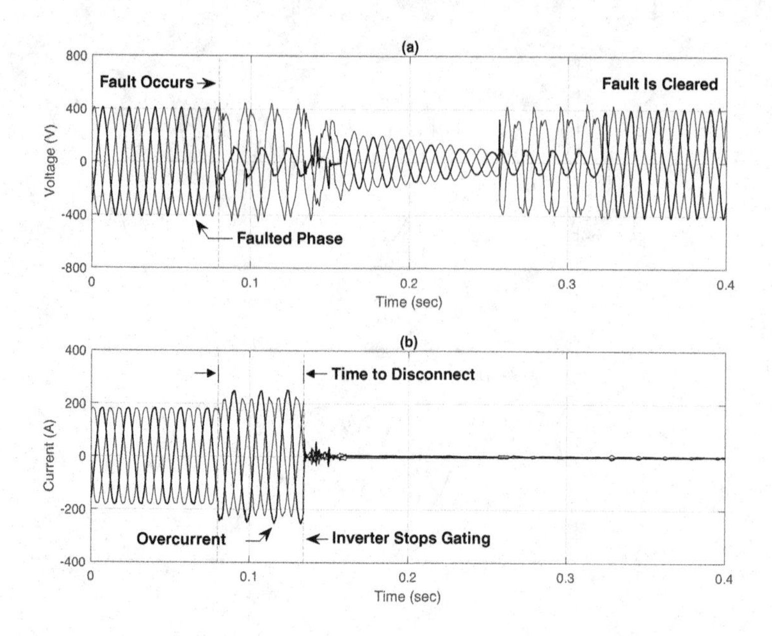

Figure 4.17 Calculating the time to disconnect for a PV inverter in response to a fault: (a) voltage waveform measurements; (b) current waveform measurements.

and each one may manifest its unique signatures in voltage and/or current waveform measurements during a variety of fault scenarios. For example, see the analysis of faults in circuit breakers in [212, 223], surge arresters in [215, 224], switching devices in [225, 226], and inverters and other power electronics devices in [227–229].

4.3.6 Impact of Faults on DERs

While it is important to study different types of faults and their root causes and characteristics, it is important also to study how different equipment and devices may *respond* to faults. This is particularly critical when it comes to inverter-based resources, such as most DERs. Examples of inverter-based resources include PVs, wind turbines, stationary batteries, and grid-connected electric vehicles.

Example 4.14 The voltage and current waveforms of a three-phase 480 V solar PV inverter during a fault are shown in Figure 4.17. The faulted phase is marked on the figure. The fault occurs at 0.0802 seconds, as marked by the first vertical dashed line. It immediately creates a sudden drop in voltage on the faulted phase. The fault also causes a surge in current at the PV unit, which quickly reaches as high as 140% of the pre-fault current. This ultimately causes the inverter's protection system to act. The inverter stops gating and abruptly cuts off current at all three phases at around 0.1341 seconds, as marked by the second vertical dashed line. At this point, the PV unit trips off-line. The *time to disconnect*, also known as the trip time or run-on time [230, 231], is about 54 msec. The fault is later cleared after a few cycles, yet the PV unit stays disconnected for the *next three minutes*, not shown here.

For the scenario in Example 4.14, the inverter was *unable to ride through the fault* so as to resume normal operation as soon as the fault was cleared. The inverter simply *ceased production*. This can cause severe problems in circumstances where there is a high penetration of DERs, as shown in the next example.

Example 4.15 On October 9, 2017, a wildfire caused two transmission system faults in Southern California. The first fault occurred at 12:12 PM on a 220 kV transmission line. The second fault occurred at 12:14 PM on a 500 kV transmission line. Despite the fact that these faults were *cleared normally* by the transmission system's protection devices, they both resulted in reduction in solar PV generation across the region. The overall setup is shown in Figure 4.18(a). The first fault resulted in tripping 682 MW of PV resources. The second fault resulted in tripping 937 MW of PV resources. Note that no PV resource was de-energized to clear a fault because the faults had no direct relevance to the PV resources. Instead, these PV resources ceased output as a response to the fault on the system in order to protect their own inverters. It took several minutes for the PV resources to gradually restore service, as shown in Figure 4.18(b). Losing these PV resources affected the power system frequency; see Section 2.9 in Chapter 2. In particular, the second fault that resulted in over 900 MW solar PV resource loss caused a *frequency excursion* in the Western Interconnection with system frequency reaching as low as 59.878 Hz within 3.3 seconds after the fault; see Figure 4.18(c). Frequency recovered to nominal in about 100 seconds. The voltage waveform that was measured at one of the PV inverters is shown in Figure 4.18(d). Notice that the transient overvoltage took place on one phase for only a fraction of a cycle; yet it caused the PV inverter to trip because the measured voltage exceeded the overvoltage protective setting for the inverter.

After studying the above and other similar incidents, it was concluded that a large percentage of the existing grid-connected inverters are configured to trip using *instantaneous overvoltage protection*; they do *not* filter out voltage transients. Therefore, any instantaneous, sub-cycle transient overvoltage may trip the inverter off-line, making these resources susceptible to tripping on transients caused by faults or major switching actions in the bulk power system. As another example, it has been observed that some momentary system-wide *phase jumps* in the voltage waveforms that are caused by certain faults can be mistaken by inverters as a severe frequency violation event. This too can cause inverter-based resources to cease production; see [233].

The North American Electric Reliability Corporation (NERC) recommends setting the voltage protection settings and the frequency protection settings of inverters based on their actual physical equipment limitations, so that, as much as possible, the inverters can *ride through the fault* and continue injecting current to the grid; see [232, 233]. Also see the similar issues on inverter-based wind generation resources in [234].

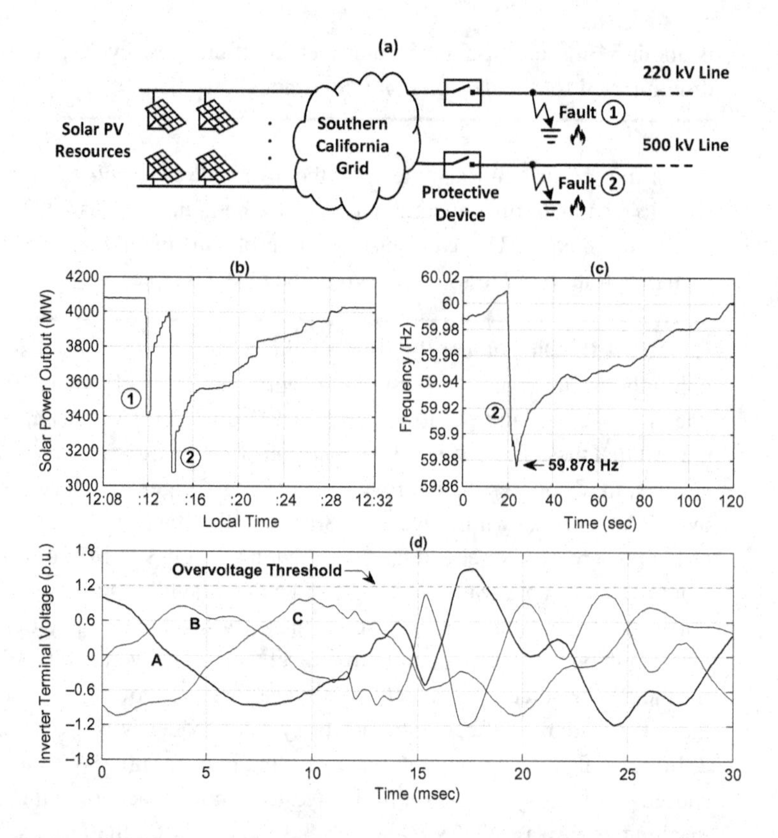

Figure 4.18 Impact of faults on solar PV resources during a wildfire based on Example 4.15: (a) an overview of the key components in the system; (b) the loss of solar generation that occurred during the two transmission line faults; (c) the impact of the second event on the power system frequency of the Western Interconnection; (d) the voltage waveform measurements at one of the solar PV resources [232].

4.4 Features and Statistical Analysis of Waveform Events

The analysis in Section 4.3 showed that a lot can be learned from each event in waveform measurements. However, given the enormous number of events that may take place in the power system every day, we need a way to translate the waveform measurements during events into *useful information* that can help diagnose issues, discover hidden patterns and unknown correlations, and make recommendations.

The key to achieving this goal is to define *quantitative features* that can characterize each event and allow conducting signature evaluation, event classification, pattern recognition, and statistical analysis. The basic idea in selecting the features is the same as what we discussed in Section 3.7.2 in Chapter 3 in the context of phasor measurements. However, there are also some quantitative features that are somewhat specific to waveform measurements.

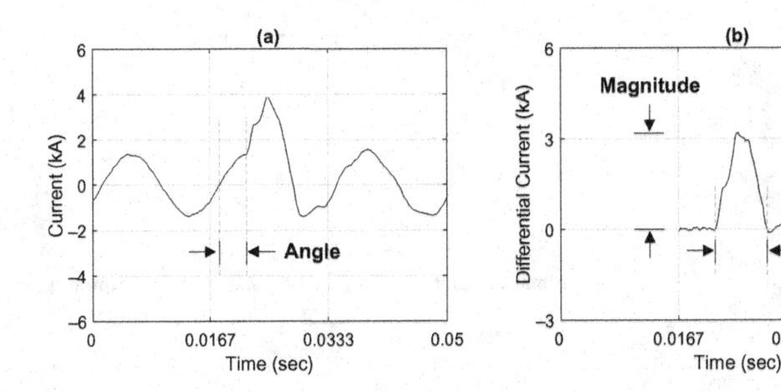

Figure 4.19 Obtaining the angle, magnitude, and duration of an event: (a) the current waveform measurement; (b) the differential waveform during the event cycle.

In this section, we discuss several of these quantitative features. Some of them are generic and can be defined for almost any event in waveform measurements. Some can be defined only for certain types of events, such as certain faults.

4.4.1 Angle, Magnitude, and Duration

These basic features can be obtained for most events. They are shown in Figure 4.19 for the case of a current waveform measurement during a self-clearing fault. The *angle* of the event is the angular difference between the moment when the event starts and the moment of the most recent zero-crossing point prior to the event. The zero-crossing point serves as the reference point. It is often easier to obtain the angle of the event directly from the original waveform measurement, as shown in Figure 4.19(a). However, when it comes to obtaining the *magnitude* and *duration* of the event, it might be better to obtain them from the differential waveform, as shown in Figure 4.19(b). For the event in Figure 4.19, the angle is 82°, the magnitude is 3.187 kA, and the duration is 7.25 msec. The duration is less than half a cycle.

The angles for certain events are meant to be at certain values or within certain ranges. In Example 4.13, energizing the capacitor should happen at or near voltage zero (zero-crossing point). Therefore, we can measure the switching angle, relative to the moment when the voltage is zero, over several days in order to evaluate the operation of the synchronous closing control mechanism. If the percentage of the switching event angles that fall outside of a given range, such as $[-45°, 45°]$, exceeds a certain threshold, then we may suspect that the performance of the synchronous closing control is poor and that the switch controller must be reprogrammed [221].

Even for some events where the angle *seems arbitrary*, a simple statistical analysis may reveal an underlying pattern with respect to the event angle which may help us with identifying the root cause of the event.

Example 4.16 The study in [235] examined the event angle in voltage waveform measurements for 73 faults in overhead power lines. The cause for each fault was already known from the utility records. The first 42 events were caused by *animal contact.*

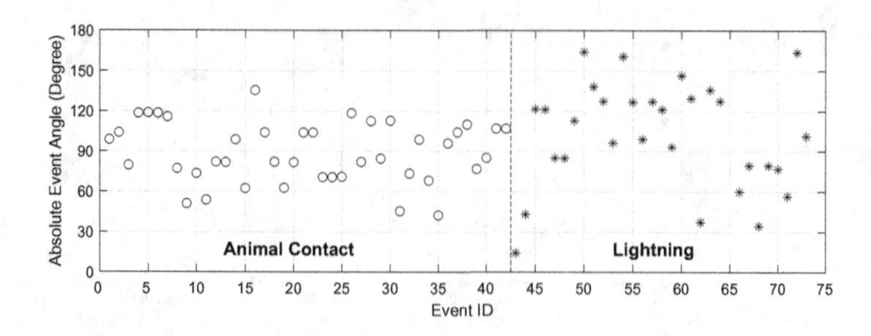

Figure 4.20 The absolute value of the event angle for several instances of recorded faults in overhead power lines, caused by animal contact or lightning [235].

The other 31 events were caused by *lightning*. The absolute value of the event angles for these 73 events are shown in Figure 4.20. We can see that the majority of the events that are caused by animal contact have an absolute angle between 60° and 120°. That means these events almost always occurred around the *positive or negative peaks* of the voltage waveform. Note that, when the voltage is at its absolute peak, the voltage gradient between the animal and the overhead power line is at its maximum. Conversely, we do not see any such pattern for the events that are caused by lightning. Therefore, we may conclude that if the angle for an event does *not* fall within the aforementioned range, then it is unlikely that the fault was caused by an animal contact.

Magnitude and duration of events in waveform measurements can also be used to characterize the events and identify their root causes.

Example 4.17 Figure 4.21(a) shows the sub-cycle blips for over 30 self-clearing single-phase faults in underground cables. The curves in this figure are derived by obtaining the differential current waveform during each fault. Four generic equations can be used to represent all these sub-cycle faults [217]:

$$1: \quad i_f(t) = 6.17\sin(695.4t - 0.126), \qquad 0 \leq t \leq 0.0047 \text{ sec},$$

$$2: \quad i_f(t) = 3.029\sin(451.37t - 0.02726), \qquad 0 \leq t \leq 0.0071 \text{ sec},$$

$$3: \quad i_f(t) = 1.275\sin(458.036t + 0.04495), \qquad 0 \leq t \leq 0.0068 \text{ sec},$$

$$4: \quad i_f(t) = 0.6846\sin(612.708t + 0.213), \qquad 0 \leq t \leq 0.0048 \text{ sec}.$$

They are shown in Figure 4.21(b). These four classes can be labeled as: (1) very high magnitude with a duration of about 1/4 of a cycle; (2) high magnitude with a duration of about 1/2 of a cycle; (3) small magnitude with a duration of about 1/2 of a cycle; and (4) small magnitude with a duration of about 1/4 of a cycle. These labels can be used to determine the cause of the fault. For instance, faults with high magnitude but short duration, under label 1, may be the result of moisture entering the cable splice; which means the cable failure is likely in its early stage. Conversely, faults with long duration

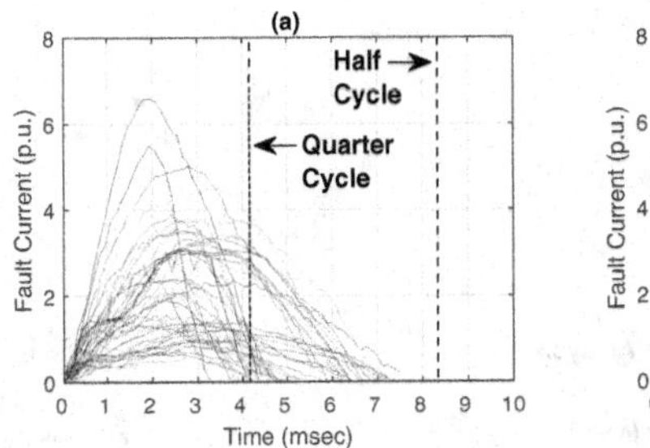

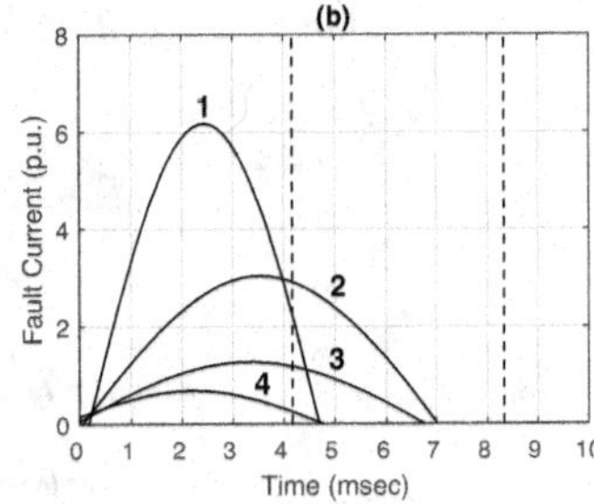

Figure 4.21 Clustering a group of sub-cycle self-clearing fault current waveforms: (a) raw measurements [217]; (b) four clusters based on four represented curves.

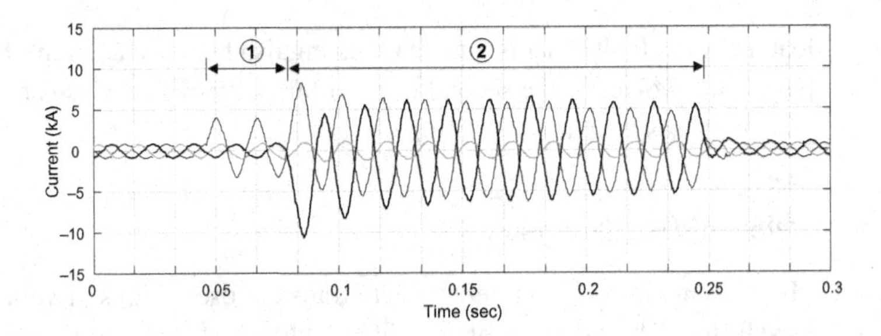

Figure 4.22 An event that begins as single-phase but evolves into two phases [236].

but small magnitude, under label 3, may be caused by insulation breakdown; which means the cable failure is perhaps in the mid stage, moving toward a more sustained fault [217].

The duration of an event may not be the same when it is calculated based on the current waveform versus when it is calculated based on the voltage waveform. For instance, consider the two faults in Figures 4.11(a) and (b) in Section 4.3.1. In both cases, the distortions in the current waveform almost disappeared in less than one cycle; however, the distortions in the voltage waveform continued for several cycles.

4.4.2 Number of Affected Phases

An event in waveform measurements may affect one phase (A, B, or C), two phases (A and B, A and C, or B and C), or all three phases (A and B and C). For example, most ground faults occur on only one phase. But most switching events take place on all three phases. Accordingly, the number of affected phases can be used as another feature for the analysis and identification of events.

Some events may begin as single-phase but *evolve* into other phases. An example is shown in Figure 4.22. This fault is initially single-phase, during the period that is marked as ①. It then evolves into a second phase, during the period that is marked as ②, before it is cleared by a protective device. Such evolving behavior by itself is a

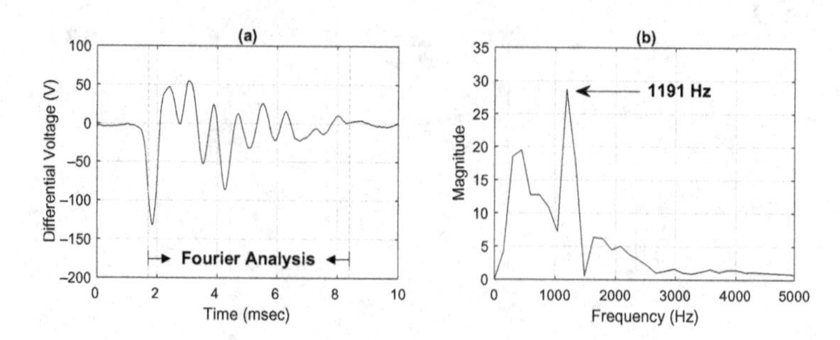

Figure 4.23 Obtaining the dominant frequency of the transient oscillations in Example 4.18: (a) differential waveform; (b) frequency spectrum based on Fourier analysis.

feature of the fault. That is, one may distinguish between an event that starts from one phase and evolves into a second phase after a few cycles, versus an event that occurs on two phases from the beginning.

4.4.3 Transient Oscillations

Many events in power systems create transient oscillations in voltage and/or current waveforms. The duration of transient oscillations may vary from a few microseconds to several milliseconds. Transient oscillations in waveform measurements are described by their *magnitude*, *duration*, and *dominant frequency*. The frequency of oscillations in waveform measurements can be obtained by using modal analysis methods that we discussed in Section 2.6.3 in Chapter 2.

Example 4.18 Consider the ringing event in Example 4.5. The differential voltage waveform for this event is shown in Figure 4.23(a). Once we apply Fourier analysis to the waveform between the two vertical dashed lines, the frequency spectrum is obtained as in Figure 4.23(b). The dominant frequency is 1191 Hz.

The frequencies for waveform transient oscillations are classified as [237]:

- Low Frequency: Below 5 kHz;
- Medium Frequency: Between 5 kHz and 500 kHz;
- High Frequency: Above 500 kHz.

Low-frequency transient oscillations take place frequently in sub-transmission and distribution systems. They are caused by different types of events. For instance, the transient oscillations during the capacitor switching event in Example 4.13 had a dominant frequency of 300 Hz, and the transient oscillations during the fault-induced capacitor ringing event in Example 4.18 had a dominant frequency of 1.2 kHz. Oscillatory transients with dominant frequencies less than 300 Hz are often associated with ferroresonance and transformer energization.

Medium-frequency transient oscillations take place in certain events, such as back-to-back capacitor energization. This happens when a capacitor bank is energized in close electrical proximity to another capacitor bank that is in service. Another cause for medium-frequency transient oscillations is cable switching.

High-frequency transient oscillations are often the result of a local response of an equipment or device to an impulsive transient disturbance in the system.

Note that in order to capture high-frequency transient oscillations, the waveform sensor must have a very high reporting rate. For instance, to capture the transient oscillations with a frequency of 500 kHz, the reporting rate of the waveform sensor must be at least 1 MHz. To support such a high reporting rate, the sampling rate of the sensor must be even higher in order to allow any pre-processing of the measured data within each reporting interval. See Section 2.3 in Chapter 2 for more discussions on reporting rates and sampling rates.

4.4.4 Transient Impulses

An impulsive transient is a sudden change in the waveform of voltage, current, or both, that is unidirectional in polarity, i.e., it is primarily either positive or negative. The most common cause of impulsive transients is lightning; see Figure 4.7 in Section 4.2. However, certain faults may also create transient impulses. For example, the four transient impulses that we saw in Figure 4.12(a) in Section 4.2 were due to a fault in the termination point of an underground cable. Notice that they appeared only in the voltage waveform and only at the start of the event; three of these four transient impulses had negative polarity, and one had positive polarity.

Transient impulses are normally characterized by their rise and decay times. For example, when an impulsive transient in a voltage waveform is described as having a 1.2/50 waveshape, 1.2 expresses a measure of the rise time in microseconds and 50 expresses a measure of the decay time in microseconds [237].

4.4.5 Fault-Specific Features

Some features in waveform measurements are specific to certain faults. For example, recall from Section 4.3.3 that failures in transformer tap changers may cause momentary zero current on a faulted phase. At its early stages, this type of fault may occur only occasionally and may last for only a fraction of a cycle. However, over time, the zero-current incidents may take place more frequently and may last longer, ultimately leading to a major failure. Therefore, we may evaluate the *trend* in the characteristics of the zero-current incidents in order to predict the fault.

Example 4.19 In a period of one week before a transformer was taken out of service for repair, about 40 zero-current events were detected, all on the same phase. The duration of these zero-current disturbances during this period are shown in Figure 4.24(a). We can see that the duration was initially at the sub-cycle level during the first two days; however, it subsequently increased to several cycles. On the day before the transformer was taken out for repair, half of the zero-current events lasted more than

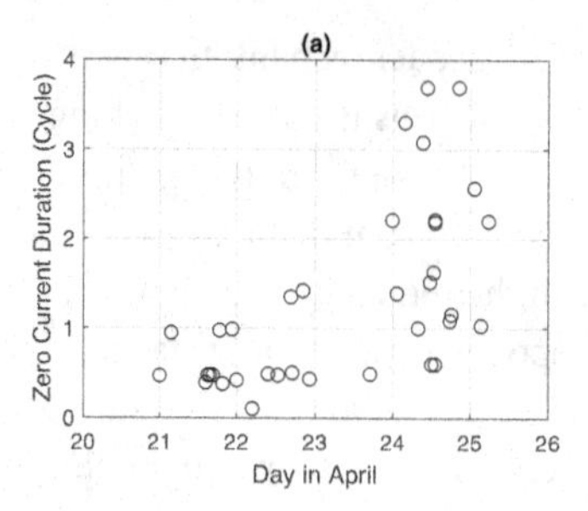

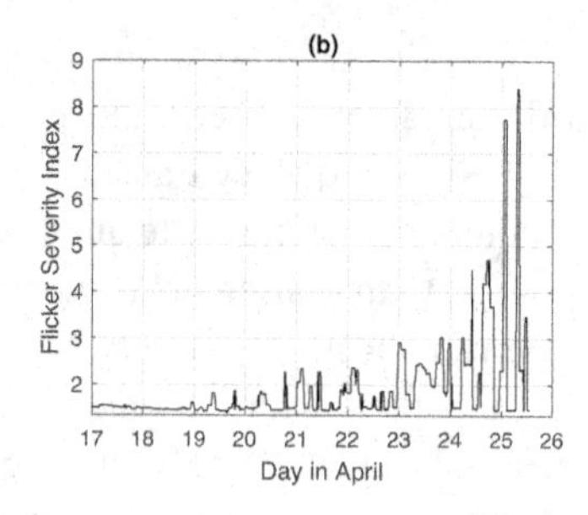

Figure 4.24 Trends to predict transformer tap changer failure [215]: (a) duration of the zero-current events (see Example 4.19); (b) flicker severity index (see Example 4.20).

two cycles. The trend in this figure can be used as an indication that the tap changer is likely to fail in the near future.

Not all incipient failures in transformer tap changers create zero-current events. However, there is an alternative feature that we can check to potentially predict tap changer failure. It is called the *long-term flicker severity*, denoted by P_{lt}, which is an index to evaluate the severity of flicker in voltage waveforms. It is obtained by taking the average of another index, called the *short-term flicker severity*, denoted by P_{st}, over intervals of two hours. This latter index itself is obtained by conducting a statistical analysis of flicker harmonics in the voltage waveform. The instructions to calculate P_{st} and P_{lt} are given in [238, 239].

Example 4.20 Again, consider the transformer failure in Example 4.19. Figure 4.24(b) shows the flicker severity index P_{lt} that was reported on the faulted phase voltage during the same week as in Example 4.19, plus a few days earlier. We can see that, at about two days before the first zero-current event occurs, we start seeing some minor abnormalities in the voltage flicker severity. These abnormalities grow drastically over the next few days. Similar trends are reported in other case studies of transformer tap changer failures; e.g., see [220, 236].

Another feature that can be measured during faults is *fault impedance*. It can be obtained from the faulted phase voltage waveform together with either the faulted phase current waveform or the neutral current waveform. There are advantages in using the neutral current waveform, if it is available, because it does not contain the load current; therefore, it can provide a better estimation of the fault impedance. We may obtain the fault impedance as follows:

$$Z_{\text{fault}} = \frac{V\angle\theta}{I_n\angle\phi_n}, \tag{4.21}$$

where $V\angle\theta$ denotes the phasor estimation of the *fundamental component* of the faulted phase voltage, and $I_n\angle\phi_n$ denotes the phasor estimation of the *fundamental component* of neutral current. Here, the faulted phase corresponds to the phase that

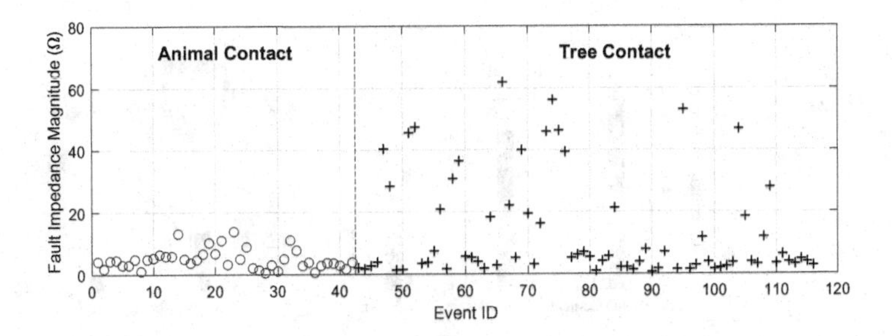

Figure 4.25 Fault impedance for several instances of recorded faults in overhead power lines, caused by animal contact and tree contact [235].

experiences the maximum voltage drop during the fault. Also, if the fault lasts for more than one cycle, then Z_{fault} can be calculated for each cycle. In that case, the minimum of the impedance value across multiple fault cycles is considered as the impedance of the fault because it corresponds to the worst fault condition, which causes the highest fault current [235]. See Exercise 4.14 for an example on how to obtain the fault impedance from the voltage and current waveform measurements. The information on the magnitude of fault impedance may help identify the cause, the location, or other characteristics of the event [240].

Example 4.21 Figure 4.25 shows the measured fault impedance for 116 recorded faults in overhead power lines. The first 42 events were caused by *animal contact.* The other 74 events were caused by *tree contact.* We can see that the fault impedance during animal contacts were always below 20 Ω. Conversely, the fault impedance during tree contacts were above 20 Ω in about one quarter of such events. Therefore, we can conclude that if the fault impedance is above 20 Ω, then it is very unlikely that the fault was caused by an animal contact.

Regarding the faults in Example 4.21 that were caused by tree contact, it is further observed that if the fault impedance magnitude is less than 20 Ω, then the event is likely caused by *tree leaves.* If the fault impedance magnitude is greater than 20 Ω, then the event is likely caused by *tree branches*; cf. [241, 242].

4.4.6 Changes in Steady-State Characteristics

Certain events in waveform measurements may create changes in the steady-state characteristics of the waveform. Here, we are comparing the steady-state characteristics *immediately before* and *immediately after* the event. For instance, we saw in Example 4.13 that energizing the capacitor significantly changed the phase angle difference between the voltage waveform and the current waveform.

Other steady-state characteristics may also be considered. For example, one may examine whether there is a considerable increase or a considerable decrease in the strengths of harmonics and the THD value in the current or voltage waveform

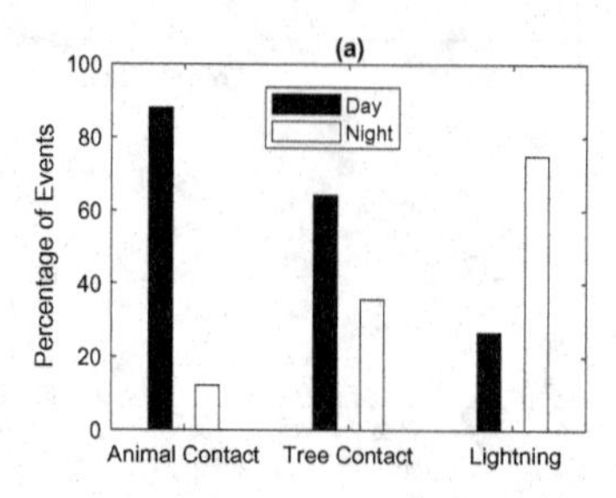

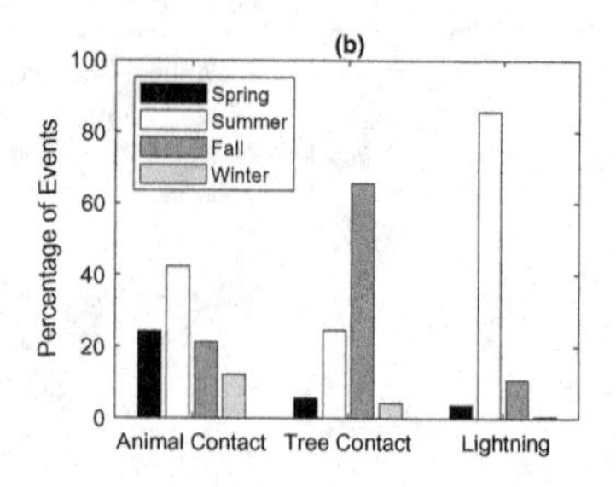

Figure 4.26 Comparing the percentages of different causes of faults at overhead power lines, evaluated based on (a) time of day; (b) season [243].

measurements *before* and *after* the event. Obtaining the pre-event THD is often straightforward; however, one has to wait until the waveform reaches a steady state after the event before the post-event THD is calculated.

4.4.7 Time, Season, and Location

The time-stamp of the event can sometimes be very informative. For example, Figure 4.26(a) shows the percentage of overhead line faults of different causes that occurred during the day versus during the night. We can see that the faults that are caused by animal contact are much more likely to happen during the day, 88% versus 12%. Combine this with what we learned about the angle of such faults in Figure 4.20, and about the impedance of such faults in Figure 4.25, and we already have several quantitative features to statistically characterize this type of fault.

Some events are highly seasonal. For example, Figure 4.26(b) shows the percentage of overhead line faults of different causes that occurred during spring, summer, fall, and winter. The faults that are caused by lightning are much more likely to happen in summer, 86% versus 14% in any other season. Of course, such seasonal correlation can change significantly depending on geographical location.

Figure 4.27 shows the percentage of all faults that are caused by animal contact versus the percentage of all faults that are caused by tree contact across the substations of two neighboring operation centers in North Carolina in the United States over a period of five years. The substations of Operation Center 1 are between 300 and 400 miles away from the substations of Operation Center 2. We can see that the two operation centers experience generally similar percentages of faults that are caused by animal contact. However, the percentages of the faults that are caused by tree contact are much higher across the substations of Operation Center 2. Better vegetation management might be needed in this operation center.

4.4.8 Other Features

There may exist many other types of quantitative factors that can be used to characterize certain types of events in waveform measurements. One example is the *weather*

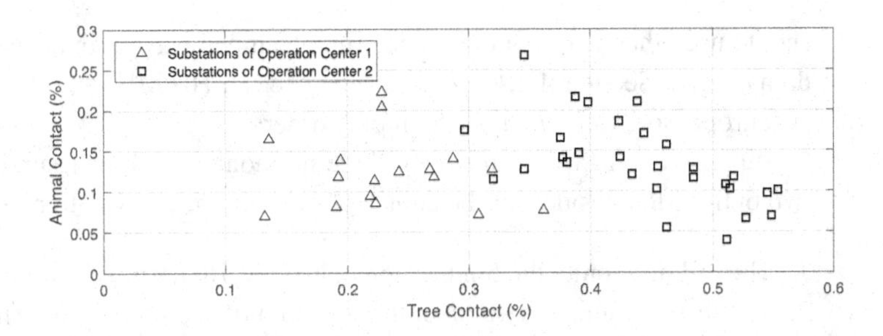

Figure 4.27 Comparing the statistics of the root-causes of the faults on overhead lines across several substations of two neighboring operation centers [244].

condition, i.e., whether the event happens during clear sky, cloudy sky, rain, thunderstorm, wind, snow, ice, or heat. Another example is the *loading condition*, such as whether the event happens during peak-load hours.

Features from Other Types of Sensors

Correlation between waveform sensors with some other sensors, such as equipment sensors, could also be used as quantitative features to evaluate certain waveform events. For instance, incipient failures in transformer tap changers may show *strong correlations* with the measurements at equipment sensors that monitor the chemicals and temperature of transformers. We will discuss these and other types of equipment sensors and measurements in Chapter 7.

4.5 Harmonic Synchrophasors

From (4.1), a voltage or current waveform with steady-state distortion can be represented in frequency-domain using its fundamental phasor $X_1 \angle \phi_1$ together with the following harmonic phasors:

$$X_2 \angle \phi_2, \; X_3 \angle \phi_3, \; X_4 \angle \phi_4, \; \ldots . \tag{4.22}$$

The frequencies for the above harmonic phasors are $2f, 3f, 4f$, etc. If the harmonic phasors are *time-stamped*, then they become *harmonic synchrophasors.*

The instrument to measure harmonic synchrophasors is called the Harmonic Phasor Measurement Unit (H-PMU). Unlike PMUs, H-PMUs are still a developing concept and they have not yet been widely adopted by the power industry.

One of the first protoypes for H-PMUs was proposed in [245], where synchronization is achieved using GPS receivers and the harmonic phasors are obtained using Fourier analysis. Recent efforts have improved the accuracy in measuring harmonic synchrophasors; see [246–249].

H-PMUs are *not* intended to stream the phasors for *all* harmonics. Providing such a heavy load of data is not easy, and it is not necessary either. In fact, if we really want to fully reconstruct the voltage or current waveforms from all harmonics, then we should use other sensors that provide time-stamped waveform measurements in time-domain; see Section 4.6.1. Thus, in practice, an H-PMU may stream the harmonic synchrophasors for *only a few* harmonic orders.

Suppose an H-PMU can stream three harmonic synchrophasors. We have at least two options to consider for the choice of the harmonic orders to be streamed:

1. The 3rd harmonic; the 5th harmonic; and the 7th harmonic.
2. The most dominant harmonic; the second most dominant harmonic; and the third most dominant harmonic.

In the first option, we provide a steady stream of three harmonic phasors that are often used in most applications. In the second option, we may switch across different harmonics depending on which harmonics are most dominant at any time. Of course, under the second option, the H-PMU should indicate the harmonic orders that it is streaming at any given time. The first option is likely more appropriate for steady-state analysis, such as in harmonic state estimation. The second option is likely more appropriate for the analysis of events and faults.

It should be noted that some of the concepts that we learned in Chapter 3 for PMUs are also applicable to H-PMUs, such as wrapping around the phase angle, the impact of ROCOF on measruing phase angle, and the definition of TVE.

4.5.1 Harmonic State Estimation

The magnitude of the harmonic component is typically much less than the magnitude of the fundamental component. From this, and also because the measurement range of an instrument is often adjusted for the range of fundamental values, harmonics cannot be measured precisely in all locations unless proper instrumentation equipment is installed specifically for the purpose of measuring harmonics. Therefore, it would be expensive and laborious to directly measure harmonics at every location in the entire power system. Instead, it is more desirable to measure harmonics rather precisely but only at certain locations, and then to use such measurements to estimate the harmonics in other locations.

The problem formulation for *harmonic state estimation* is generally similar to that of the fundamental state estimation in Section 3.8 in Chapter 3. However, one key difference is in the choice of the states of the system. While the system states in fundamental state estimation are often the magnitude and phase of the nodal voltage phasors at each bus, in harmonic state estimation, the states may rather consist of the magnitude and phase angle of the current injection phasors at each bus, or at each candidate bus that could be a source of harmonics. This is because one of the main purposes of harmonic state estimation is to identify the sources of major harmonics in the system. As for the measurements, they may include both nodal harmonic voltage phasors and branch harmonic current phasors.

Note that harmonic state estimation must be done separately for each harmonic order. For example, if our goal is to identify the sources of the 3rd, the 5th, and the 7th harmonics, then we should formulate and solve three separate harmonic state estimation problems, one for each harmonic order.

Basic Formulation

For each harmonic order h, let $\mathbf{z}\{h\}$ denote the vector of all nodal voltage and branch current harmonic phasor measurements. Also, let $\mathbf{x}\{h\}$ denote the vector of system states, i.e., the harmonic phasors for harmonic current injection at certain candidate buses; or otherwise all buses. Similar to (3.76) in Chapter 3, we can express the relationship between the measurements and the state variables as follows:

$$\mathbf{z}\{h\} = \mathbf{A}\{h\}\mathbf{x}\{h\} + \epsilon\{h\}, \tag{4.23}$$

where $A\{h\}$ is a matrix that relates $\mathbf{z}\{h\}$ and $\mathbf{x}\{h\}$; and $\epsilon\{h\}$ is the vector of measurement errors. The size and entries of matrix $\mathbf{A}\{h\}$ depend on the network configuration, as well as the number and choices of the measurements and states. The harmonic state estimation problem at harmonic order h is formulated as

$$\min_{\mathbf{x}\{h\}} \ \|\mathbf{z}\{h\} - \mathbf{A}\{h\}\mathbf{x}\{h\}\|_2 . \tag{4.24}$$

If enough harmonic synchrophasor measurements are available, then we can solve the above problem by using the standard LS method:

$$\mathbf{x}\{h\} = (\mathbf{A}\{h\}^T\mathbf{A}\{h\})^{-1}\mathbf{A}\{h\}^T\mathbf{z}\{h\}. \tag{4.25}$$

Example 4.22 Consider the 4-bus transmission network in Example 3.20 in Chapter 3. Suppose the steady-state current waveform on the transmission line between bus 2 and bus 4 is measured, as shown in Figure 4.28(a). Fourier analysis reveals the presence of a considerable 5th harmonic. Suppose the source of the harmonic is unknown. There are two candidate locations, bus 2 and bus 4, which are both load buses. Harmonic state estimation must be done on harmonic order $h = 5$ in order to obtain the source of harmonics. The network model is shown in Figure 4.28(b). In order to do harmonic state estimation, all power plants are represented by their internal impedances.

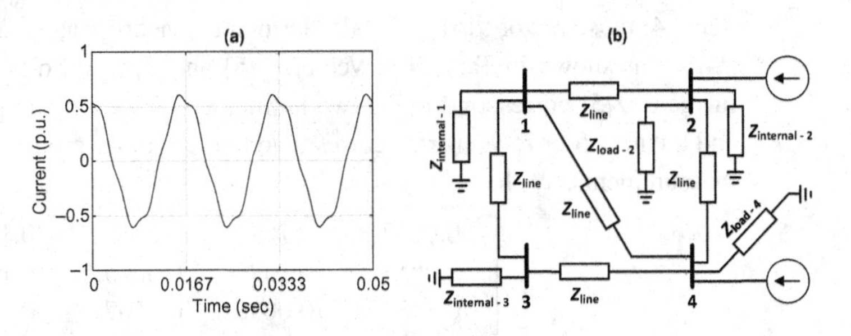

Figure 4.28 The measurements and the network model corresponding to the harmonic state estimation problem in Example 4.22: (a) steady-state current waveform on the transmission line between buses 2 and 4; (b) the network model at the 5th harmonic.

Table 4.1 The true and the measured 5th harmonic synchrophasors in Example 4.22.

Bus #	True Voltage (p.u.)	Measured Voltage (p.u.)
1	$0.034352\angle 67.5606°$	–
2	$0.034154\angle 67.3670°$	–
3	$0.034417\angle 67.6243°$	$0.033539\angle 67.0418°$
4	$0.035034\angle 67.6777°$	$0.034312\angle 66.6236°$
Line #	**True Current (p.u.)**	**Measured Current (p.u.)**
1,2	$0.011465\angle 8.3390°$	$0.011709\angle 8.6784°$
1,3	$0.003802\angle -171.6580°$	–
1,4	$0.034291\angle 164.1274°$	$0.035615\angle 164.3268°$
2,4	$0.044994\angle 170.1260°$	$0.046402\angle 170.3159°$
3,4	$0.030863\angle 161.2316°$	$0.030204\angle 161.8703°$

Two harmonic current sources are added to the network model at the two candidate harmonic source locations. All impedances must be represented based on the 5th harmonic. For instance, recall from Example 3.20 in Chapter 3 that the admittance of each transmission line is $0.5 - j10$ p.u. Accordingly, for the network model under the 5th harmonic, we have:

$$Z_{\text{line}} = 1/\left(0.5 - j10 \times 5\right) = 0.0002 + j0.02 \text{ p.u.} \tag{4.26}$$

The internal impedance at each of the three power plants is

$$Z_{\text{internal}} = j0.25 \times 5 = 1.25 \text{ p.u.} \tag{4.27}$$

The load impedances at load buses 2 and 4 are

$$\begin{aligned} Z_{\text{load},2} &= 0.5710 + j0.1903 \times 5 = 0.5710 + j0.9515 \text{ p.u.} \\ Z_{\text{load},4} &= 0.2239 + j0.0564 \times 5 = 0.2239 + j0.2820 \text{ p.u.} \end{aligned} \tag{4.28}$$

The state variables are the two *unknown* injected harmonic current phasors at buses 2 and 4; thus, vector $\mathbf{x}\{5\}$ is 2×1. Harmonic synchrophasors are measured using H-PMUs, as shown in Table 4.1. Vectors $\mathbf{z}\{5\}$ and $\boldsymbol{\epsilon}\{5\}$ are 6×1. The first two rows in vector $\mathbf{z}\{5\}$ correspond to the two harmonic voltage synchrophasor measurements, and the next four rows in $\mathbf{z}\{5\}$ correspond to the four harmonic current synchrophasor measurements. We have:

$$\mathbf{A} = \begin{bmatrix} 0.0647 + j0.1572 & 0.0655 + j0.1591 \\ 0.0657 + j0.1576 & 0.0665 + j0.1620 \\ -0.3136 + j0.0098 & 0.0567 + j0.0083 \\ 0.0822 + j0.0453 & -0.1649 + j0.0469 \\ 0.3958 + j0.0354 & -0.2216 + j0.0386 \\ -0.0218 + j0.0485 & -0.1461 + j0.0497 \end{bmatrix}. \tag{4.29}$$

From (4.25), the states of the system at buses 2 and 4 are obtained as

$$0.00331\angle{-173.71^\circ} \quad \text{and} \quad 0.20017\angle{-0.34^\circ}. \tag{4.30}$$

We can see that the magnitude of the estimated injected current at bus 4 is almost 200 times more than the magnitude of the estimated injected current at bus 2. Therefore, we can conclude that the source of the harmonic is at bus 4.

Sparsity

If only a small number of H-PMUs are available, then (4.24) can be under-determined. In particular, if the number of harmonic synchrophasor measurements is less than the number of candidate buses, then matrix $\mathbf{A}\{h\}^T\mathbf{A}\{h\}$ can be singular; which leads to unbounded estimation error. Even if the number of harmonic synchrophasor measurements is just enough for this matrix to be theoretically non-singular, it can still be ill-conditioned in case of low redundancy; which can create numerical issues in solving the harmonic state estimation problem. However, the good news is that, in practice, there are only a small number of simultaneous large harmonic sources among the candidate buses. Therefore, for any harmonic order, the solution space for the harmonic state estimation can be *sparse*, in particular in power distribution systems. Therefore, we can use the concept of *sparse recovery* from signal processing, cf. [250], and reformulate problem (4.24) as in [251]:

$$\begin{aligned} \min_{\mathbf{x}\{h\}} \quad & \|\mathbf{x}\{h\}\|_0 \\ \text{s.t.} \quad & \|\mathbf{z}\{h\} - \mathbf{A}\{h\}\mathbf{x}\{h\}\|_1 \le \varepsilon, \end{aligned} \tag{4.31}$$

where $\|\{h\}\|_0$ is the l_0 norm, which equals the number of non-zero entries; and $\|\{h\}\|_1$ is the l_1 norm, which equals the summation of the absolute values of all entries. Basically, in (4.31), we want to select the *minimum number of harmonic source locations* among all the candidate locations, subject to maintaining a small level of residue in the measurements. While (4.31) is a convex optimization problem [103], it is not easy to solve this problem due to computational issues that arise because of the l_0 norm in the objective function. Therefore, in practice, one may replace the l_0 norm in the objective function with l_1 norm to obtain an approximate solution. Details about sparse recovery in power system state estimation are available in [154, 251].

4.5.2 Topology Identification

Harmonic synchrophasors can also be used in parameter estimation. For example, we may use harmonic synchrophasors to estimate the status of a switch in order to solve the topology identification problem, also see Section 3.8.3 in Chapter 3. Of course, whether a switch is open or closed is not by itself related to harmonics; however, the presence or the absence of certain harmonics may help us determine the status of the switch and therefore identify the network topology.

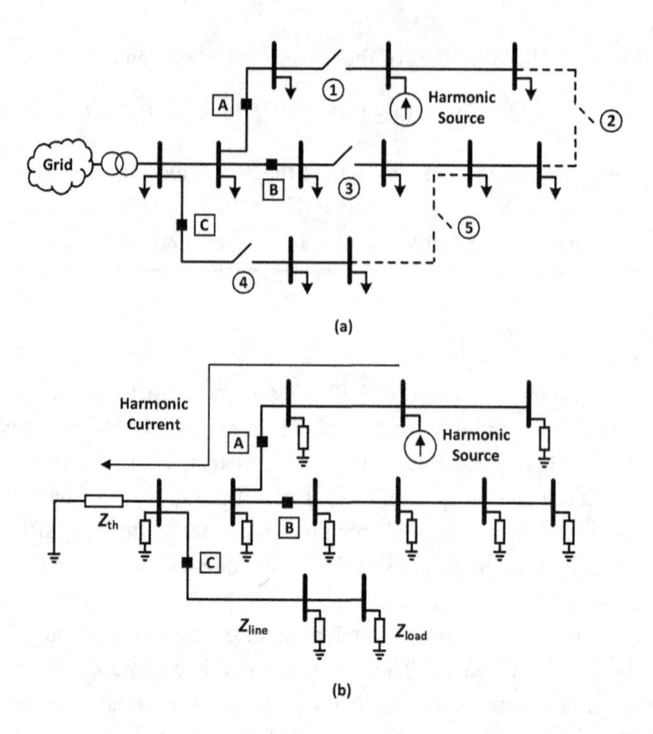

Figure 4.29 Application of H-PMUs in topology identification: (a) a power distribution network with 11 buses, three normally closed switches, two normally open switches, and three H-PMUs; (b) the network model at the harmonic when the normally closed switches are closed and the normally open switches are open.

Consider the distribution network in Figure 4.29(a). The network includes five switches. Switches ①, ③, and ④ are the normally closed switches. Switches ② and ⑤ are the normally open switches. There is a single source of harmonic on the top lateral. The location of the harmonic source is known. There are three H-PMUs at the beginning of each lateral, which are denoted by A, B, and C. They measure the harmonic current synchrophasors on the three laterals.

Suppose switches ①, ③, and ④ are closed and switches ② and ⑤ are open. In that case, the network model at the harmonic is obtained as shown in Figure 4.29(b), where the only source is the harmonic source. All other sources are removed. This includes the voltage source in the Thevenin equivalent circuit model of the distribution substation. All loads are modeled by their impedance. We can make the practical assumption that the impedance of the distribution lines is negligible. All impedances are adjusted based on the order of the harmonic, i.e., similar to Section 4.5.1.

In practice, the magnitude of Z_{th} is much less than the magnitude of Z_{load}. Therefore, the harmonic current almost entirely flows through the substation and not through the loads; see the arrow in Figure 4.29(b). As a result, only H-PMU A may measure a harmonic current phasor with a considerable magnitude.

The above analysis can be used to identify the status of switches. For example, if considerable harmonic current is reported by H-PMU A but not by H-PMUs B and C, then we may conclude that switch ① is closed and switch ② is open. If considerable harmonic current is reported by H-PMU C but not by H-PMUs A and B, then we may conclude that switches ① and ③ are open and switches ②, ④, and ⑤ are closed. Note that, in this analysis, we assume that all network topologies are *radial*; that means there is no loop in the distribution network. See [252] for more details on how harmonic synchrophasors can be used to identify the network topology in power distribution systems.

Probing Interpretation

The harmonic component that is generated by the harmonic source in the above analysis can be seen as a *probing* signal that is monitored by the three H-PMUs in order to identify the network topology. Such a harmonic source could be simply a nonlinear load that already exists in the system at a known location. However, if needed, the harmonic source could also be intentionally positioned in the network such that it can help with identifying the network topology. In this latter setup, the harmonic could be generated only when the operator is unsure about the network topology. We will discuss the concept of probing in greater detail in Chapter 6.

4.5.3 Differential Harmonic Synchrophasors

The concepts of phasor differential and differential synchrophasors that we covered in Section 3.5 in Chapter 3 can also be defined for harmonic synchrophasors. PD corresponding to harmonic order h is obtained as

$$\Delta X_h = X_{h,\text{before}} - X_{h,\text{after}}, \tag{4.32}$$

where $X_{h,\text{before}}$ and $X_{h,\text{after}}$ denote the harmonic phasor measurements of order h that are obtained *before* and *after* the event, respectively. Once such PDs are obtained at multiple H-PMUs, they can be time synchronized in order to create *differential harmonic synchrophasors* of harmonic order h. Differential harmonic synchrophasors capture how an event may create, eliminate, or change the magnitude or phase angle of steady-state waveform distortions in voltage or current.

Similar to differential synchrophasors, we can use differential harmonic synchrophasors to identify the location of an event. This is particularly useful if the event mainly affects the harmonics, as opposed to the fundamental. An example for such event is a *high-impedance fault* (HIF) that can be caused, for instance, by a tree with a power line; see Section 4.4.5. Due to its high impedance, an HIF may not create a major change in the steady-state fundamental voltage or current phasors that are measured by PMUs. Thus, if we use the methodology in Section 3.5.4 in Chapter 3, it may not accurately identify the location of the fault. However, an HIF does often create a considerable *3rd* harmonic [253], which can cause a major change in the steady-state harmonic current and voltage phasors that are measured by H-PMUs. Therefore, we may still use the method in Section 3.5.4, but we should apply it to differential harmonic synchrophasors of the 3rd harmonic, instead of the fundamental differential synchrophasors; see [254].

4.6 Synchronized Waveform Measurements

Once the measurements from multiple waveform sensors are time-synchronized, such as by using GPS as we discussed in Section 3.2 in Chapter 2, they can collectively form *synchronized waveform measurements*, also known as *synchrowaveform* measurements. In this regard, we can extend the concept of PMUs and H-PMUs and introduce a new class of sensors, called Waveform Measurement Units (WMUs), which report time-stamped voltage and current waveform measurements; cf. [255]. Unlike the measurements from PMUs and H-PMUs that are in frequency-domain, the measurements from WMUs are in time-domain.

The measurements that are provided by WMUs in time-domain during *steady-state* conditions practically provide the same level of information as the measurements that are provided by PMUs and H-PMUs in frequency-domain. However, the story is different when it comes to monitoring *transient* conditions. Since both PMUs and H-PMUs work based on applying Fourier analysis to a window of waveform measurements, they are inherently incapable of providing the same level of information about the transient behavior in the voltage and current waveforms that are provided by WMUs in time-domain. Therefore, the measurements from WMUs are likely to be most appreciated when they are used to analyze events and transient conditions across the power system, e.g., see Figure 4.30 [256]. Nevertheless, in the future, it might be valuable for WMUs to provide continuous streaming of the synchronized waveform measurements, similar to the continuous streaming of the synchronized phasor measurements in PMUs. WMUs that can provide such continuous synchrowaveform measurements may also be referred to as Continuous Point-on-Wave (CPoW) sensors.

4.6.1 Relative Waveform Difference

Consider two synchronized voltage waveform measurements $v_1(t)$ and $v_2(t)$ that are reported by two WMUs. We can define the Relative Voltage Waveform Difference (RVWD) for these two synchronized waveform measurements as

$$\text{RVWD}(t) = v_1(t) - v_2(t), \tag{4.33}$$

which itself is a waveform; see Figure 4.31(a). We can see that the event creates a major signature in RVWD. It involves a spike followed by damping oscillations. RVWD can also be analyzed in frequency domain, as shown in Figure 4.31(b). Two dominant frequencies are identified, at 60 Hz, which is the fundamental frequency, and at 440 Hz, which is the frequency of the damping oscillations. Both the time-domain signature in Figure 4.31(a) and the frequency-domain signature in Figure 4.31(b) can be used to detect and characterize the event.

Relationship with RPAD

The concept of RVWD is related to the concept of RPAD in Chapter 3. Let $V_1\angle\theta_1$ and $V_2\angle\theta_2$ denote the phasor representation of the synchronized voltage waveforms

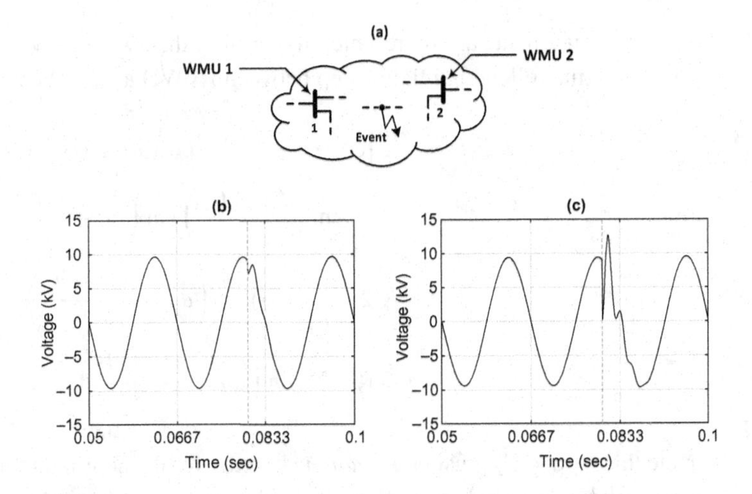

Figure 4.30 An illustrative example for synchronized waveform measurements: (a) two WMUs simultaneously capture the voltage waveform during an event; (b) voltage waveform that is captured by WMU 1; (c) voltage waveform that is captured by WMU 2.

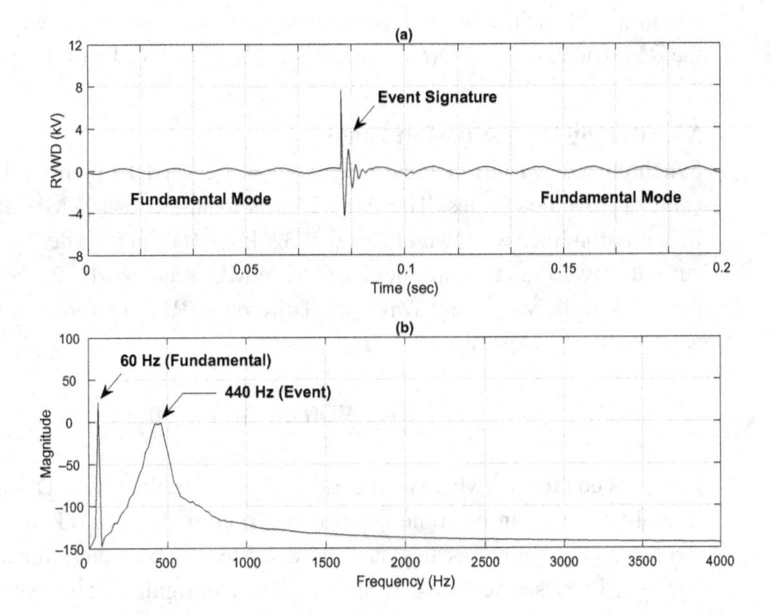

Figure 4.31 Analysis of RVWD across WMU 1 and WMU 2 in Figure 4.30: (a) analysis in time-domain; (b) analysis in frequency-domain.

$v_1(t)$ and $v_2(t)$ at WMU 1 and WMU 2, respectively. At steady state, and assuming that voltage waveforms are purely sinusoidal, we have:

$$v_1(t) = \sqrt{2}V_1 \cos(\omega t + \theta_1), \tag{4.34}$$

$$v_2(t) = \sqrt{2}V_2 \cos(\omega t + \theta_2). \tag{4.35}$$

If the measurements are represented in per unit, then we have $V_1 = V_2 = 1$ p.u. We can derive the following relationship between RVWD and RPAD:

$$\begin{aligned} \mathrm{RVWD}(t) &= \sqrt{2}\left[\cos(\omega t + \theta_1) - \cos(\omega t + \theta_2)\right] \\ &= -2\sqrt{2}\sin\left(\frac{\theta_1 - \theta_2}{2}\right)\sin\left(\omega t + \frac{\theta_1 + \theta_2}{2}\right) \\ &\approx -\sqrt{2}\,(\theta_1 - \theta_2)\sin\left(\omega t + \frac{\theta_1 + \theta_2}{2}\right) \\ &= \sqrt{2}\,\mathrm{RPAD}\ \sin\left(\omega t + \frac{\theta_1 + \theta_2}{2}\right). \end{aligned} \tag{4.36}$$

Note that θ_1 and θ_2 must be in *radians* in order for the approximation in the third line to be valid; accordingly, RPAD must also be expressed in radians.

The expression in (4.36) can help explain the steady-state sinusoidal oscillations before and after the event in Figure 4.31(a). The frequency of such sinusoidal oscillations is ω, which is the fundamental angular frequency of the power system. The magnitude of such sinusoidal oscillations is $\sqrt{2}$ RPAD. In other words, at steady state, the RMS value of RVWD(t) is approximately equal to RPAD.

Monitoring Transmission Lines

Synchronized waveform measurements can be used to monitor long underground cables or overhead lines. The basic idea is to install one WMU at each end of the transmission line, as shown in Figure 4.32. Let $i_1(t)$ and $i_2(t)$ denote the synchronized current waveforms that are measured by WMU 1 and WMU 2, respectively. We can define the Relative Current Waveform Difference (RCWD) for these two synchronized waveform measurements as:

$$\mathrm{RCWD}(t) = i_1(t) - i_2(t). \tag{4.37}$$

If there is no fault anywhere on the transmission line, then RCWD is theoretically zero. In reality, there can be some small nonzero current in RCWD due to measurement errors or other minor issues such as discrete spectral interference for the case of overhead lines; see Section 4.7. Nevertheless, any significant increase in the magnitude of RCWD can be seen as an indication for the presence of *fault current*, which is denoted by $i_f(t)$ in Figure 4.32. Recall from Section 4.3 that the fault current can be due to various faults or even incipient faults, such as moisture penetration into underground cables and tree contact with overhead lines.

It should be noted that the above application of WMUs in transmission line monitoring is similar to the concept of *differential protection* in differential power system relays, where time-synchronization between the two measurement points are often achieved directly by using fiber-optic communications; e.g., see [257].

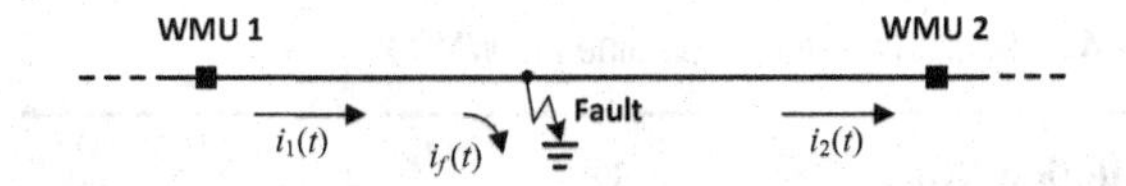

Figure 4.32 Identifying and characterizing a fault in an underground cable or overhead power line by obtaining RCWD across two WMUs.

4.6.2 Modal Analysis of Synchronized Transient Waveforms

Measurements captured by WMUs can also be analyzed in frequency-domain. Of course, if our concern is to examine the steady-state synchronized waveform measurements, then we should use PMUs or H-PMUs. However, if our concern is to examine the transient behavior across the synchronized waveform measurements, then we can apply the modal analysis (see Section 2.6.3 in Chapter 2) to the waveform measurements that are captured by each WMU.

Example 4.23 Again, consider the two synchronized voltage waveform measurements in Figures 4.30(b) and (c). The moment when the event occurs is marked on both waveforms using vertical dashed lines. We can apply the modal analysis to each waveform immediately after the event occurs. The five most dominant oscillatory modes are obtained as shown in Tables 4.2 and 4.3, respectively. Both waveform measurements include a dominant fundamental mode, at frequency 60 Hz with no damping. Each waveform measurement also includes several damping oscillatory modes. The frequencies of the damping oscillatory modes are not the same at WMU 1 and WMU 2. The frequency of the most dominant damping oscillatory mode at WMU 1 is 441.91 Hz, and at WMU 2 is 434.72 Hz. These frequencies coincide with the inter-harmonics between the 7th and the 8th harmonics. Note that all these damping oscillatory modes are transient and caused by the event. They dissipate quickly, as we saw in Figures 4.30(b) and (c).

The modal analysis of the synchronized voltage and current waveform measurements that are captured by multiple WMUs during a transient oscillation event can be utilized to identify the location of the event. There are ongoing efforts to address this emerging research area; e.g., see [256, 258].

Table 4.2 Oscillation modes obtained at WMU 1

Oscillation Mode	1	2	3	4	5
Frequency (Hz)	60	355.27	441.91	556.85	792.92
Damping Factor (Hz)	0	−1.98	−1.49	−2.08	−2.44
Amplitude (kV)	4.842	0.325	1.189	0.493	0.153
Phase Angle (°)	17.94	−176.48	151.64	−97.87	−55.75

Table 4.3 Oscillation modes obtained at WMU 2

Oscillation Mode	1	2	3	4	5
Frequency (Hz)	60	345.43	434.72	526.47	748.12
Damping Factor (Hz)	0	−1.72	−1.67	−1.90	−2.40
Amplitude (kV)	4.803	0.912	6.615	2.816	0.800
Phase Angle (°)	16.14	173.96	175.94	−38.90	13.88

4.7 Accuracy in Waveform Measurements

The basic metrics to evaluate the accuracy of waveform sensors are similar to those that we saw in Section 2.3 in Chapter 2. For instance, for the waveform sensor in [259], the accuracy is presented as 0.33% of the full scale (`FS`), where `FS` is 600 V. Accordingly, the accuracy for this sensor is ±1.98 V.

4.7.1 Impact of Noise and Interference

When it comes to waveform sensors, it can sometimes be very difficult to distinguish abnormal signatures from noise and interference. For example, consider the three-phase voltage waveform measurements in Figure 4.33 that are obtained at an overhead transmission line conductor. Each of the three sub-figures shows the waveform on one phase. We can observe that the measurements appear as noisy in all three phases. For instance, let us zoom in at the two areas that are marked as ① and ②. They are shown in Figures 4.34(a) and (b), respectively.

First, consider the fluctuations in Figure 4.34(a). Are these fluctuations due to measurement noise in a potentially defected sensor? Or does the voltage at the line conductor actually fluctuate? For the example in this figure, the answer is the latter. These fluctuations are due to Discrete Spectral Interference (DSI), which is caused by radio broadcasting. An overhead transmission line is practically a long wire *antenna*; and therefore, it acts as a *receiver* for long-wave radio transmissions. Sometimes, more than 100 radio stations can be identified in the waveform measurement data at an overhead transmission line. For the example in Figure 4.34(a), the biggest permanent source of DSI on the waveform measurements was the 225 kHz carrier wave of a 1 MW transmitter at a radio station that was situated 200 miles away from the measurement site [261]. Broadcasting of the radio waves is variable and it depends on many factors, cf. [262]. DSIs that are caused by radio broadcasting can often be recognized based on their frequency. DSIs can be removed by using adequate low-pass filters.

Next, consider the impulse in Figure 4.34(b). Notice that the impulse lasts for only a few *microseconds*. Therefore, capturing this impulse requires a waveform sensor that has a very high time resolution. Broadly speaking, this type of impulses may

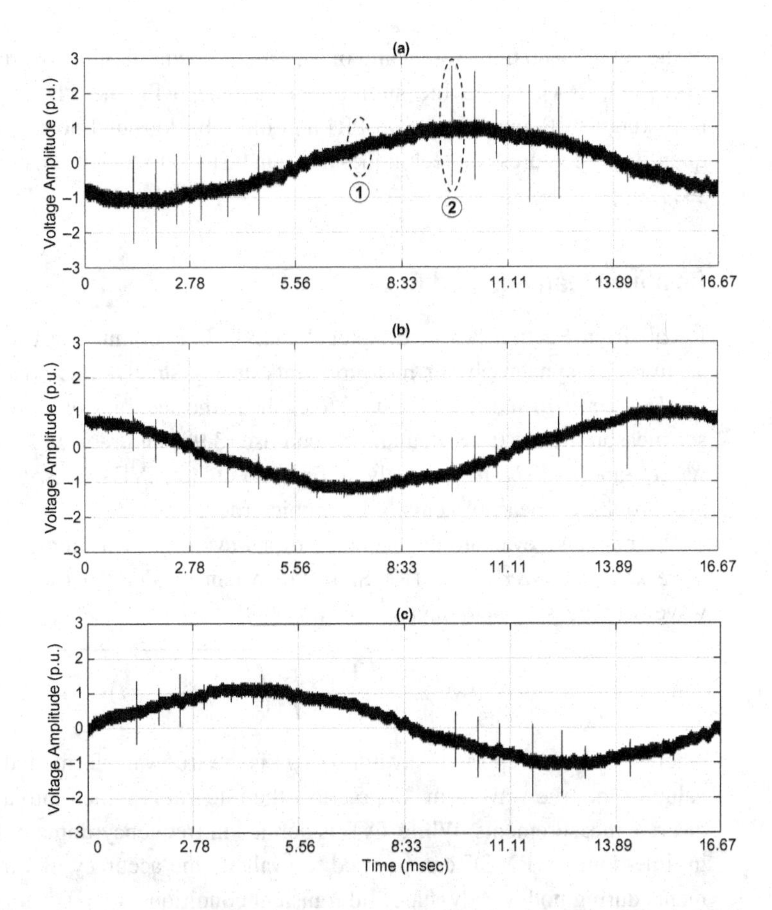

Figure 4.33 Voltage waveform measurement at an overhead transmission line during one cycle: (a) Phase A; (b) Phase B; (c) Phase C [260].

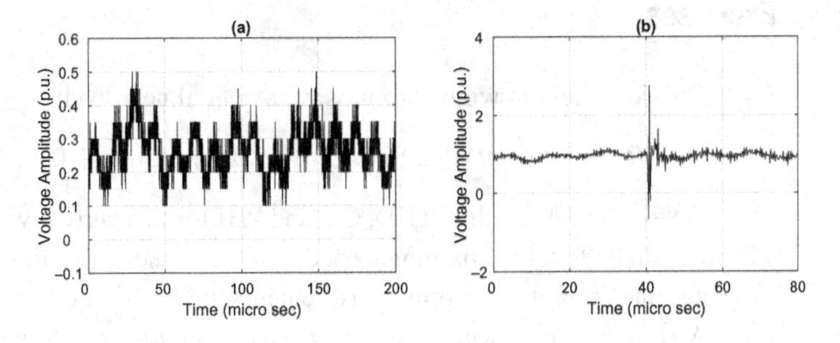

Figure 4.34 Close-up view of the two marked areas in the waveform measurement in Figure 4.33(a): (a) high-frequency fluctuations in ①; (b) impluse in ②.

have both benign or alarming causes. It could be a benign signature if it is due to Random Pulses Interference (RPI), which could be caused by transmission line corona, switching operations, etc. However, it could also indicate an incipient fault if it is due to internal Partial Discharge (PD). PDs are known as indicators of degradation of the cable insulation system and of rupture or downfall of the overhead transmission line [263, 264]. Given the similarities between RPI and PD, it is often difficult to distinguish PD patterns from PRI and other background noises. There are certain quantitative features and techniques that can be used to resolve this issue; for example, see [261].

4.7.2 Relative Mean Squared Error

Recall from Section 3.9 in Chapter 3 that TVE is commonly used to evaluate the accuracy of synchrophasor measurements during steady-state conditions. A similar *total vector error* index can be used to evaluate the accuracy of harmonic synchrophasor measurements. In particular, one can use (3.98) and replace $X\angle\theta$ and $\hat{X}\angle\hat{\theta}$ with $X_h\angle\theta_h$ and $\hat{X}_h\angle\hat{\theta}_h$, respectively, in order to obtain TVE corresponding to harmonic synchrophasor measurements of harmonic order h.

We may also evaluate the accuracy of a waveform measurement by using the *relative mean squared error* (RMSE), which can be defined over each cycle of the waveform measurement data as follows [248]:

$$\text{RMSE} = \frac{1}{x_{\text{rms}}}\sqrt{\frac{1}{T}\int_0^T \left(\hat{x}(t) - x(t)\right)^2 dt}, \tag{4.38}$$

where $\hat{x}(t)$ is the reported waveform, $x(t)$ is the true waveform, and x_{rms} is the RMS value of the true waveform. In practice, the integral is turned into a summation over discrete measurements. While TVE is defined in frequency-domain, RMSE is defined in time-domain. RMSE can be used to evaluate the accuracy of waveform measurements during both steady-state and transient conditions.

Exercises

4.1 Consider the following current wave at a nonlinear load:

$$i(t) = 5\cos(\omega t) + 2.8\cos(3\omega t) + 0.5\cos(5\omega t) + 0.7\cos(7\omega t). \tag{4.39}$$

Obtain the RMS value, THD, CF, and PHI for this current wave.

4.2 Obtain and plot the harmonic spectrum, up to the 25th harmonic, for both voltage and current waveform measurements in file `E4-2.csv`.

4.3 A periodic signal with period $T = 2\pi/\omega$ is *half-wave symmetric* if the second half of its waveform at each period is exactly the opposite of the first half, i.e., $x(t + T/2) = -x(t)$ at any time t. Show that a half-waved symmetric signal comprises only the odd-numbered harmonics.

4.4 The voltage and current measurements at a power electronics load are as in Figure 4.35 [265]. The RMS value of the current wave is 9.062 A, 9.350 A, and 10.283 A on Phases A, B, and C, respectively. The fundamental component of the current wave is 5.339 A, 5.536 A, and 6.267 A on Phases A, B, and C,

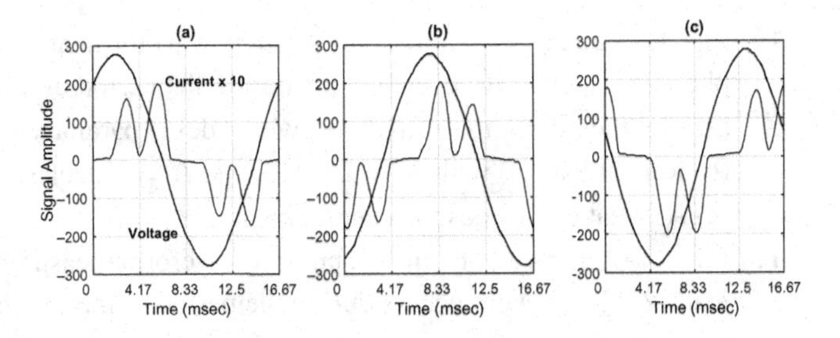

Figure 4.35 Voltage and current waveforms at a three-phase power electronics load in Exercise 4.4: (a) Phase A; (b) Phase B; (c) Phase C.

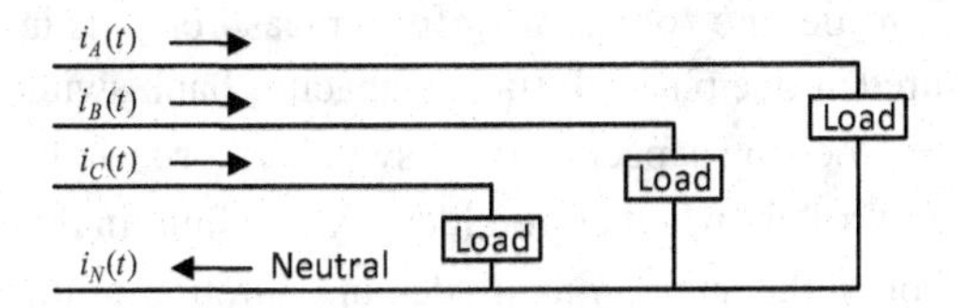

Figure 4.36 Voltage and current waveforms at a three-phase power electronics load in Exercise 4.8: (a) Phase A; (b) Phase B; (c) Phase C.

respectively. Obtain the THD and CF for each phase of the current wave. Note that you do *not* need the raw measurements to answer this question.

4.5 Calculate the *per-cycle* RMS values of the voltage signal in Example 4.3. Each cycle takes 16.667 msec, as in the fundamental component.

4.6 Consider the distorted voltage waveform in Figure 4.6(b) in Example 4.5. How many positive or negative zero-crossing points do you see in this waveform within the time period between 0.0733 seconds and 0.0767 seconds?

4.7 File `E4-7.csv` contains voltage waveform measurements for a voltage wave with notching. Obtain the notch depth, notch width, and notch area.

4.8 Consider a three-phase load, as in Figure 4.36. Suppose the load is *balanced.* That is, $i_B(t)$ is identical to $i_A(t)$, but it is 120° behind $i_A(t)$; and $i_C(t)$ is identical to $i_A(t)$, but it is 120° ahead of $i_A(t)$.

(a) Show that harmonic currents of orders which are *not* a multiple of *three* cancel each other out in the neutral current. As an example, plot the neutral current if the load current is as in Load Scenario 1 in file `E4-8.csv`.

(b) Show that harmonic currents of orders which *are* a multiple of *three* are added up arithmetically in the neutral current. As an example, plot the neutral current if the load current is as in Load Scenario 2 in file `E4-8.csv`.

4.9 Consider the voltage waveform measurements in file `E4-9.csv` [218].

(a) Obtain and plot the per-cycle RMS value profile.

(b) Obtain and plot the per-cycle THD profile.

(c) Capture and plot all waveform events when RMS value drops below 95% rated voltage or THD exceeds 5%. Let $C_{\text{before}} = 4$ and $C_{\text{after}} = 5$.

4.10 File `E4-10.csv` contains the current waveform measurements at a single-phase load [218]. An event occurs during the period of measurements.
(a) Plot the differential waveform with a delay parameter of $N = 1$ cycle.
(b) Repeat Part (a) with $N = 2$ cycles and $N = 3$ cycles.
(c) At what cycle does the event take place?

4.11 File `E4-11.csv` contains the current waveform measurements at a three-phase load [218]. An event occurs during the period of measurements.
(a) Plot the neutral current waveform.
(b) At what cycle does the event occur?
(c) Is this a single-phase, a two-phase, or a three-phase event?

4.12 Consider the voltage waveform measurements in file `E4-12.csv` that are captured at one phase during a capacitor bank switching event [221].
(a) Does the capacitor bank switch on or switch off?
(b) Explain whether you identify any fault in this switching operation.

4.13 Obtain the angle, magnitude, and duration of the incipient fault that is captured by the current waveform measurements in file `E4-13.csv`.

4.14 File `E4-14.csv` contains the three-phase voltage and current waveform measurements during a fault [218]. The fault takes place only on one phase.
(a) Identify the faulted-phase, whether it is Phase A, Phase B, or Phase C.
(b) Obtain the fault impedance on the faulted phase at each fault cycle.

4.15 A Lissajous curve is a graph that is constructed by plotting the voltage waveform versus the current waveform; e.g., see [266, 267]. Plot the Lissajous curve for *each phase* for the three-phase voltage and current waveform measurements in Exercise 4.14. Use the Lissajous curves that you will plot to identify the faulted phase; whether it is Phase A, Phase B, or Phase C.

4.16 Revise (3.1) and (3.2) in Chapter 3 to estimate harmonic synchrophasors.

4.17 Consider the non-sinusoidal voltage wave in file `E4-17.csv`.
(a) Use the expressions in (3.1) and (3.2) to obtain the magnitude and phase angle for the phasor corresponding to the fundamental component.
(b) Use the expressions that result from Exercise 4.16 to obtain the harmonic synchrophasors for the 3rd, the 5th, and the 7th harmonics.
(c) Combine the results from Parts (a) and (b) to reconstruct the original voltage wave as a summation of its fundamental and harmonic components.

4.18 File `E4-18.csv` contains the three-phase voltage and current waveform measurements that are obtained at a three-phase load.
(a) Obtain and plot the instantaneous power at each phase.
(b) How much is the average power per cycle that is delivered to the load?

4.19 Suppose we want to remotely evaluate the tap changing operation of a transformer on a power distribution feeder to identify a potential incipient fault, similar to the type of incipient fault that we discussed in Example 4.12. Is it better to place a waveform sensor in the *upstream* of the transformer, i.e., at location 1 in Figure 4.37(a), or in the *downstream* of the transformer, i.e., at location 2 in Figure 4.37(a)? Justify your answer.

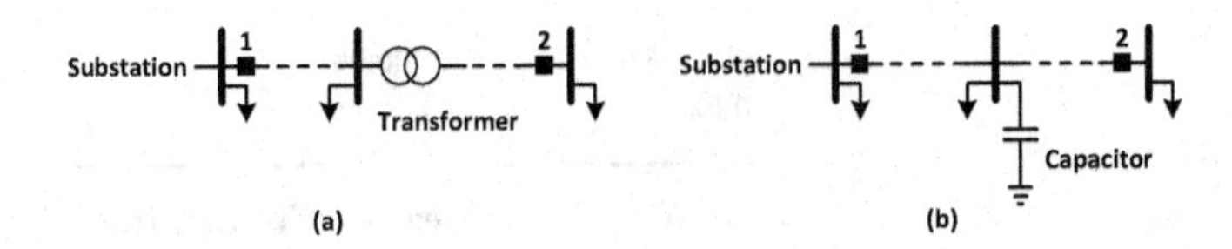

Figure 4.37 Sensor location for asset monitoring: (a) Exercise 4.19; (b) Exercise 4.20.

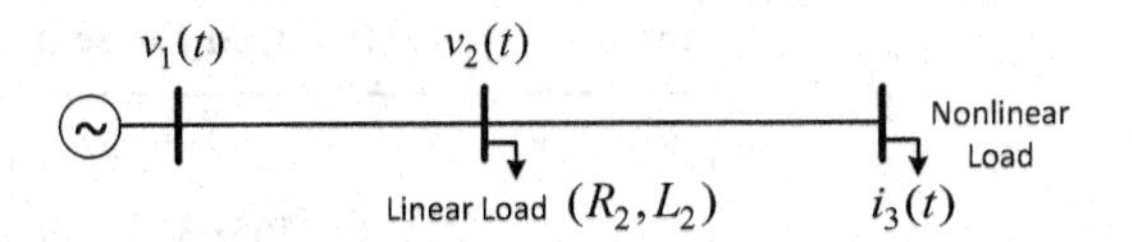

Figure 4.38 The network setup in Exercise 4.21.

4.20 Repeat Exercise 4.19 for the case of evaluating the operation of a capacitor bank, based on the two locations that are marked on Figure 4.37(b). Our goal here is to identify a potential incipient fault in the capacitor bank, similar to the type of incipient fault that we discussed in Example 4.13.

4.21 Two waveform sensors are installed at the 3-bus network in Figure 4.38 to measure voltage $v_1(t)$ and current $i_3(t)$. The measurements are given in file `E4-21.csv`. The linear load at bus 2 is a 70 Ω resistor in series with a 90 mH inductor. Each power line has 1 Ω resistance and 5 mH inductance. Estimate and plot the waveform for voltage $v_2(t)$. As a hint, you can first obtain the fundamental phasors and the dominant harmonic phasors of voltage $v_1(t)$ and current $i_3(t)$; then you can solve the circuit in frequendy domain at each frequency mode separately. After that, you can combine the results at different frequencies in order to reconstruct the voltage waveform $v_2(t)$.

4.22 Consider the power distribution network in Figure 4.39. The network includes six switches. Switches ①, ③, ④, and ⑤ are the normally closed switches. Switches ② and ⑥ are the normally open switches. There is one source of the 3rd harmonic and also one source of the 5th harmonic in this network. The locations of the harmonic sources are known. There are three H-PMUs at the beginning of each lateral, which are denoted by A, B, and C. They measure the harmonic current synchrophasors.

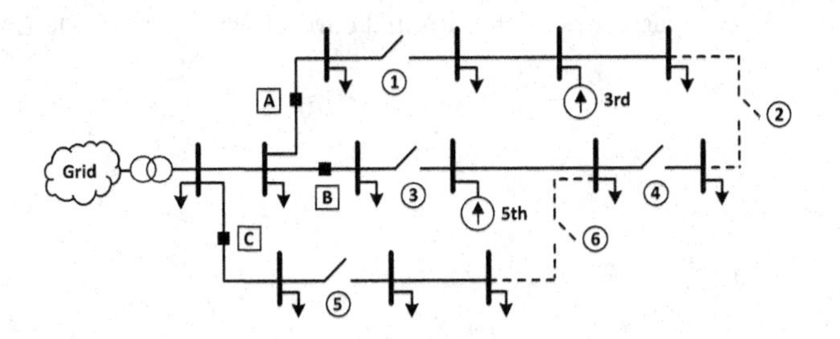

Figure 4.39 The power distribution network that is considered in Exercise 4.22.

Table 4.4 The 5th harmonic synchrophasors in Exercise 4.23.

Bus #	**Measured Voltage (p.u.)**
3	$0.033122\angle 67.025117°$
4	$0.033436\angle 66.278581°$
Line #	**Measured Current (p.u.)**
1,2	$0.062529\angle 178.101217°$
1,4	$0.017765\angle 31.562689°$
2,4	$0.078087\angle 5.274117°$
3,4	$0.009945\angle 112.685782°$

(a) How many topology configurations are possible in this network that are *radial* and include *all* the buses in the network?

(b) Suppose H-PMU A records the presence of the 3rd harmonic, H-PMU B does not record any harmonic, and H-PMU C records the presence of the 5th harmonic. What are the possible radial topology configurations?

4.23 Consider the harmonic state estimation problem in Example 4.22. Suppose the 5th harmonic phasor measurements are as in Table 4.4. Identify the bus number for the location of the harmonic source.

4.24 File `E4-24.csv` contains the synchronized voltage waveform measurements from two WMUs. The measurements are in per unit.

(a) Obtain and plot RPAD between the two voltage waves.

(b) Obtain and plot RVWD between the two voltage waves.

(c) Obtain the RMS value of the RVWD signal. Is it equal to $\sqrt{2}$ RPAD?

4.25 Consider the current waveform in file `E4-25.csv`.

(a) Obtain the harmonic phasors up to the 30th harmonic.

(b) Use (4.1) to reconstruct the original current waveform based on the harmonic phasors that are estimated in Part (a). Try the first 10, the first 20, and all 30 estimated harmonic phasors, respectively.

(c) Use (4.38) to obtain the RMSE for each of the three current waveforms that are reconstructed in Part (b) to evaluate their reconstruction accuracy in comparison with the original current waveform in file `E4-25.csv`.

5 Power and Energy Measurements and Their Applications

Measuring power and energy is another fundamental aspect in monitoring power systems. In fact, many components in the power grid are currently monitored based only on measuring their power and energy consumption or generation, such as the load of most utility customers. Therefore, it is important to know how power and energy are measured and how such measurements can be used in various smart grid applications, either when they are the only available type of measurements or when they are available together with other types of measurements.

Furthermore, there are some smart grid components that are characterized only (or primarily) based on their power and energy characteristics. There are also some smart grid applications that mostly use only the power and energy measurements, even if the voltage and current measurements are also available.

In this chapter, we discuss the fundamentals of measuring power and energy as well as several applications of power and energy measurements.

5.1 Measuring Power

The instrument to measure electric power is the *wattmeter*; see Figure 5.1. Recall from Section 1.2.4 in Chapter 1 that the *instantaneous power* is obtained by multiplying voltage and current. In an analog wattmeter, this multiplication is done implicitly by using a current coil that is connected in series to the circuit, the top coil in Figure 5.1(a), and a potential coil that is connected in parallel to the circuit, the bottom coil in Figure 5.1(a). Both coils create proportional deflections in a mechanical power indicator, such as a needle. In a digital wattmeter, the multiplication is done rather explicitly, by measuring voltage using a voltmeter subsystem and measuring current using an ammeter subsystem, then conducting the multiplication. The symbol for wattmeters may not always show its internal sub-systems, such as in Figure 5.1(b).

Depending on their current rating and voltage rating, wattmeters may use CTs and/or PTs to step down current and/or voltage, respectively.

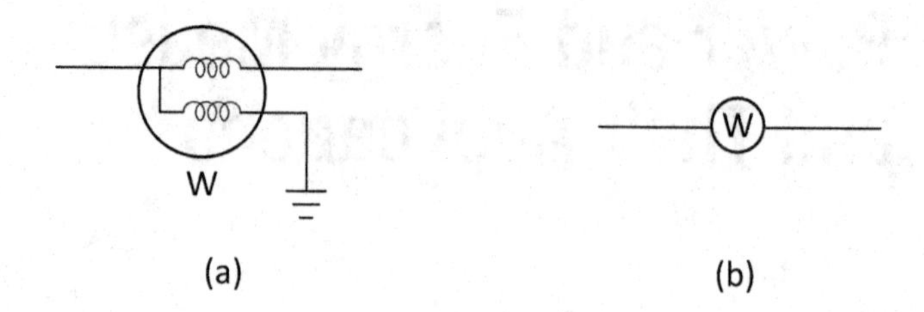

Figure 5.1 Two symbols for wattmeters (a) with and (b) without showing its subsystems.

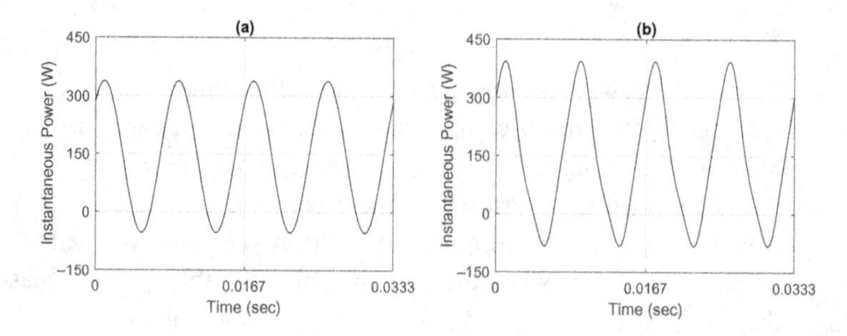

Figure 5.2 Instantaneous power waveforms in Example 5.1 that are obtained: (a) based on the voltage and current waveforms in Example 3.1 in Chapter 3; (b) based on the voltage and current waveforms in Example 4.1 in Chapter 4.

Example 5.1 Consider the voltage and current waveforms in Example 3.1 in Chapter 3. The instantaneous power that is delivered to the load is obtained as

$$\begin{aligned} p(t) &= v(t)\, i(t) \\ &= 391.2 \cos(\omega t) \cos(\omega t - 0.7532). \end{aligned} \tag{5.1}$$

The above instantaneous power waveform is shown in Figure 5.2(a). It is purely sinusoidal, and its frequency is 120 Hz, which is twice the frequency of $v(t)$ and $i(t)$. Next, consider the voltage and current waveforms in Example 4.1 in Chapter 4. The instantaneous power that is delivered to the load is obtained as

$$\begin{aligned} p(t) &= v(t)\, i(t) \\ &= 384 \cos(\omega t) \cos(\omega t - 0.7532) \\ &\quad + 64.8 \cos(\omega t) \cos(3\omega t - 0.4323) \\ &\quad + 36 \cos(\omega t) \cos(5\omega t + 3.3058). \end{aligned} \tag{5.2}$$

The above instantaneous power waveform is shown in Figure 5.2(b). It is non-sinusoidal but periodic, and its frequency is again 120 Hz. Note that, in both examples, $p(t)$ has negative values at certain sub-intervals. This is due to the partly inductive nature of the motor load in these two examples.

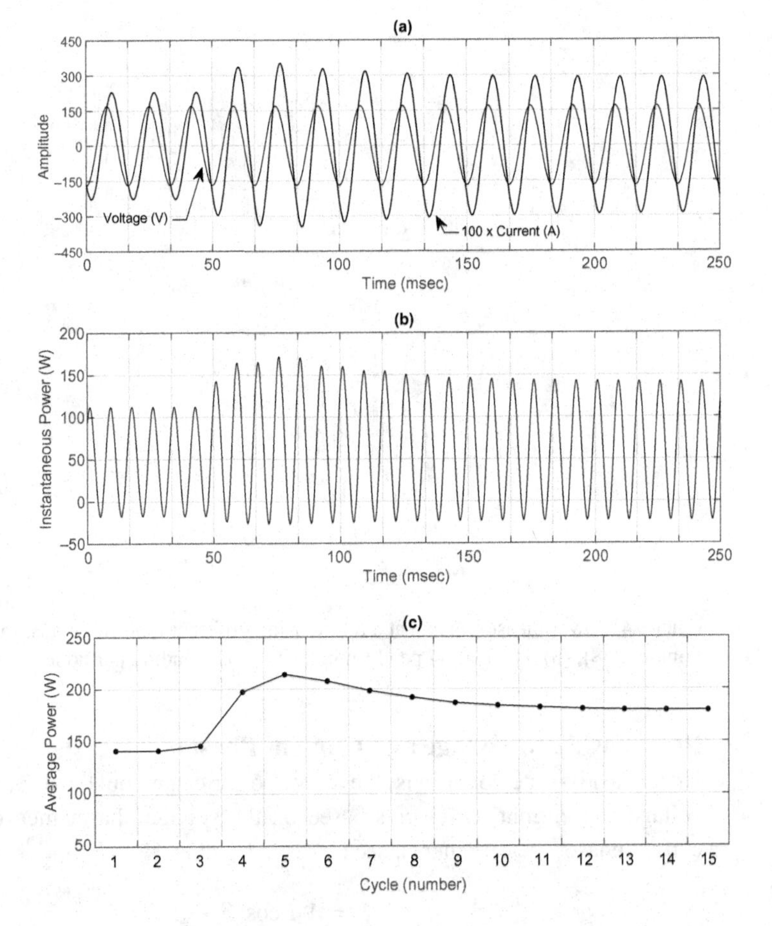

Figure 5.3 Changes in average power over time: (a) voltage waveform and current waveform; (b) instantaneous power; (c) average power.

5.1.1 Active Power

Measuring instantaneous power is rarely of practical use. A wattmeter rather reports the *average* of the instantaneous power across one or multiple cycles. In other words, a wattmeter measures active power; see (1.21) in Chapter 1:

$$P = \frac{1}{T} \int_0^T p(t), \tag{5.3}$$

where $T = 1/f$. In Example 5.1, we have $P = 142.7$ W for the instantaneous power in (5.1); and $P = 140.1$ W for the instantaneous power in (5.2).

In practice, the average power itself varies over time. This happens due to the changes in current and/or voltage waveforms. An example is shown in Figure 5.3. Here, the voltage waveform does not change, but there are changes in the current waveform. The instantaneous power and the average power change accordingly.

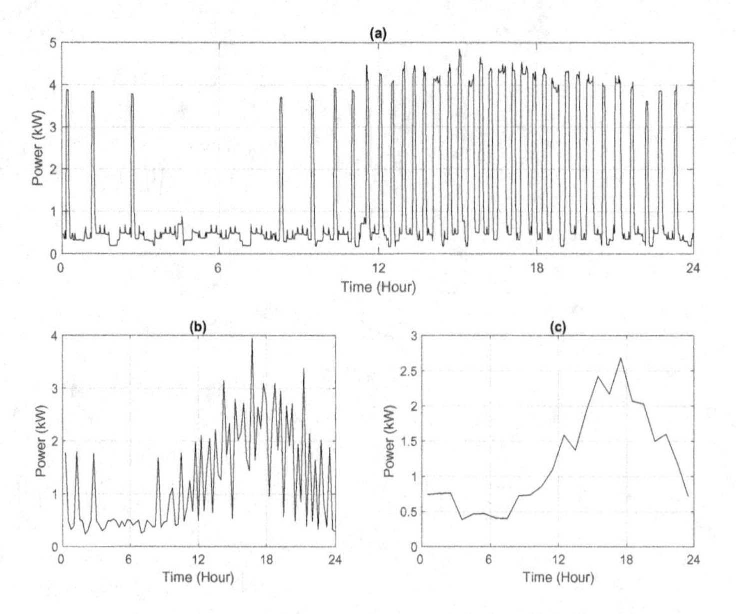

Figure 5.4 Power measurements at a house using different reporting rates: (a) one reading per minute [268]; (b) one reading per 15 minutes; (c) one reading per hour.

Relationship to Voltage and Current Phasors

Under steady-state conditions, the active power can be obtained for purely sinusoidal voltage and current waveforms based on the voltage phasor measurements and the current phasor measurements, see (1.22) in Chapter 1:

$$P = VI\cos(\theta - \phi), \tag{5.4}$$

where the voltage phasor measurement is denoted by $V\angle\theta$ and the current phasor measurement is denoted by $I\angle\phi$. Recall from Section 1.2.2 in Chapter 1 that V and I denote the RMS values for voltage and current, respectively.

We can also obtain the average power for voltage and current waveforms that contain steady-state harmonic distortions based on the voltage and current harmonic phasor measurements; see Section 4.1.1 in Chapter 4:

$$P = \sum_{h=1}^{\infty} V_h I_h \cos(\theta_h - \phi_h), \tag{5.5}$$

where at each harmonic h, the voltage harmonic phasor measurement is denoted by $V_h\angle\theta_h$ and the current harmonic phasor measurement is denoted by $I_h\angle\phi_h$.

Reporting Rate

The reporting interval of wattmeters is often much longer than one AC cycle; therefore, active power is *averaged* across several cycles. Power measurements are typically reported once every few seconds to once every few minutes.

Example 5.2 Figure 5.4 shows the power consumption measurements based on different reporting rates at a house during a day. Clearly, a higher reporting rate is more informative and can show more details about power usage.

5.1.2 Power Profile

It is informative to examine the daily profiles based on power measurements at different grid components. Several examples are shown in Figure 5.5. They include a wide range of *consumption*, *generation*, *storage*, and *delivery* scenarios.

Figure 5.5(a) shows the daily profile for power consumption at a single appliance, which is an oven. The appliance operates only once in late afternoon. Figure 5.5(b) shows the daily profile for power consumption at a house. This power profile is essentially the aggregation of the power profiles for all appliances, the air-conditioning cycles, the refrigeration cycle, etc. Figure 5.5(c) shows the daily profile for power consumption at a commercial building on a weekday. Notice that the load increases significantly during business hours. Figure 5.5(d) shows the daily profile for power generation at a solar power generation unit during a partially cloudy day. The power generation takes place during the day; however, it varies due to the variations in solar irradiance and the movement of the clouds. Figure 5.5(e) shows the daily profile for a combination of power consumption and power generation at a commercial building with behind-the-meter solar power generation. The building acts as a load at night and during the evening. However, it acts as a generator during the day due to its excessive behind-the-meter solar power generation. Figure 5.5(f) shows the daily profile for power generation at a wind turbine. Power generation depends on weather conditions, in particular the direction and speed of wind. As a result, the amount of power generation varies during the day due to varying wind speed, wind direction, etc. Figure 5.5(g) shows the daily profile based on the charge and discharge cycles of a grid-connected battery energy storage system. Finally, Figure 5.5(h) shows the daily profile for the daily power flow at a power distribution line. The amount of power flow changes due to the changes in the load of the distribution feeder.

Load Factor

Given the power profile of a customer, the customer's *load factor* is obtained as

$$\text{Load Factor} = \frac{\text{Maximum of the Power Profile}}{\text{Average of the Power Profile}}. \tag{5.6}$$

For example, if the peak load and the average load of a customer during a day is 4 kW and 1.6 kW, respectively, then the customer's load factor is 2.5. Load factor indicates how balanced (over time) the load power profile is during a given time-period of interest. Load factor of 1 means a flat load profile.

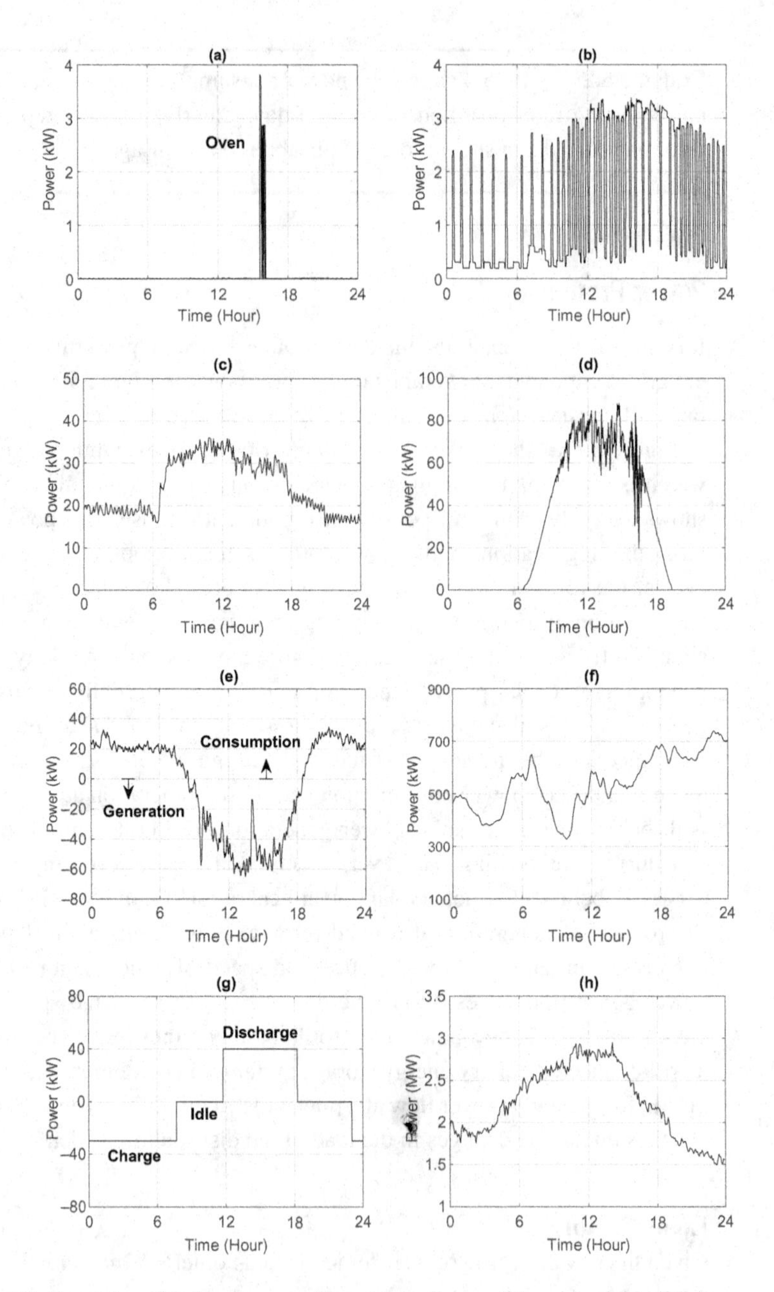

Figure 5.5 Examples of daily power profiles: (a) appliance; (b) residential building; (c) commercial building; (d) solar power generation; (e) commercial building with behind-the-meter solar power generation; (f) wind power generation; (g) battery energy-storage charge and discharge; (h) power flow at a power line [88, 90, 268, 269].

5.2 Measuring Reactive Power and Power Factor

The instrument to measure reactive power is a *varmeter*. Given the measurements of active power and reactive power, one can calculate power factor. However, to increase accuracy, a separate electromechanical instrument, called a *power factor meter*, is used to measure power factor. These instrumentation distinctions are of concern mainly in electromechanical instruments. A single digital sensor is often capable of serving as a wattmeter, a varmeter, and a power factor meter.

Relationship to Voltage and Current Phasors

Both reactive power and power factor are primarily defined based on purely sinusoidal voltage and current waveforms under steady-state conditions; see Section 5.2.3 for the non-sinusoidal case. Recall from (1.25) in Chapter 1 that reactive power can be obtained based on the voltage and current phasors as

$$Q = VI\sin(\theta - \phi). \tag{5.7}$$

We can also obtain power factor as

$$\text{PF} = \cos(\theta - \phi). \tag{5.8}$$

If the voltage or current waveforms are distorted, i.e., if they are *not* purely sinusoidal, then the common approach is to obtain reactive power and power factor based on the *fundamental* components of the voltage and current waveforms:

$$Q = V_1 I_1 \sin(\theta_1 - \phi_1) \tag{5.9}$$

and

$$\text{PF} = \cos(\theta_1 - \phi_1), \tag{5.10}$$

where $V_1 \angle \theta_1$ is the phasor for the fundamental component of the voltage waveform; and $I_1 \angle \phi_1$ is the phasor for the fundamental component of the current waveform. There is also an alternative approach to work directly with the distorted voltage or current waveform that we will see in Section 5.2.3.

In Example 5.1, we have $Q = 133.8$ VAR and PF = 0.7295 for the instantaneous power in (5.1). As for the instantaneous power in (5.2), we can obtain Q and PF based on the fundamental component of the distorted current waveform. From (4.3) in Chapter 4, the fundamental component is $1.60\sqrt{2}\cos(\omega t - 0.7532)$. Therefore, we can obtain $Q = 131.3$ VAR and PF = 0.7295.

5.2.1 Reactive Power and Power Factor Profiles

The daily profiles of reactive power and power factor can often reveal useful information about the operation of the power system. For instance, recall from Example 2.9 in Chapter 2 that the daily RMS voltage and current profiles at a power distribution feeder can identify the events that are induced by the operation of loads and the grid

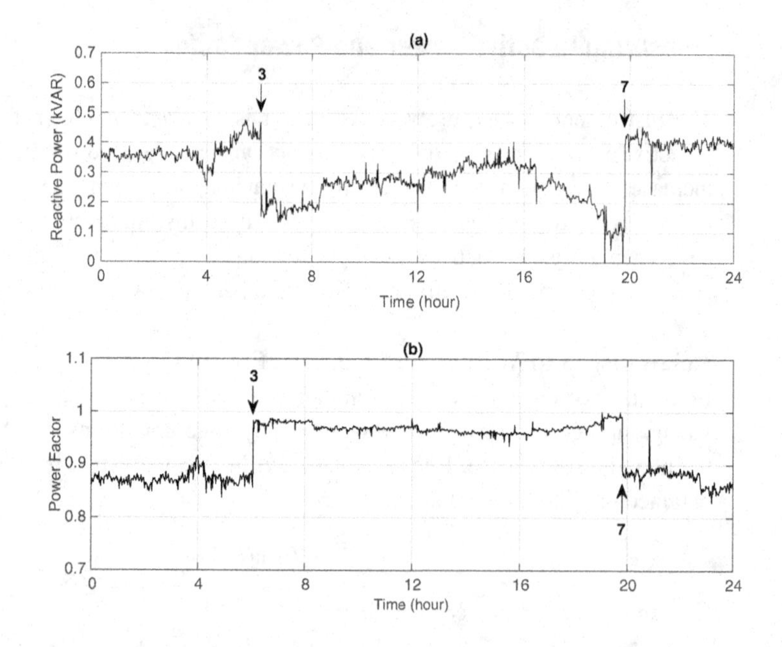

Figure 5.6 Reactive power and power factor profiles in Example 5.3: (a) Reactive power; (b) power factor. The event numbers are the same as those in Example 2.9 in Chapter 2.

equipment. Similar information can also be extracted from the daily reactive power profile or daily power factor profile.

Example 5.3 Consider the daily reactive power and power factor profiles in Figure 5.6 [152] that are measured on the same day and at the same power distribution feeder as in Example 2.9 in Chapter 2. The reporting rate is once per minute. We had previously concluded in Example 2.9 that events 3 and 7 are likely caused by some local issues on the understudy feeder. The exact causes can now be identified based on the measurements in Figure 5.6. At event 3, there is a sudden decrease in reactive power consumption and a sudden increase in power factor. At event 7, there is a sudden increase in reactive power consumption and a sudden decrease in power factor. These events are in fact due to *switching on* and *switching off* of a major capacitor bank on this feeder, respectively. The capacitor bank is a local solution to support reactive power and regulate voltage.

5.2.2 Apparent Power

Apparent power can be measured in different ways depending on the type of measurements that are available. It can be measured from RMS voltage and RMS current, i.e., by using a voltmeter and an ammeter. Note that we have:

$$S = V_{\text{rms}}\, I_{\text{rms}}. \tag{5.11}$$

It can also be measured by using active power measurements and reactive power measurements, i.e., by using a wattmeter and a varmeter. Note that we have:

$$S = \sqrt{P^2 + Q^2}. \tag{5.12}$$

Apparent power can also be measured by a wattmeter and a power factor meter as well as by a varmeter and a power factor meter. Note that we have:

$$S = P/\text{PF} = Q/\sqrt{1 - \text{PF}^2}. \tag{5.13}$$

Of course, any one of the above options is sufficient to obtain apparent power. If multiple measurements are available, for example if voltage, current, active power, and reactive power are measured separately, then we have *redundancy*. In that case, we can simply use the average of the results from different methods.

5.2.3 True Power Factor

If the voltage or current waveforms are distorted, then power factor in (5.10) is often referred to as the *displacement power factor* (DPF). However, we can also measure the *true power factor* (TPF), which is obtained as follows:

$$\text{TPF} = \frac{P}{S}, \tag{5.14}$$

where P is the average power in (5.3) and S is the apparent power in (5.11).

Example 5.4 Again, consider the voltage and current waveforms in Example 4.1 in Chapter 4. Recall that the current waveform is distorted. We have:

$$V_{\text{rms}} = 120 \text{ V}, \quad I_{\text{rms}} = 1.63 \text{ A}, \quad S = 195.6 \text{ VA}. \tag{5.15}$$

The average power in this example is $P = 140.1$ W. Therefore, we have:

$$\text{TPF} = \frac{140.1}{195.6} = 0.7163. \tag{5.16}$$

We can see that TPF is smaller than DPF, which is 0.7295 in this example.

In practice, harmonic distortion is very small in voltage compared to in current. Therefore, the contribution of the harmonics to the delivery of active power is not significant. Thus, it is reasonable to approximate $V_{\text{rms}} \approx V_1$ and $P \approx V_1 I_1 \cos(\theta_1 - \phi_1)$. From these, together with (4.6) in Chapter 4, we can derive:

$$\text{TPF} = \frac{P}{S} \approx \frac{V_1 I_1 \cos(\theta_1 - \phi_1)}{V_1 I_{\text{rms}}} = \frac{\text{DPF}}{\sqrt{1 + \text{THD}^2}}, \tag{5.17}$$

where THD is associated with the current waveform. TPF is always *less than or equal* to DPF. If THD = 0, i.e., if there is no distortion, then TPF = DPF.

Distortion Power

There are different ways to define reactive power in presence of non-sinusoidal voltage and current signals. For example, six different definitions for reactive power in this context are surveyed in [270, Section 2.4]. One option is to follow the definition of active power in (5.5) and define reactive power as

$$Q = \sum_{h=1}^{\infty} V_h I_h \sin(\theta_h - \phi_h). \tag{5.18}$$

The triangular equality does not hold among S, P, and Q. In fact, we can show that S^2 is always greater than or equal to $P^2 + Q^2$; see Exercise 5.5. Therefore, a quantity named *distortion power*, denoted by D, is defined as follows:

$$D^2 = S^2 - P^2 - Q^2, \tag{5.19}$$

which yields the equation $S^2 = P^2 + Q^2 + D^2$. Of course, if the voltage and current waveforms are purely sinusoidal, then $D = 0$ and $S^2 = P^2 + Q^2$.

5.3 Measuring Energy

The instrument to measure electric energy is a *watthour meter*. Energy is measured over a certain time interval by taking the integral of instantaneous power on said interval. The time interval can be one hour, one day, or one month. For the interval between time t_1 and time t_2, energy E is calculated as the integral of instantaneous power $p(t)$ from $t = t_1$ to $t = t_2$, as shown below:

$$E = \int_{t=t_1}^{t_2} p(t)dt. \tag{5.20}$$

In a traditional analog watthour meter, the above integral is taken implicitly by counting the rotations of a metal disc, which is made to rotate at a speed proportional to the power passing through the meter. In a digital watthour meter, the integral is calculated explicitly, by measuring power and using discrete summation while taking into account the sampling rate for accurate computation.

The unit of measuring E is watt hour. For the two scenarios in Example 5.1, the energy that is delivered to the motor load over a one-minute interval is obtained as 2.38 Wh and 2.34 Wh, respectively. The energy that is delivered over a one-hour interval is obtained as 142.8 Wh and 140.4 Wh, respectively.

Reporting Interval

The reporting interval for measuring energy could be fixed, or it could vary. This is because the energy measurements can be reported in fixed intervals, or they can be reported in fixed increments. Both methods are explained next.

Table 5.1 Minute-by-minute energy usage in watts at a building.

Minute	Energy Usage	Minute	Energy Usage
1	383	11	160
2	502	12	267
3	630	13	404
4	446	14	281
5	661	15	320
6	378	16	549
7	269	17	648
8	330	18	805
9	298	19	718
10	122	20	829

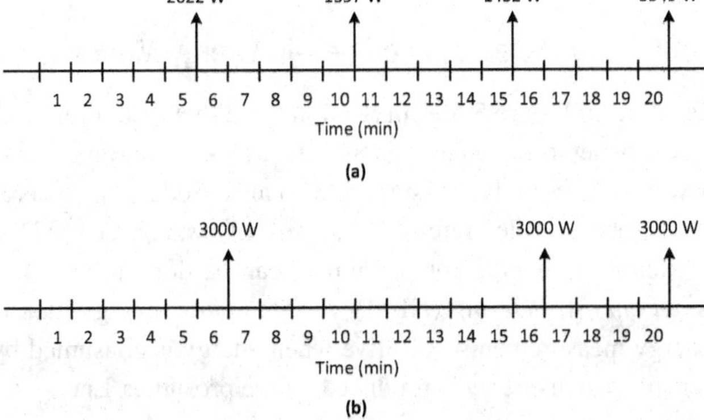

Figure 5.7 Reporting of energy measurements in Table 5.1 based on (a) fixed intervals: one reading every five minutes; (b) fixed increments: one reading every 3000 W.

5.3.1 Fixed Intervals

It is typical for energy measurements to be reported in fixed intervals. For instance, suppose the minute-by-minute energy usage of a building is as shown in Table 5.1. Suppose the watthour meter reports the measurements once every five minutes. Figure 5.7(a) shows the reported measurements. The location of each arrow indicates when each measurement is reported. The number above each arrow indicates the energy measurement that is reported by the watthour meter. The amount that is reported can vary, but the reporting is done at fixed intervals.

5.3.2 Fixed Increments

Some watthour meters operate as *pulse meters* and report energy usage in fixed increments. Each pulse corresponds to a certain amount of energy in kWh. Accordingly, one can calculate the amount of energy usage by counting the number of pulses. For example, if one pulse is equivalent to 0.5 kWh, then generating six pulses per minute indicates that the energy usage is 3 kWh per minute.

Again, consider the minute-by-minute energy usage in Table 5.1. Suppose the watthour meter reports the measurements once every 3000 W, i.e., at fixed increments. Figure 5.7(b) shows the reported measurements. As in the previous case in Section 5.3.1, the location of each arrow indicates when each measurement is reported. The number above each arrow indicates the energy measurement that is reported by the watthour meter. The amount that is reported is always fixed, but the reporting is done at varying intervals. Notice that, based on Table 5.1, the summation of the energy usage during the first six minutes is 3000 W. The summation of the energy usage during the next 10 minutes is also 3000 W. The summation of the energy usage during the last four minutes is also 3000 W.

5.3.3 Net Energy Metering and Feed-In Energy Metering

Recall from Figure 5.5(e) in Section 5.1.2 that a building with behind-the-meter solar power generation sometimes acts as a power consumer and at other times acts as a power producer. The consumers who also produce and share surplus energy with the power grid are often referred to as *prosumers*; e.g., see [271–273].

Measuring energy for prosumers can be done in two different ways. One option is *net energy metering* (NEM). In this option, energy measurement is *bidirectional*. Energy measurement is positive when energy is consumed by the prosumer; and it is negative when energy is produced by the prosumer. Energy measurements from NEM can be used to bill prosumers. In this scenario, the prosumer is credited by the utility for its excess energy generation at *the same price* that it is charged for its energy consumption; e.g., see [274].

Another option is *feed-in energy metering* (FIEM). In this option, energy measurement is *unidirectional*; energy consumption is measured separately, and energy generation is also measured separately. This option often requires using two watthour meters, one to measure energy consumption and one to measure energy generation. The second watthour meter is installed at the local source of energy generation. Energy measurements from FIEM can be used to bill prosumers based on Feed-in Tariffs (FIT). In this billing scenario, the prosumer is credited by the utility for its excess energy generation at a price that is *different* from the price that the prosumer is charged for its energy consumption; e.g., see [275].

Example 5.5 Consider a prosumer which consumes 298.7 kWh energy during one day. On that same day, this prosumer generates 384.1 kWh solar energy. First, suppose the prosumer is charged at 12 ¢/kWh for its energy usage and it is credited at the same price for its energy generation. By using net energy metering, this prosumer is credited $(384.1 - 298.7) \times 0.12 = \$10.25$. Next, suppose the prosumer is charged at 12 ¢/kWh for its energy usage and it is credited at 27 ¢/kWh for its energy generation [276]. By using feed-in energy metering, this prosumer is credited $384.1 \times 0.27 - 298.7 \times 0.12 =$

$67.86. In this example, an FIT pays *more* than the retail electricity rate for renewable power generation; which is intended to provide incentives to consumers to install solar panels.

5.4 Smart Meters and Their Applications

Electric meters are watthour meters that are used by utilities to measure the amount of electric energy consumed by a customer, such as a residence or a business, for billing purposes. Traditionally, electric meters are read manually by the utility personnel, once per billing period, such as one every month.

Smart meters are the next generation of electric meters in the era of smart grids. They are digital energy meters that record customer energy usage much more frequently, such as once every 15 minutes, and provide near real-time automated meter reading (AMR) to the utility, through a *two-way communications* infrastructure.

Smart meters record not only the energy usage but also the *time* and *date* of energy usage. This is in sharp contrast to traditional mechanical meters that record only the incremental electricity usage and do not record the time or date of usage. In addition to measuring energy usage, some smart meters also measure RMS voltage, RMS current, power factor, and power quality [277].

Measurements from smart meters provide the consumer with greater clarity of consumption behavior, and the utility with enhanced system monitoring and the ability to implement different billing mechanisms, such as time-of-use pricing.

Smart meters are the most widely deployed smart grid technology to date. About 100 million smart meters are already installed in the United States [278].

5.4.1 Price-Based Demand Response

Demand response (DR) programs provide opportunities for consumers to *reduce* their electricity usage during peak periods, or *shift* part of their usage to off-peak periods, *in response* to time-of-use rates or other forms of financial incentives. DR programs can be divided into two categories, *price-based* and *incentive-based*. Smart metering is a key enabling technology for DR programs in both categories.

In this section, we discuss the use of smart meters to facilitate price-based DR. We will discuss incentive-based DR in Section 5.4.2.

Time-of-Use Pricing

Traditionally, utility customers have been charged with *flat rates*. If a customer is charged with a flat rate, then all its usage during a given period of time, such as a monthly billing cycle, is charged with the same rate. For instance, if the flat rate is 12 ¢/kWh, and a customer's electricity usage is 914 kWh in one month, then this customer is charged $914 \times 0.12 = \$109.68$. It does *not* matter at what time during

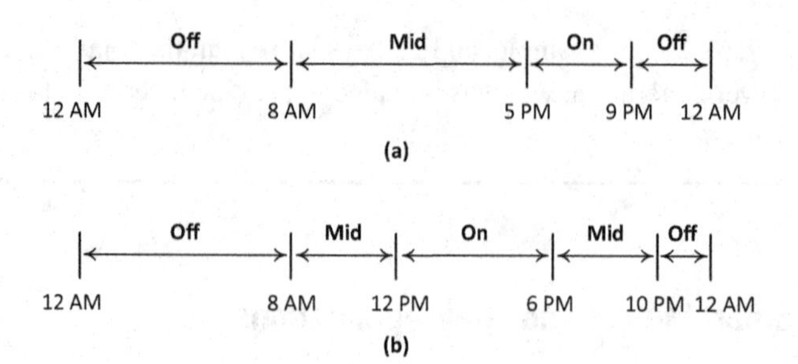

Figure 5.8 An example for on-peak, mid-peak, and off-peak hours during: (a) winter; (b) summer. The choice of these hours depends on the overall climate in the region.

the day or night the energy was consumed. It also does *not* matter whether the energy consumption occurred during weekdays or weekends.

However, flat rates do not reflect the *variations* in the cost of electricity generation, such as the higher cost of generation during *peak demand hours* [279].

Smart meters can keep track of the energy usage during different times of a day and different days of a week. Therefore, they can facilitate billing customers based on time-of-use (ToU) pricing. Under a ToU rate plan, price of electricity varies depending on the time of day, day of week, and season. Prices are higher during peak demand hours and lower during low demand hours.

Here is an example for ToU prices for a utility in California [280]:

- On-Peak Hours: 10.33 ¢/kWh
- Mid-Peak Hours: 8.28 ¢/kWh
- Off-Peak Hours: 7.27 ¢/kWh.

Notice that the price of electricity is 42% higher during on-peak hours compared to off-peak hours. Figures 5.8(a) and (b) show the on-peak, mid-peak, and off-peak hours for this utility during winter and during summer, respectively.

Example 5.6 Suppose the hourly electricity usage of a utility customer is reported by a smart meter during a day in summer, as shown in Table 5.2. Suppose the customer is charged based on the ToU rates that we saw in the above bullet points. The energy usage of this customer during the on-peak hours is 12.173 kWh, which is the summation of its hourly usages during hours 13, 14, 15, 16, 17, and 18. The energy usage during the mid-peak and off-peak hours are 10.608 kWh and 6.321 kWh, respectively. Accordingly, the total energy usage charge to this customer during this day is obtained as \$2.60 dollars, where

$$\begin{aligned} \text{On-Peak Charge} &= 0.1033 \times 12.1734 = \$1.26, \\ \text{Mid-Peak Charge} &= 0.0828 \times 10.6084 = \$0.88, \\ \text{Off-Peak Charge} &= 0.0727 \times 6.3208 = \$0.46. \end{aligned} \tag{5.21}$$

Table 5.2 Hourly energy usage in kWh of a utility customer in Example 5.6.

Hour	Energy Usage	Hour	Energy Usage	Hour	Energy Usage
1	0.7428	9	0.7240	17	2.1726
2	0.7575	10	0.7372	18	2.6853
3	0.7650	11	0.8557	19	2.0673
4	0.3867	12	1.0984	20	2.0270
5	0.4673	13	1.5828	21	1.4991
6	0.4711	14	1.3717	22	1.5997
7	0.4069	15	1.9402	23	1.2005
8	0.4048	16	2.4208	24	0.7182

Time-Shiftable Loads

ToU pricing encourages customers to shift their energy use from on-peak hours to off-peak or mid-peak hours. If customers have energy usage that can be shifted from peak hours to off-peak hours, then they can reduce their energy bill.

Example 5.7 An electric vehicle (EV) is parked at home and plugged in to an EV Charger from 4:00 PM till 6:00 AM. This EV needs 75 kWh to be charged. The rate of charge is 7.7 kW and the charge efficiency is 95%. Accordingly, this EV needs to be charged for $(75/0.95)/7.7 = 10.25$ hours, i.e., 10 hours and 15 minutes. If the customer is charged for its energy usage at a flat rate, then it is natural for the customer to start charging as soon as the EV is plugged in. In this scenario, the EV is charged from 4:00 PM till 2:15 AM; see Figure 5.9(a). However, if the customer is charged for its energy usage based on ToU rates, then the customer can *schedule* charging the EV to lower its cost. Suppose the ToU rates are 10.33 ¢/kWh, 8.28 ¢/kWh, and 7.27 ¢/kWh during the on-peak, mid-peak, and off-peak hours, respectively. Suppose the on-peak, mid-peak, and off-peak hours are as in Figure 5.8(a), which indicates operation during the winter. The energy usage cost to charge this EV is *minimized* if the charging is scheduled in two intervals, first from 4:00 PM till 5:15 PM, and then from 9:00 PM till 6:00 AM; see Figure 5.9(b). In this scenario, the EV is charged for one hour during the mid-peak period, for 15 minutes during the on-peak period, and for nine hours during the off-peak period. This schedule results in

$$0.1033 \times 0.25 + 0.0828 \times 1 + 0.0727 \times 9 = 76¢ \qquad (5.22)$$

as the total energy usage cost to charge this EV. In contrast, if the EV is charged as soon as it is plugged in, then the total energy usage cost of charging becomes:

$$0.1033 \times 4 + 0.0828 \times 1 + 0.0727 \times 5.25 = 88¢. \qquad (5.23)$$

An EV is a *time-shiftable load*, also known as a *deferrable load*. A time-shiftable load requires a certain amount of energy within a given time period; however, the timing of operation is flexible. Other examples of time-shiftable loads include various home appliances such as washing machines, dryers, and dishwashers [281, 282], water

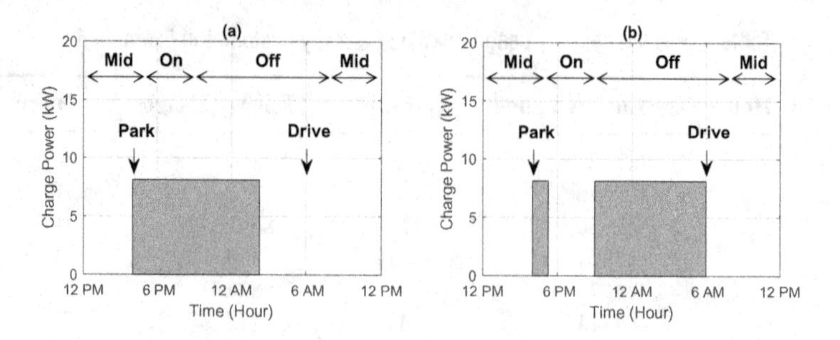

Figure 5.9 EV charging load in Example 5.7: (a) charging starts as soon as the EV is plugged in; (b) charging is scheduled to minimize the energy usage cost of charging.

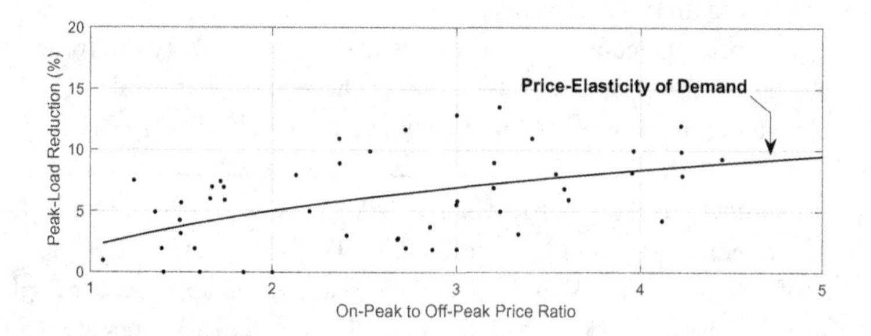

Figure 5.10 Price-elasticity of demand in price-based DR, based on the study in [290].

heaters [283], industrial equipment in process control and manufacturing [284, 285], batch processes in data centers and computer servers [286], irrigation pumps [287], and swimming pools [288]. Some time-shiftable loads are *interruptable*, meaning that their operation can be interrupted and then later resumed; such as in the case of charging EVs as we saw in Example 5.7. Some time-shiftable loads are *uninterruptable*, such as some industrial processes, e.g., see [289].

Price-Elasticity of Demand

The detailed measurements from smart meters can help us evaluate the *impacts* of ToU prices in reducing the peak load. An example is shown in Figure 5.10 based on the analysis in [290], which is based on different pilot projects. The key parameter here is the *ratio* between the price during on-peak hours and the price during off-peak hours. In general, a higher ratio is more forceful to encourage customers to shift their load to off-peak hours; thus contributing to reducing the overall peak load in the power system. The curve is the result of an approximation curve fitting to the points in the figure. This curve shows the *price-elasticity* of the demand. It shows how much load shifting, and thus peak-load reduction, can be achieved by imposing different levels of the price ratio between the on-peak and off-peak hours.

Other Pricing Methods

Smart meters can also facilitate using other types of pricing methods in price-based DR. One example is *peak-load pricing*, which requires customers to pay peak-load charges, also known as *demand charges*, based upon their *highest* amount of *power consumption* during any given time interval, typically 15 minutes, during the billing period. Another example is *real-time pricing*, which requires the customers to be charged for their electricity usage based on the price of electricity in wholesale electricity. See Section 7.3.2 in Chapter 7 for a related discussion.

5.4.2 Incentive-Based Demand Response

In incentive-based demand response, customers receive financial incentives for their participation in the DR program. Their participation is rather explicit because they are expected to reduce their load *upon receiving a notification* from the utility. In other words, customers are paid to be available to reduce their demand when needed, i.e., when a *demand response event* occurs. DR events occur occasionally, such as 5–10 times a year. They take place at times when wholesale electricity market prices are high or when system reliability is jeopardized. The amount of incentive payments to a customer depends on how much the customer is capable of reducing its load when a DR event occurs.

Baseline Calculation

Once a customer that is enrolled to an incentive-based DR program receives the notification for a DR event, it must curtail its load accordingly. However, it is not easy to calculate *how much curtailment* the customer actually makes *in response* to the DR event. The process of making such a calculation is called baseline calculation. The key in baseline calculation is to compare the load of the customer with a *baseline load*, also known as the "business as usual" load, which is the load that the customer *would have* in case the customer *had not* responded to the DR event.

An illustrative example is shown in Figure 5.11. The DR event occurs at 10:25 AM. It lasts for four hours, ending at 2:25 PM. It is clear that the customer did respond to the DR event. However, due to its inherent volatility, the customer's load fluctuates during the DR event. To evaluate the customer's performance during the DR event, we need to figure out how much of the customer's new load was the result of its curtailment efforts, and how much was due to its normal load variations. Therefore, it is necessary to establish a baseline load profile for the duration of the DR event, as shown in the figure. Here, ΔP is meant to indicate the actual curtailment in the customer's load during the DR event.

Baseline Window

Baseline calculation is done often by examining the load of the customer over the past few days, i.e., over the *baseline window*. In this regard, the baseline window is defined as the window of time prior to the DR event, typically a number of days, over which the customer load data is collected in order to establish the baseline. A common

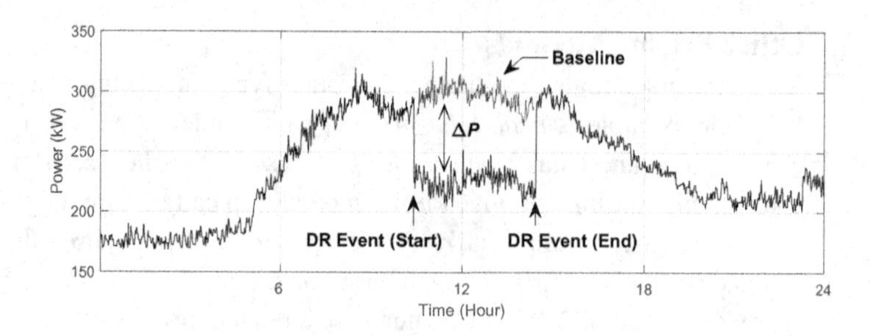

Figure 5.11 Baseline calculation in incentive-based demand response programs.

choice in practice for the baseline window is the previous 10 (non-event) business days [291]. Using a 10-day time window is considered an appropriate choice because it is short enough to account for near-term trends and long enough to limit opportunities for gaming the system [292].

High 5 of 10 Method

A class of baseline calculation methods look only at a few *highest load days* within the baseline window. For example, in the "High 5 of 10" method, baseline calculation is done based only on the five days, out of the 10 days in the baseline window, that have the highest average load for the corresponding duration of the DR event. The other five days are *excluded* from baseline calculation [291].

Example 5.8 Consider a customer that is enrolled in an incentive-based DR program. Suppose a DR event occurs at 1:15 PM and lasts for 90 minutes. It ends at 2:45 PM. Table 5.3 provides the average power usage of this customer during 15-minute intervals from 1:15 PM till 2:45 PM over the past 15 days. After we exclude the four weekends and also the day of the previous DR event, we obtain the baseline window which includes the following ten days: 1, 2, 3, 4, 8, 9, 10, 11, 14, 15. Next, consider the average power usage across all of the six intervals, i.e., the number on the second to the last column. The following five days have the highest average power usage within the baseline window: 1, 9, 10, 11, and 14. Accordingly, we can obtain the baseline as

$$
\begin{aligned}
&\text{Interval 1: } (308 + 307 + 307 + 314 + 303)/5 = 308 \text{ kW},\\
&\text{Interval 2: } (309 + 303 + 314 + 318 + 306)/5 = 310 \text{ kW},\\
&\text{Interval 3: } (302 + 305 + 314 + 308 + 304)/5 = 307 \text{ kW},\\
&\text{Interval 4: } (296 + 302 + 308 + 304 + 309)/5 = 304 \text{ kW},\\
&\text{Interval 5: } (297 + 305 + 306 + 300 + 309)/5 = 303 \text{ kW},\\
&\text{Interval 6: } (301 + 304 + 307 + 302 + 307)/5 = 304 \text{ kW}.
\end{aligned}
\tag{5.24}
$$

Table 5.3 Average power usage over the past 15 days in Example 5.8.

Day	Day of Week	Interval 1	2	3	4	5	6	Average Usage	DR Event
1	Tue	308	309	302	296	297	301	302	No
2	Wed	293	297	289	291	281	278	288	No
3	Thu	282	277	281	283	285	284	282	No
4	Fri	301	300	296	291	288	290	294	No
5	Sat	165	165	164	161	158	156	162	No
6	Sun	144	141	136	135	135	136	138	No
7	Mon	295	295	294	289	268	272	286	Yes
8	Tue	292	289	286	295	298	296	293	No
9	Wed	307	303	305	302	305	304	304	No
10	Thu	307	314	314	308	306	307	309	No
11	Fri	314	318	308	304	300	302	308	No
12	Sat	173	169	170	170	165	166	169	No
13	Sun	148	145	143	146	145	143	145	No
14	Mon	303	306	304	309	309	307	306	No
15	Tue	291	307	305	310	288	297	300	No

Suppose the load of this customer during the DR event period is:

$$\begin{aligned} &\text{Interval 1: 289 kW,} \quad \text{Interval 2: 291 kW,} \quad \text{Interval 3: 290 kW,} \\ &\text{Interval 4: 288 kW,} \quad \text{Interval 5: 285 kW,} \quad \text{Interval 6: 287 kW.} \end{aligned} \tag{5.25}$$

By subtracting (5.25) from (5.24), we estimate that the customer curtailed 19 kW, 19 kW, 17 kW, 16 kW, 18 kW, and 17 kW at intervals 1–6, respectively.

Other Baseline Calculation Methods

It is generally preferred to have a baseline calculation method that is simple and therefore *easy to understand by the customer*. However, given the challenges in calculating the baseline load profile, there are also many advanced methods that use statistical and *machine learning* techniques to calculate baseline load profiles, e.g., see [293–296]. Of course, in all these methods, ultimately the key to success is having access to detailed power and energy usage data, i.e., the type of measurements which are provided by smart meters.

5.4.3 Energy Usage Clustering

The load profiles that are obtained by smart meters can be used by utilities to perform customer energy usage clustering, also known as *usage segmentation*, i.e., classifying

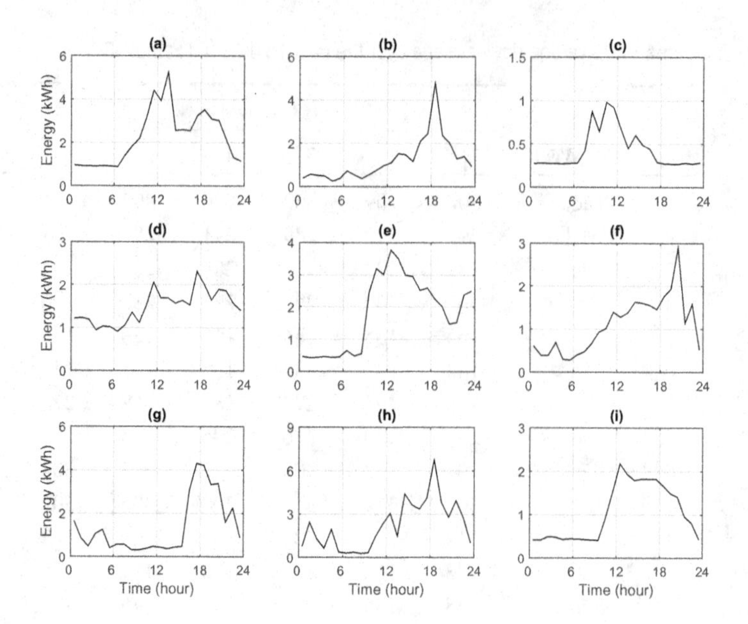

Figure 5.12 Load profiles of nine customers measured by smart meters on the same day [268]. These load profiles can be divided into two classes as in (5.28).

customers based on their load profiles [297, 298]. The results can be used, for example, to calculate baselines in *demand response* programs, see Section 5.4.1; or to enhance accuracy in *load forecasting*, e.g., see [299–301].

k-Means Clustering

Classification of load profiles can be done by using techniques such as k-means clustering. As an example, consider the $n = 9$ hourly load profiles in Figure 5.12 that are reported by nine smart meters on the same day. We seek to divide these nine load profiles into $k = 2$ clusters, based on their *similarities*.

The first step is to introduce adequate *features* to quantitatively represent each load profile. Suppose we define two features for each load profile:

- Feature 1: Peak-to-average magnitude divided by 3.
- Feature 2: Peak-time hour divided by 24.

The features are normalized by 3 and 24 in order to take values around the same range. For instance, for the first load profile, the peak-to-average magnitude is 2.3472 and the peak hour is 14; thus, Feature 1 is 0.7824 and Feature 2 is 0.5833. The features of all nine load profiles in Figure 5.12 are given in Table 5.4.

Let $\mathbf{x}_i$ denote the 2×1 vector of features for load profile i, that is:

$$\mathbf{x}_1 = \begin{bmatrix} 0.7824 \\ 0.5833 \end{bmatrix}, \quad \ldots, \quad \mathbf{x}_9 = \begin{bmatrix} 0.6970 \\ 0.5417 \end{bmatrix}. \tag{5.26}$$

Table 5.4 Features of the load profiles in Figure 5.12.

Load Profile #	Feature 1	Feature 2
1	0.7824	0.5833
2	1.3252	0.7917
3	0.7616	0.4583
4	0.5233	0.7500
5	0.6947	0.5417
6	0.8845	0.8750
7	1.0406	0.7500
8	1.0064	0.7917
9	0.6970	0.5417

In k-means clustering, we seek to partition $n = 9$ load profiles into $k = 2$ sets, denoted by S_1 and S_2, so as to *minimize* the within-cluster sum-of-squares:

$$\sum_{i \in S_1} \|\mathrm{x}_i - \mu_{S_1}\|^2 + \sum_{i \in S_2} \|\mathrm{x}_i - \mu_{S_2}\|^2, \tag{5.27}$$

where μ_{S_1} and μ_{S_2} denote the *average* vectors of set S_1 and set S_2, respectively. That is, notation μ_{S_1} is the 2×1 vector of the average of the features among the load profiles that are members of set S_1. Notation μ_{S_2} is the 2×1 vector of the average of the features among the load profiles that are members of set S_2.

Once we apply the k-means clustering method to the features in Table 5.4, we can classify the nine load profiles into the following two sets:

$$\mathrm{S}_1 = \{1,3,4,5,9\}, \qquad \mathrm{S}_2 = \{2,6,7,8\}. \tag{5.28}$$

The corresponding average vectors for the two clusters are obtained as

$$\mu_{S_1} = \begin{bmatrix} 0.6918 \\ 0.5750 \end{bmatrix}, \quad \mu_{S_2} = \begin{bmatrix} 1.0642 \\ 0.8021 \end{bmatrix}. \tag{5.29}$$

The within-cluster sum-of-squares is 0.0881 and 0.1125, respectively.

The proper choice of features is critical in order to have an effective clustering. In fact, *feature extraction* and *feature selection* often require statistical characterization of the load profiles and also considering external factors; e.g., see [302].

The above classification method that is based on clustering is considered an *unsupervised* learning method in the field of machine learning. Here, we do *not* need to first manually label a few training samples into set S_1 or set S_2. In contrast, in classification methods that are based on *supervised* learning; we must do prior manual labeling. An example for supervised learning is classification based on *support vector machines*; see Section 3.7.2 in Chapter 3 for more details.

Baseline Calculation by Using Load Clustering

Recall from Section 5.4.2 that the baseline for a customer that participates in an incentive-based DR program is to look at the recent load of that same customer. Alternatively, one can calculate the baseline by looking at the recent load of not only the customer in question itself but also the *other similar customers*. This can be done by using the customer classification method that we learned in this section. For instance, suppose we would like to calculate the baseline for Customer #1. From (5.28), the load profile of this customer is classified to belong to set S_1. Accordingly, we may want to calculate the baseline for Customer #1 by looking at the recent load of not only Customer #1, but also Customers #3, #4, #5, and #9, since they belong to the same class of loads. In this option, we can select the five days with the highest average power usage based on the load of all five of these customers [291].

5.4.4 Other Applications of Smart Meter Measurements

Besides what we discussed above, smart meter measurements may support many other smart grid applications. Next, we briefly discuss some of those applications.

Remote Service Connection and Disconnection

In addition to the ability to read the smart meter measurements remotely, utilities can also remotely connect or disconnect service through smart meters, without the need to send their crew to the customer location [303].

Outage Notification and Outage Management

Many smart metering systems offer a "last gasp" message transmission capability to tell the utility that the customer has lost power. Therefore, unlike in traditional metering systems, the customer no longer needs to call the utility to report the outage. The outage notification is sent by the smart meter when it detects a zero voltage event lasting more than a programmed period of time [304]. The outage notifications that are sent by the group of affected smart meters can also help the utility's Outage Management System (OMS) to understand and efficiently respond to outage conditions, such as by helping to identify the approximate location of the fault that may have caused the outage [305].

Electricity Theft Detection

Non-technical losses, i.e., losses due to electricity theft, account for billions of dollars of revenue loss for utilities around the world as individuals may tamper with the electric meters to slow or stop the accumulation of energy usage [306]. The measurements from smart meters can be used to detect electricity theft. For example, we may detect electricity theft by detecting abnormal energy consumption patterns in the load profile of customers [307]. Another option is to examine the energy losses, i.e., to check the balance between the energy that is *supplied* by the utility transformers and the total energy that is *consumed* by the customers that are served by that transformer as reported by the smart meters [308].

Identifying Customers with DR Potential

Smart meters are necessary to facilitate various demand response programs, as we discussed in Sections 5.4.1 and 5.4.2. They can also help *identify* the customers that are a good fit to participate in demand response. This can be done by evaluating *variability in usage*, *sensitivity of usage to temperature*, *occupancy status*, and *inter-temporal usage dynamics*; see [309–311]. It should be noted that, besides smart meters, other types of sensors can also help with the identification and participation of customers in DR programs, such as different types of occupancy sensors; see Section 7.2 in Chapter 7.

Distribution System State Estimation

The measurements by smart meters can help with solving the state estimation problem in distribution systems; see [312–315], and also Section 5.9.3.

Topology and Phase Identification

We previously discussed the problem of topology identification in power distribution systems in Section 4.5.2 in Chapter 4. We also previously discussed the problem of phase identification in power distribution systems in Section 2.8.3 in Chapter 2, and Section 3.6.4 in Chapter 3. Smart meter measurements can help with both applications. For example, see the analysis in [316], where the authors inferred the topology of the distribution system from smart meter energy measurements. We will discuss the application of smart meter measurements in phase identification in Section 5.9.2.

Load Modeling and Load Forecasting

Smart meter measurements can be used in load modeling, as we will discuss in Section 5.7.1. They can also be used in load forecasting. Load forecasting at the customer level can be aggregated to help with the typical system-wide operation needs of the utility; they can also be used for the operation of a specific substation or a specific distribution feeder [317].

Customer Reports

Smart meters provide consumers with greater clarity on their electricity bills and also their own consumption behavior. Customer reports can break down the usage of the customer during different days and different on-peak, mid-peak, and off-peak hours. Customers may access their detailed load profile through the utility's website. This can help them identify opportunities to reduce their electricity cost, such as by coordinating the operation of their *programmable communicating thermostats* and *smart appliances* with the data from smart meters [318].

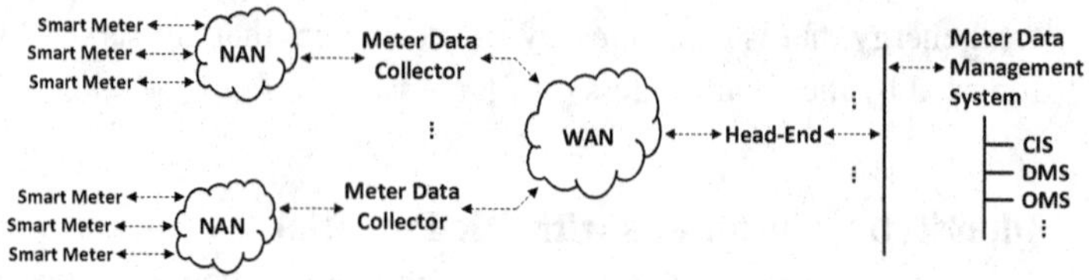

Figure 5.13 The high-level view of AMI and its main components and their interactions. Each dashed double-sided arrow indicates two-way communications.

5.5 Advanced Metering Infrastructure

Advanced metering infrastructure (AMI) is an integrated system of three main components: smart meters, communications networks, and data management systems. These components are shown in Figure 5.13. We have already discussed smart meters and their applications in the previous sections. In this section, we discuss communications networks and data management systems.

5.5.1 AMI Communications Networks

The AMI communication system typically includes two basic layers [319]:

- A *neighborhood area network* (NAN) that provides the last-mile communication system between the smart meters to meter data collectors.
- A *wide area network* (WAN) that serves as the backhaul communication system between the meter data collectors in the field and the head-end system.

Utilities have different geographical conditions, load densities, and legacy communication facilities. Therefore, the communication system to be implemented in each layer of the AMI system can be different for each utility [319].

In Europe, *power line communication* (PLC) is the common choice for the communication technology in a NAN. PLC is a *wired* (as opposed to *wireless*) communication technology that reuses power lines as the media for the purpose of data transmission. We will discuss PLC further in Section 6.5 in Chapter 6.

In North America, *wireless* communications are the more popular choice for the communication technology in a NAN. Two different architectures are often considered. In the *mesh* architecture, the NAN is in the form of a *wireless mesh network* (WMN). In this architecture, the smart meters that are near the meter data collector communicate with the meter data collector directly. However, smart meters that are away from the meter data collector *use other meters as repeaters* to communicate with the meter data collector. This is due to the short-range communication capability in WMNs, i.e., from 1 to 5 miles. In the *star* architecture, all smart meters communicate with the meter data collector directly. Different *radio frequency* (RF) technologies are used. RF technologies provide longer-range communication capability, i.e., from 5 to 15 miles [319].

Other NAN technologies may include fiber-optic communications, wireless broadband communications, and satellite communications (in remote areas).

The second layer of communications, i.e., the WANs, offer a mix of different technologies, such as fiber, microwave, public/private cellular networks, and satellite links [320]. In recent years, cellular communications have been particularly popular in WANs around the world, due to their immediate availability from cellular phone service providers. By leveraging the third-party cellular networks, this option requires a relatively low capital investment by the utility.

5.5.2 AMI Data Management Systems

Meter Data Management System (MDMS) is a database software application that interfaces with the AMI head-ends to collect, store, and analyze the smart meter readings [321]. MDMS also interfaces with all smart meter applications and other smart grid information systems, such as the Consumer Information System (CIS), which also includes billing; Geographic Information System (GIS); Demand Response Management System (DRMS); Distribution Management System (DMS); Distribution Automation System (DAS); Outage Management System (OMS); Fault Location Isolation and Service Restoration (FLISR); Power Quality Management System (PQMS); and Load Forecasting System (LFS) [319].

5.6 Disaggregation and Sub-Metering

5.6.1 Load Disaggregation

Load disaggregation refers to the problem of extracting the power or energy usage of individual *appliances* from their aggregated power and energy usage measurements, such as from the whole-building power or energy usage measurements [322]. For instance, consider the load profile of a residential customer as shown in Figure 5.14, over a period of 30 minutes. At any point in time, the power consumption of this customer is a *summation* of the power consumption of multiple appliances. For example, at point ①, the total load is the summation of the load of the oven, the load of the stove burner, and some other background loads. As another example, at point ②, the total load is the summation of the load of the dishwasher, the load of the refrigerator, and some other background loads. In load disaggregation, we seek to identify the load of each of these major appliances.

Load disaggregation is a *non-intrusive* technique to monitor appliance-level loads because this technique does *not* require placing sensors on individual appliances in the customer's property [323]. The alternative is *sub-metering*, which is an intrusive technique. We will discuss sub-metering in Section 5.6.3.

Applications of load disaggregation may include demand response [324], load forecasting [301], appliance and equipment health monitoring [325], and household appliance marketing by the utility or third-party businesses [326, 327].

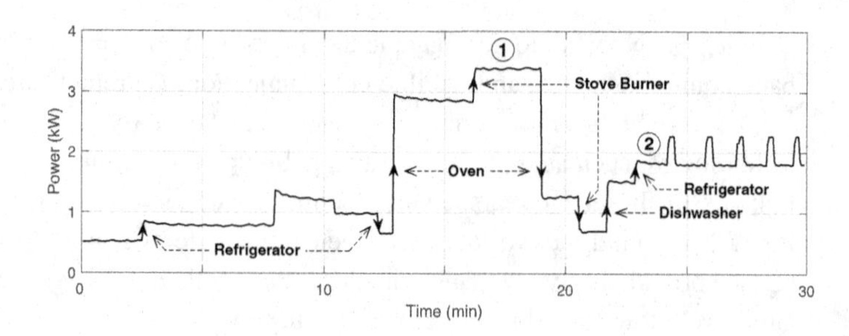

Figure 5.14 Total power consumption of a house over a period of 30 minutes. The arrows indicate the step changes in the total load due to individual appliance operation [323].

There are different methods to solve the load disaggregation problem. Let us first look at an elementary analysis in Example 5.9.

Example 5.9 Consider a residential customer whose total power usage is measured by a wattmeter. The power usage of this customer's largest loads are as follows:

- Water Heater: 4500 W
- Central Air Conditioning: 3250 W
- Clothes Dryer: 2300
- Dishwasher: 1700 W
- Oven: 1200 W
- Clothes Washer: 800 W.

If none of the above major appliances and equipment is on, then the power usage of this residential customer is somewhere between 200 W and 500 W. Based on this information, we want to disaggregate the total load of this customer. First, assume that the total power usage of this customer is measured at 13.2 kW. In that case, we can conclude that all the major loads, except for the clothes washer, are on; and the load of the remaining (non-major) appliances is 250 W:

$$4500\text{ W} + 3250\text{ W} + 2300\text{ W} + 1700\text{ W} + 1200\text{ W} + 250\text{ W} = 13.2\text{ kW}. \quad (5.30)$$

However, if the total power usage is measured at 11.7 kW, then there are *two possibilities*. The first possibility is that all the major loads, except for the clothes dryer, are on; and the load of the remaining (non-major) appliances is 250 W:

$$4500\text{ W} + 3250\text{ W} + 1700\text{ W} + 1200\text{ W} + 800\text{ W} + 250\text{ W} = 11.7\text{ kW}. \quad (5.31)$$

Another possibility is that all the major loads, except the dishwasher and clothes washer, are on; and the load of the remaining (non-major) appliances is 450 W:

$$4500\text{ W} + 3250\text{ W} + 2300\text{ W} + 1200\text{ W} + 450\text{ W} = 11.7\text{ kW}. \quad (5.32)$$

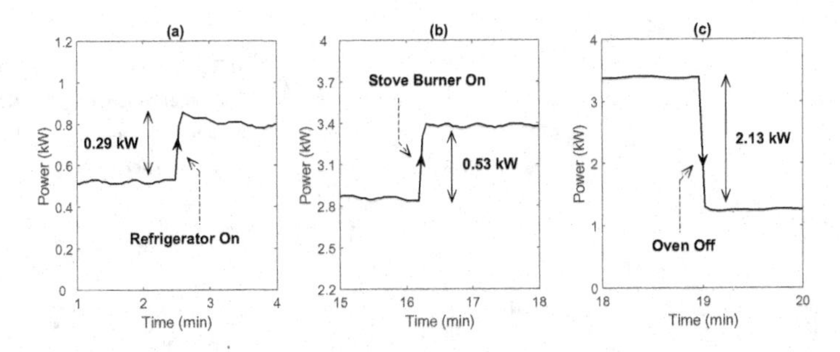

Figure 5.15 Three example load switching events in the power profile in Figure 5.14: (a) refrigerator switches on; (b) stove burner switches on; (c) oven switches off.

The basic idea that we examined in Example 5.9 has been considered in the literature, in form of a combinatorial optimization to identify *which combination of appliances* can match the total load; e.g., see the binary optimization problem formulation in [323]. However, this approach is not always effective and it can be prone to uncertain solutions, as we saw in the second case in Example 5.9.

Analysis of Load Signatures

Another approach is to identify the *signatures* of different types of appliances, when they *switch on* or when they *switch off*. Accordingly, in this approach, the focus is on *load switching events*. For instance, again consider the load profile in Figure 5.14. If we focus on the *sharp edges* in the total load, they correspond to the switching events of the individual appliances. A *positive* sharp edge indicates that a major appliance switches on; and a *negative* sharp edge indicates that a major appliance switches off. Three examples of such events are shown in Figure 5.15. When the refrigerator switches on, the total load suddenly increases by 0.29 kW; see Figure 5.15(a). When the stove burner switches on, the total load suddenly increases by 0.53 kW; see Figure 5.15(b). When the oven element switches off, then the total load suddenly decreases by 2.13 kW; see Figure 5.15(c).

Once we identify which appliance switches on at each positive sharp edge and which appliance switches off at each negative sharp edge, we can identify all the appliances that are on at any time, thus solving the load aggregation problem.

The analysis of load signatures during load switching events is in principle similar to the analysis of other types of events that we have studied throughout this book. For example, we can *detect* a switching event in power measurements by using similar methods that we learned in Section 2.7.2 in Chapter 2.

Features: Active and Reactive Power

The *changes* in the total active power and the *changes* in the total reactive power are among the most common features that are looked at when load disaggregation is done based on the analysis of load signatures. Figure 5.16 shows the scatter plot for these two features for several appliances in a house [328]. The points that correspond to the switching of the major appliances are marked from 1 to 8. It should be noted that, all

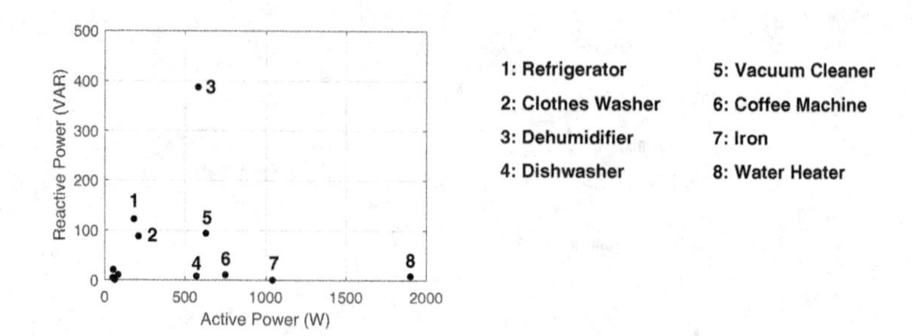

Figure 5.16 Scatter plot of the changes caused in active power and reactive power of an example household due to switching on various appliances [328].

the points in this figure are based on switching *on* events. All the major loads that are marked with numbers are either mainly resistive or both resistive and inductive.

We can see in Figure 5.16 that, while the smaller loads are difficult to distinguish, the larger loads have distinct signatures in the active and reactive power measurements that can help identify them. Notice that the dehumidifier and the dishwasher have similar active power consumption, but they have different reactive power consumption that can help distinguish them. Also, the clothes washer and the vacuum cleaner have similar reactive power consumption, but they have different active power consumption that can help distinguish them.

Features: Instantaneous Power and Harmonics

Instantaneous power waveform can serve as another feature in load disaggregation. A few examples are shown in Figure 5.17 during one cycle of the AC power signal at 60 Hz. Recall from Section 1.2.4 in Chapter 1 that the frequency of the instantaneous power waveform is twice the frequency of the voltage and current waveforms. That is why in Figure 5.17 we see two cycles of the instantaneous power waveform within $1/60$ Hz $=$ 16.667 msec. The shape of the instantaneous power waveform is very different among the three types of loads that are shown in this figure. If the instantaneous power waveform measurements are available, then we can detect when these load types switch on, as soon as we notice *the presence* of these specific instantaneous power waveforms; and detect when these load types switch off, as soon as we notice *the absence* of these specific instantaneous power waveforms.

If instantaneous power waveform measurements are *not* available, then we may still use harmonics in current measurements. The induction cooker in Figure 5.17 generates the 89th and 91st harmonics in the current; and the television in Figure 5.17 generates the 3rd and 5th harmonics in the current.

These additional features can be used to better distinguish different load types that have similar active and reactive power consumption. For instance, it is quite possible that two load types have more or less similar active power and reactive power signatures; but they have very different harmonic signatures.

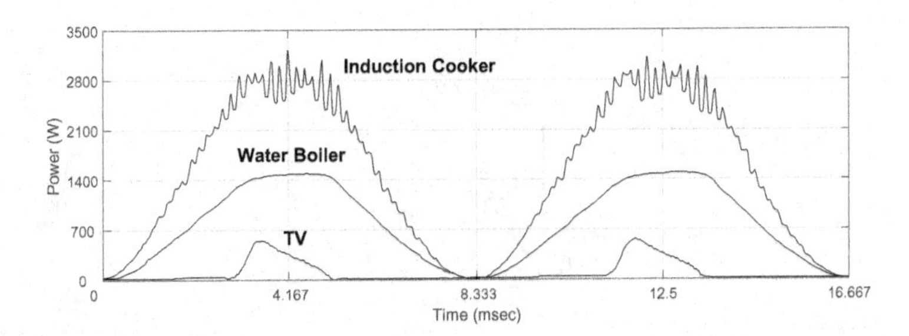

Figure 5.17 Application of instantaneous power waveform in load disaggregation [329].

Features: Time, Day, and Other External Factors

The performance of load disaggregation may improve by considering various other features. Time and day might add a factor of *likelihood* to the operation of certain appliances. For example, an oven is less likely to be operated at 3:00 AM versus at 6:00 PM. One can also define features based on the human behavior in a household. For example, the *sequence of operation* among appliances can be considered, such as the sequence between the clothes washer and the clothes dryer; or some likely sequential operation of certain kitchen appliances [330].

5.6.2 Net Load Disaggregation

The increasing penetration of *behind-the-meter* renewable power generation resources, such as rooftop Photo-Voltaic (PV) resources in residential customers, has direct impact on the problem of load disaggregation. The reason is that the utility's meters measure the *net aggregate* of the customer's load minus the customer's generation. This is shown in Figure 5.18(a). What the utility measures at its meter is

$$P_{\text{Measured}} = P_{\text{Load}} - P_{\text{PV}}. \tag{5.33}$$

This new paradigm transforms the traditional load disaggregation problem that we saw in Section 5.6.1 into the new problem of *net load disaggregation.*

Apart from the initial goal of identifying the appliance-level load of a customer, net load metering can also help with estimating the customer's behind-the-meter renewable power generation. This by itself is an important task, in particular, when it comes to estimating behind-the-meter solar power generation. This task is often referred to as *solar generation disaggregation* [331, 332].

Figure 5.18(a) shows a commercial building with a behind-the-meter solar power generation unit. Figure 5.18(b) shows the net aggregated power consumption measurements at this building. As in (5.33), the utility measures the customer's load minus the customer's solar power generation. The net load of the customer *reduces during the day* due to solar power generation. The power consumption profile for the disaggregated load and the power generation profile for the disaggregated solar power generation are shown in Figures 5.18(c) and (d), respectively.

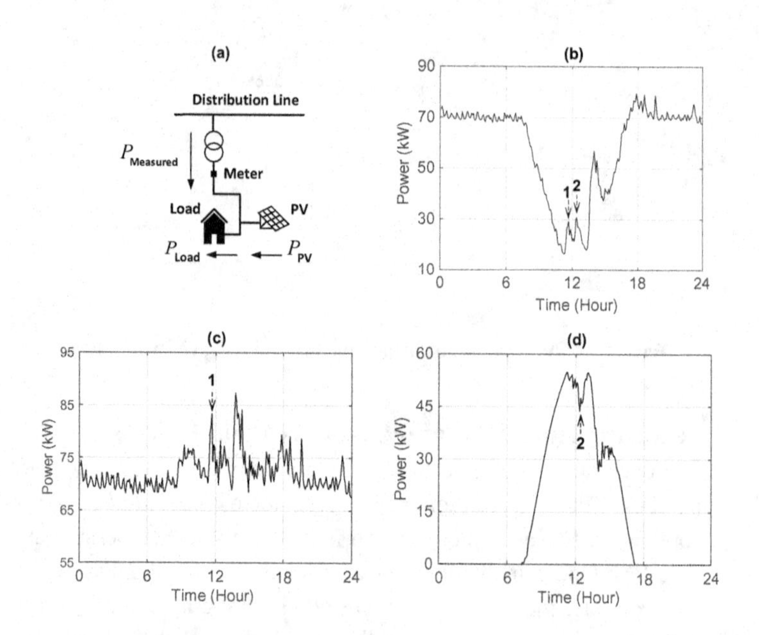

Figure 5.18 Solar generation disaggregation: (a) the layout of the behind-the-meter system; (b) net aggregated power consumption measurements; (c) disaggregated power consumption profile; (d) disaggregated solar power generation profile.

Two events in form of sudden increases in the net load of the customer are marked in Figure 5.18(b) with numbers 1 and 2. These two events have two different causes. The first event was caused due to a *sudden increase* in the load of the customer, as marked in Figure 5.18(c). The second event was caused due to a *sudden decrease* in solar power generation, as marked in Figure 5.18(d). The sudden decrease in solar power generation could be due to cloud passing. Therefore, the variations in solar power generation can create their own events in the net load, which may not be easy to distinguish from the signature of loads.

Proxy Measurements

One option that is sometimes considered in solar generation disaggregation, such as in [332, 333], is to use *proxy measurements*. A PV proxy is a PV system that is in a *close-by location*, and its power generation is measured directly. The solar power generation at the PV proxy can be used to estimate the solar power generation portion of a net load at the location where we need net load disaggregation.

For example, consider again the scenario in Figure 5.18. Another PV system is located about *two miles away* from this customer, as shown in Figure 5.19(a), and its solar power generation is measured directly. Therefore, this other PV system can be used as a PV proxy. Figure 5.19(b) shows the solar power generation at the original PV system and also at the solar power generation at the proxy PV system. Overall, the two power generation profiles are similar, because the two locations are close to each other. Two points, denoted by ① and ②, are marked on this figure at the times when

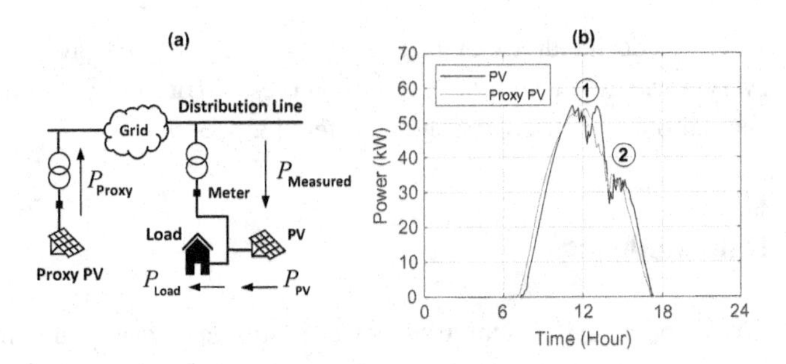

Figure 5.19 Application of proxy PV in net load disaggregation: (a) proxy PV is located two miles away from the PV of interest; (b) comparing the solar power generation of the proxy PV and the PV of interest.

solar power generation is volatile. There are some considerable differences at ①, but the two solar power generation profiles are mostly similar at ②.

As an approximate method, we can disaggregate the net load as

$$P_{\text{PV}} \approx P_{\text{Proxy}}, \qquad P_{\text{Load}} \approx P_{\text{Measured}} - P_{\text{Proxy}}. \tag{5.34}$$

The accuracy of the above approximate method may improve by factoring in some physical characteristics of the two PV systems, taking into account the information about clouds movement in the region [334], or training a model to use a proper mixture of measurements from *multiple* PV proxies.

Other Methods

Different methods have been proposed in the literature to solve the solar generation disaggregation problem. Some methods use fundamental physical models and take into account relationships between location, weather, physical characteristics, and solar irradiance; e.g., see [331]. Some other methods use data-driven and statistical models, such as extracting features to discribe load profiles with PVs and load profiles without PVs; e.g., see [335]. Some methods also use a combination of physical models and statistical models, such as in [336].

5.6.3 Sub-Metering

The discussions in Sections 5.6.1 and 5.6.2 are concerned with *non-intrusive* disaggregation. They are called non-intrusive because they do *not* require placing sensors on individual appliances, individual equipment, or individual PV units. The alternative to non-intrusive disaggregation is *sub-metering*. In sub-metering, we install a separate meter to measure power or energy usage or power or energy generation of a subsystem or a component of interest; e.g., see [268, 337].

Sub-metering has recently received a technology boost due to the advent of *Internet-of-Things* (IoT). IoT technologies have lowered the cost of sensor installation and data collection. They can help enhance energy efficiency and facilitate participation in demand response. An IoT-based sub-metering system in a building may include hundreds of meters at every lighting fixture, every power outlet, and every variable-air-volume valve, compressor, or other subsystems of the heating, ventilation, and air conditioning system [338–340].

5.7 Load Modeling

Load modeling is essential to power system analysis, planning, and operation. The purpose of load modeling is to understand the behavior of the load, such as in response to changes in voltage or frequency. Load modeling can be done at the transmission level to model the aggregate load at different locations on an interconnected power system, e.g., see [341]; or it can be done at the distribution level to model the load of a power distribution feeder, the load of a single customer, or even the load of a single appliance, e.g., see [342–344].

Load modeling can be *component-based*, where the knowledge of physical behavior of the load components, such as the physical parameters of a motor load, are used to model the functioning of load devices [345, 346]. Load modeling can also be *measurement-based*, where the measurements from various sensors are used to capture the behavior of the load. Measurement-based load modeling has several advantages over component-based load modeling. For example, it can be applied to a load, even when we have no prior knowledge about the physical behavior or physical parameters of the load. Moreover, measurements-based load modeling can update the model over time in order to capture the changes in the load.

In this section, we discuss only measurement-based load modeling. A detailed review of different load modeling techniques is available in [347].

Measurement-based load models are either *static* or *dynamic*. We discuss static load models in Section 5.7.1 and dynamic load models in Section 5.7.2.

5.7.1 Static Load Model

ZIP Model

A popular measurement-based load model is the ZIP model. This model represents the relationship between (active and reactive) power consumption and voltage in a polynomial equation that combines *constant impedance* (Z), *constant current* (I), and *constant power* (P) components of the load:

$$P = P_0 \left[\alpha_Z \left(\frac{V}{V_0} \right)^2 + \alpha_I \left(\frac{V}{V_0} \right) + \alpha_P \right], \tag{5.35}$$

$$Q = Q_0 \left[\beta_Z \left(\frac{V}{V_0} \right)^2 + \beta_I \left(\frac{V}{V_0} \right) + \beta_P \right], \tag{5.36}$$

where P and Q are active power consumption and reactive power consumption at operating voltage V, respectively; and P_0 and Q_0 are active power consumption and reactive power consumption at rated voltage V_0, respectively. A ZIP model has six parameters: α_Z, α_I, and α_P are the coefficients for active power in (5.35); and β_Z, β_I, and β_P are the coefficients for reactive power in (5.36).

Suppose P_i and V_i denote the measured active power and the measured voltage at the load that we seek to model, where $i = 1, \ldots, n$, and n is the total number of measurements. We can obtain the unknown coefficients of the ZIP model for active power by solving the following *least square* (LS) optimization problem:

$$\min_{\boldsymbol{\alpha}} \ \|\mathbf{b} - \mathbf{M}\boldsymbol{\alpha}\|_2, \tag{5.37}$$

where

$$\mathbf{b} = \begin{bmatrix} P_1/P_0 \\ P_2/P_0 \\ \vdots \\ P_n/P_0 \end{bmatrix}, \quad \mathbf{M} = \begin{bmatrix} (V_1/V_0)^2 & (V_1/V_0) & 1 \\ (V_2/V_0)^2 & (V_2/V_0) & 1 \\ & \vdots & \\ (V_n/V_0)^2 & (V_n/V_0) & 1 \end{bmatrix}, \tag{5.38}$$

and the unknown coefficients are

$$\boldsymbol{\alpha} = \begin{bmatrix} \alpha_Z \\ \alpha_I \\ \alpha_P \end{bmatrix}. \tag{5.39}$$

We can solve the above LS problem by using an LS solver; such as `lsqlin` in MATLAB [102]. Alternatively, we can obtain the solution in closed-form as [103]:

$$\boldsymbol{\alpha} = (\mathbf{M}^T\mathbf{M})^{-1}\mathbf{M}^T\mathbf{b}. \tag{5.40}$$

Given Q_i and V_i as the measured reactive power and the measured voltage at the load, where $i = 1, \ldots, n$, we can obtain the coefficients β_Z, β_I, and β_P for the model in (5.36) by formulating and solving a similar LS optimization problem.

Example 5.10 Voltage, active power consumption, and reactive power consumption are measured at a load. The measurements are shown in Table 5.5, where $n = 20$. The rated voltage, the rated active power, and the rated reactive power of the load are $V_0 = 120$ V, $P_0 = 1109$ W, and $Q_0 = 487$ VAR, respectively [348]. By solving the LS optimization problem in (5.37), we can obtain:

$$\alpha_Z = 0.7294, \quad \alpha_I = 0.4279, \quad \alpha_P = -0.1568. \tag{5.41}$$

Similarly, we can obtain:

$$\beta_Z = -1.5544, \quad \beta_I = 4.487, \quad \beta_P = -1.9325. \tag{5.42}$$

If $P = P_0$ and $V = V_0$, then from (5.35), we have:

$$\alpha_Z + \alpha_I + \alpha_P = 1. \tag{5.43}$$

Table 5.5 Voltage, active power, and reactive power measurements in Example 5.10.

i	V (Volt)	P (Watt)	Q (VAR)	i	V (Volt)	P (Watt)	Q (VAR)
1	120.6	1120	491	11	124.2	1184	510
2	119.3	1097	484	12	122.5	1154	501
3	123.7	1175	507	13	115.7	1036	462
4	122.1	1146	499	14	122.4	1152	501
5	120.3	1115	489	15	115.3	1029	460
6	116.4	1048	467	16	118.1	1076	477
7	121.6	1138	496	17	121.7	1139	497
8	124.9	1196	514	18	125.1	1200	515
9	120.6	1120	491	19	126.3	1222	521
10	115.1	1025	459	20	126.4	1223	521

The above equation is sometimes added to the minimization in (5.37) as a constraint; e.g., see [348]. From (5.43), we have $\alpha_Z = 1 - \alpha_I - \alpha_P$. Thus, the number of independent optimization variables in (5.37) reduces from three to two.

Exponential Model

ZIP load model may also be represented in *exponential form* as follows:

$$P = P_0 \left(\frac{V}{V_0} \right)^{\gamma_P}, \tag{5.44}$$

$$Q = Q_0 \left(\frac{V}{V_0} \right)^{\gamma_Q}. \tag{5.45}$$

where γ_P and γ_Q are the parameters of the model. They vary between 0 and 2. If $\gamma_P = 2$, then the load model in (5.44) reduces to a constant impedance load. If $\gamma_P = 1$, then the load model in (5.44) reduces to a constant current load. If $\gamma_P = 0$, then the load model in (5.44) reduces to a constant power load.

Given V_0 and P_0, one can measure V and P to obtain

$$\gamma_P = \log(P/P_0) \,/\, \log(V/V_0). \tag{5.46}$$

If several measurements are available, then we can develop an LS optimization problem formulation to estimate γ_P, such as the following, see Exercise 5.18:

$$\min_{\gamma_P} \left\| \mathbf{b} - \mathbf{M}\gamma_P \right\|_2, \tag{5.47}$$

where

$$\mathbf{b} = \begin{bmatrix} \log(P_1/P_0) \\ \log(P_2/P_0) \\ \vdots \\ \log(P_n/P_0) \end{bmatrix}, \quad \mathbf{M} = \begin{bmatrix} \log(V_1/V_0) \\ \log(V_2/V_0) \\ \vdots \\ \log(V_n/V_0) \end{bmatrix}. \tag{5.48}$$

If V_0 and P_0 are not known, then we can estimate γ_P with two pairs of voltage and active power measurements, such as during a *voltage event*; e.g., see [343].

Frequency-Dependant Model

Power consumption of certain loads can be affected by the frequency of the power system. For example, some motor loads slow down when the frequency of the system drops; see a related discussion under the subject of *frequency response* of interconnected power systems in Section 2.9 in Chapter 2.

Both the ZIP model and the exponential model can be extended to incorporate the impact of frequency. The ZIP model can be extended to:

$$P = P_0 \left[\alpha_Z \left(\frac{V}{V_0} \right)^2 + \alpha_I \left(\frac{V}{V_0} \right) + \alpha_P \right] \left[1 + \alpha_f (f - f_0) \right], \tag{5.49}$$

$$Q = Q_0 \left[\beta_Z \left(\frac{V}{V_0} \right)^2 + \beta_I \left(\frac{V}{V_0} \right) + \beta_P \right] \left[1 + \beta_f (f - f_0) \right], \tag{5.50}$$

where α_f and β_f are the coefficients of the model, f is the operation frequency, and f_0 is the nominal frequency. Parameter α_f is positive; but parameter β_f can be positive or negative [349]. We can similarly extend the exponential model to:

$$P = P_0 \left(\frac{V}{V_0} \right)^{\gamma_P} \left[1 + \alpha_f (f - f_0) \right], \tag{5.51}$$

$$Q = Q_0 \left(\frac{V}{V_0} \right)^{\gamma_Q} \left[1 + \beta_f (f - f_0) \right]. \tag{5.52}$$

Example 5.11 Consider an agricultural pump with $\alpha_f = 5.6$ and $\beta_f = 4.2$ [349]. If the frequency drops by 0.09 Hz (as in Example 2.23 in Chapter 2), then

$$1 + \alpha_f (f - f_0) = 1 + 5.6 \times -0.09 = 0.496. \tag{5.53}$$

$$1 + \beta_f (f - f_0) = 1 + 4.2 \times -0.09 = 0.622. \tag{5.54}$$

Active power load drops by 50%, and reactive power load drops by 38%. This can contribute to the *inertial response* of the system, as we saw in Example 2.23.

5.7.2 Dynamic Load Model

Static loads are static functions of voltage and/or frequency. The power consumption of a static load at any instant depends on only the voltage and/or frequency at that *same instant*. Static loads are modeled by *algebraic equations*, such as in (5.35)–(5.36), (5.44)–(5.45), and (5.49)–(5.52).

Dynamic loads, such as induction motors, on the other hand, take some time in *transient conditions* before they reach steady-state conditions. This is because the power consumption of a dynamic load at any instant depends on not only the voltage

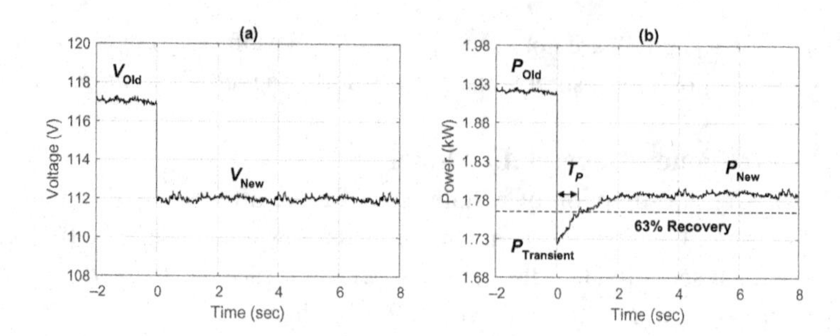

Figure 5.20 Recovery load model: (a) sudden drop in voltage; (b) power consumption.

and/or frequency at that same instant, but also the internal state variables of the load at *previous* instances. Dynamic loads are modeled as a combination of *algebraic equations* and also *differential equations* [350, 351].

Recovery Model

This model is a popular model in voltage stability analysis [352]. This model captures the *recovery response* of a load to a sudden voltage drop. That is, it captures how the active power consumption and the reactive power consumption of the load changes before it reaches steady-state conditions. This is illustrated in Figure 5.20. Voltage suddenly drops at time $t_0 = 0$ from $V_{\text{Old}} = 117$ V to $V_{\text{New}} = 112$ V; see Figure 5.20(a). In response, the active power consumption of the load drops from $P_{\text{Old}} = 1.92$ kW to $P_{\text{Transient}} = 1.72$ kW; and then it gradually increases to $P_{\text{New}} = 1.79$ kW, see Figure 5.20(b). The primary focus of the recovery model is to capture the transient response of the load.

The recovery model for active power consumption is described by the following combination of algebraic and differential equations [352]:

$$\begin{aligned} P(t) &= P_0 \left(\frac{V(t)}{V_0}\right)^{\zeta_P} + \frac{x(t)}{T_P} \\ \frac{dx(t)}{dt} &= -\frac{x(t)}{T_P} + P_0 \left(\frac{V(t)}{V_0}\right)^{\gamma_P} - P_0 \left(\frac{V(t)}{V_0}\right)^{\zeta_P}. \end{aligned} \tag{5.55}$$

Here, $x(t)$ is the state-variable of the load; γ_P is the static coefficient, which is the same parameter as in the exponential load model in (5.44) in Section 5.7.1; ζ_P is the transient recovery coefficient; and T_P is the recovery time constant. We can obtain a similar model to capture the dynamics of reactive power consumption.

We can measure V_{Old}, V_{New}, P_{Old}, $P_{\text{Transient}}$, and P_{New}. If the rated voltage V_0 and the rated active power consumption P_0 are known, then we can obtain:

$$\begin{aligned} \gamma_P &= \log\left(P_{\text{New}}/P_0\right) \big/ \log\left(V_{\text{New}}/V_0\right), \\ \zeta_P &= \log\left(P_{\text{Transient}}/P_0\right) \big/ \log\left(V_{\text{New}}/V_0\right). \end{aligned} \tag{5.56}$$

Notice that γ_P is obtained in (5.56) by placing $P(t) = P_{\text{New}}, V(t) = V_{\text{New}}$, and $dx(t)/dt = 0$ in (5.55). Also, ζ_P is obtained in (5.56) by placing $P(t) = P_{\text{Transient}}$, $V(t) = V_{\text{New}}$, and $x(t) = 0$ in the first line in (5.55).

Parameter T_P can be obtained by determined the *time instance* at which $P(t)$ in Figure 5.20(b) crosses the 63% threshold line during its recovery from $P_{\text{Transient}}$ to P_{New}. This threshold is based on the analysis of the pole of the dynamic system in (5.55) and the fact that $100 \times (1 - \exp(-1)) = 63\%$; see [352] for more details. For the example in Figure 5.20(b), we have $T_P = 0.824$.

Other Dynamic Load Models

Other dynamic load models that are used in measurement-based load modeling include: the induction motor (IM) model [353], a combination of the ZIP model and the IM model [354], and load models that are based on training neural networks [355].

5.8 State Estimation Using Power Measurements

We previously discussed the state estimation problem in Section 3.8 in Chapter 3, where we assumed that the measurements are voltage and current synchrophasors that are obtained by PMUs. Recall that the *states* of a power system are the voltage magnitudes and voltage phase angles at all buses. Therefore, when synchrophasor measurements are available, we are able to directly measure a subset of the state variables. The use of synchrophasor measurements also results in formulating a state estimation problem that is inherently *linear*; see the relationship in (3.76).

However, prior to the development of PMUs, the states of the power system could not be measured directly. In particular, the voltage phase angles could be only *inferred* from the power flow measurements [356]. Accordingly, the traditional state estimation problem is the problem of using active power and reactive power measurements to solve the power flow equations that we previously saw in Section 1.3.1 in Chapter 1, so as to estimate the voltage magnitudes and voltage phase angles at all buses.

5.8.1 Basic Nonlinear Formulation

In this section, we discuss the basic formulation of the traditional state estimation problem, which makes use of active and reactive power measurements. This problem is a nonlinear and non-convex optimization problem. We will discuss a linearized approximate formulation of the state estimation problem in Section 5.8.2. We will also briefly discuss other formulations in Section 5.8.3.

Let $\mathbf{x}$ denote the vector of all states of the power system. Let $\mathbf{z}$ denote the vector of all active and reactive power measurements, whether at buses or on power lines.

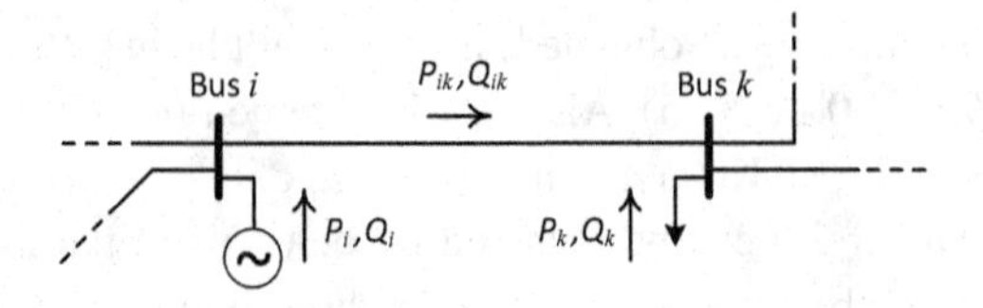

Figure 5.21 Power measurements are used in state estimation in Section 5.8.

An example is shown in Figure 5.21. The vector of state variables and the vector of measurements for the example in this figure are:

$$\mathbf{x} = \begin{bmatrix} \vdots \\ V_i \\ V_k \\ \vdots \\ \theta_i \\ \theta_k \\ \vdots \end{bmatrix}, \quad \text{and} \quad \mathbf{z} = \begin{bmatrix} \vdots \\ P_i \\ Q_i \\ P_k \\ Q_k \\ \vdots \\ P_{ik} \\ Q_{ik} \\ \vdots \end{bmatrix}, \tag{5.57}$$

respectively. Here, $V_i \angle \theta_i$ and $V_k \angle \theta_k$ denote the voltage phasors at buses i and k. Also, P_i and Q_i denote the active power and reactive power at bus i; P_k and Q_k denote the active power and reactive power at bus k; and P_{ik} and Q_{ik} denote the active power and reactive power on line (i,k). Note that unlike in the state estimation problem in Section 3.8.1 in Chapter 3, where the state variables are defined in Cartesian form, here, the state variables are defined in Polar form.

Recall from Section 1.3.1 in Chapter 1 that the following relashionships hold between the voltage phasors and nodal power injections and line power flows:

$$\begin{aligned} P_i &= \sum_{k=1}^{n} V_i V_k \left(G_{ik}^{\text{bus}} \cos(\theta_i - \theta_k) + B_{ik}^{\text{bus}} \sin(\theta_i - \theta_k) \right) \\ Q_i &= \sum_{k=1}^{n} V_i V_k \left(G_{ik}^{\text{bus}} \sin(\theta_i - \theta_k) - B_{ik}^{\text{bus}} \cos(\theta_i - \theta_k) \right) \end{aligned} \tag{5.58}$$

and

$$\begin{aligned} P_{ik} &= -V_i^2 G_{ik}^{\text{bus}} + V_i V_k \left(G_{ik}^{\text{bus}} \cos(\theta_i - \theta_k) + B_{ik}^{\text{bus}} \sin(\theta_i - \theta_k) \right) \\ Q_{ik} &= V_i^2 B_{ik}^{\text{bus}} + V_i V_k \left(G_{ik}^{\text{bus}} \sin(\theta_i - \theta_k) - B_{ik}^{\text{bus}} \cos(\theta_i - \theta_k) \right), \end{aligned} \tag{5.59}$$

where G_{ik}^{bus} and B_{ik}^{bus} denote the real part and the imaginary part of the entry in row i and column j of the Y-bus matrix; see (1.52) in Chapter 1. From (5.58) and (5.59), we can relate the measurements to the state variables as follows:

$$\mathbf{z} = \mathbf{h}(\mathbf{x}) + \epsilon, \tag{5.60}$$

where $\mathbf{h}(\mathbf{x})$ is the vector of nonlinear functions of the forms in (5.58) and (5.59); and ϵ is the vector of measurement noise. Accordingly, the state estimation problem can be formulated as the following LS optimization problem:

$$\min_{\mathbf{x}} \|\mathbf{z} - \mathbf{h}(\mathbf{x})\|_2 . \tag{5.61}$$

As we previously defined in (5.57), the number of state variables is $2n$, where n is the number of buses; and the number of measurements is m. Bus 1 can be assumed to be the *reference bus*. We assume to know the voltage magnitude at the reference bus. We also set the voltage phase angle at the reference bus to be zero, i.e., $\theta_1 = 0$. Recall from Section 3.1.2 in Chapter 3 that we can rotate the same phasors and represent them differently based on different references; therefore, we need to use a reference phase angle in order to avoid ambiguity in defining the phase angles.

The optimization problem in (5.61) is *non-convex*. There do exist some solvers to deal with non-convex least-square optimization problems, such as `lsqnonlin` in MATLAB [357]. We can also use iterative algorithms such as the Gauss–Newton method that we will discuss in the next sub-section. However, non-convex optimization problems are often difficult to solve, and there is no guarantee that an optimal solution can be obtained [103]. Therefore, in practice, one option is to solve the original nonlinear problem in (5.61), where exact optimality is not guaranteed. Another option is to solve a linearized approximate version of the problem in (5.61) using the standard LS method; see Section 5.8.2. There are also some other problem formulations that we will discuss in Section 5.8.3.

Gauss–Newton Iterations

As mentioned earlier, the optimization problem in (5.61) is non-convex and generally difficult to solve. One option is to use the Gauss–Newton method. Each iteration in the Gauss–Newton algorithm is formulated as [358]:

$$\mathbf{x} \leftarrow \mathbf{x} + \left[\mathbf{H}(\mathbf{x})^T \mathbf{H}(\mathbf{x})\right]^{-1} \mathbf{H}(\mathbf{x})^T \left(\mathbf{z} - \mathbf{h}(\mathbf{x})\right), \tag{5.62}$$

where

$$\mathbf{H} = [\partial \mathbf{h} / \partial \mathbf{x}] \tag{5.63}$$

is the measurement Jacobian. The entry in row l and column j of matrix $\mathbf{H}$ is the partial derivative of row l of function $\mathbf{h}(\mathbf{x})$ with respect to the state variable in row j of vector $\mathbf{x}$. The iterations in (5.62) start with a given initial value for $\mathbf{x}$. Note that the phase angle at the reference bus should always be kept at zero, i.e., $\theta_1 = 0$. The iterations continue until the norm of the *measurement residues* is less than a given threshold δ, i.e., until the following inequality holds:

$$\|\mathbf{z} - \mathbf{h}(\mathbf{x})\|_2 \leq \delta. \tag{5.64}$$

Note that matrix $\mathbf{H}$ has to be updated in each iteration.

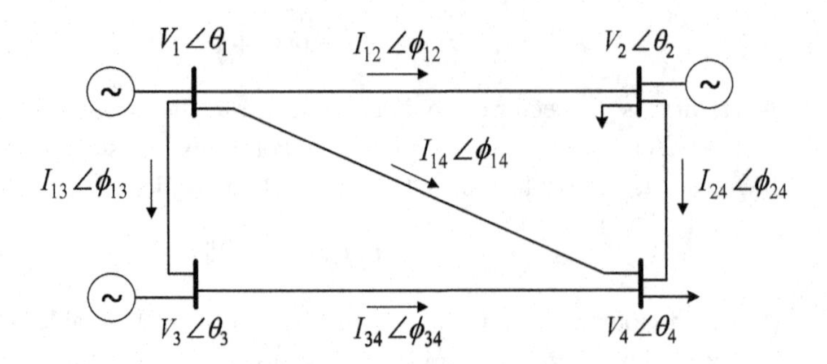

Figure 5.22 The 4-bus network in the state estimation problem in Example 5.12.

Example 5.12 Again, consider the 4-bus transmission network in Example 3.20 in Chapter 3. We have shown the same network also in Figure 5.22. Suppose we measure the power injection at each bus. The true versus measured power values are listed in Table 5.6. But 1 is the reference bus. At the reference bus, the voltage phase angle is assumed to be zero and the voltage magnitude is measured directly. The true and the measured voltage magnitudes at bus 1 are 1.0332 and 1.0297, respectively. The state variables that need to be estimated are the voltage magnitude and the voltage phase angle at buses 2, 3, and 4. The state estimation problem is formulated as in (5.61), where $\mathbf{x}$ is a 6×1 vector and $\mathbf{z}$ is a 12×1 vector. The measurement matrix $\mathbf{H}$ is 12×6. The initial value for all unknown states is set to one for magnitude and zero for phase angle. The Gauss–Newton algorithm is based on the iterations in (5.62). The iterations stop when the norm of the measurement residue is less than $\delta = 0.005$ p.u. The state estimation results are obtained as

$$\begin{aligned} V_1 \angle \theta_1 &= 1.0297 \angle 0^\circ, \\ V_2 \angle \theta_2 &= 0.9938 \angle -7.4071^\circ, \\ V_3 \angle \theta_3 &= 1.0464 \angle -1.8296^\circ, \\ V_4 \angle \theta_4 &= 0.9718 \angle -10.6214^\circ. \end{aligned} \tag{5.65}$$

The Gauss–Newton algorithm converges after seven iterations. If we rotate all phase angles by 38.8884° clockwise, they can be presented equivalently as 38.8884°, 31.4813°, 37.0588°, and 28.2670°. These new values are comparable with the phase angles that are obtained in Example 3.20 in Chapter 3.

Other Iterative Algorithms

Other iterative algorithms have also been used to solve the state estimation optimization problem in (5.61). Some of these algorithms include the Gauss–Seidel method, the Newton-Rophson method, and various decoupling methods; see the state estimation textbooks such as in [356, 359–361]. Another important topic that is often discussed in these textbooks is about the methods that can detect and discard *bad data* in the

Table 5.6 The true and measured apparent power in Example 5.12.

Bus #	True Apparent Power (p.u.)	Measured Apparent Power (p.u.)
1	$3.5622 + j0.8768$	–
2	$-0.7795 + j0.0000$	$-0.7786 + j0.0000$
3	$1.2622 + j1.0203$	$1.2600 + j1.0206$
4	$-4.0000 - j1.0000$	$-3.9927 - j0.9982$
Line #	**True Apparent Power (p.u.)**	**Measured Apparent Power (p.u.)**
1,2	$1.3420 + j0.3884$	$1.3394 + j0.3879$
1,3	$0.3369 - j0.1848$	–
1,4	$1.8832 + j0.6734$	$1.8858 + j0.6731$
2,1	$-1.3329 - j0.2060$	–
2,4	$0.5534 + j0.2060$	–
3,1	$-0.3362 + j0.1986$	–
3,4	$1.5984 + j0.8217$	$1.5975 + 0.8200$
4,1	$-1.8646 - j0.2996$	–
4,2	$-0.5517 - j0.1710$	–
4,3	$-1.5838 - j0.5294$	$-1.5850 - j0.5295$

state estimation problem, i.e., erroneous measurements, to enhance the accuracy of the state estimation solution.

Using Both Power Measurements and Phasor Measurements

The state estimation problem may also involve a combination of both power measurements and synchronized phasor measurements. Such a problem can be solved in a way similar to the above traditional state estimation problem. However, one challenge is that, while the power flow equations are formulated and solved based on a *locational* reference, where the voltage phase angle is zero at a reference bus, synchronized phasor measurements are obtained based on a *temporal reference*, where the voltage phase angle is zero for the reference waveform. Therefore, it is necessary to choose a reference bus that is equipped with a PMU.

5.8.2 Linearized Approximate Formulation

In this alternative formulation, we define the state estimation problem based on the *linearized* power flow equations in (1.65) and (1.66), instead of the original nonlinear power flow equations in (1.56) and (1.57). Accordingly, the following approximate relationships are considered between the voltage phase angles and nodal power injections, and the line power flows; see Section 1.3.1 in Chapter 1:

$$P_i = \sum_{k=1}^{n} B_{ik}^{\text{bus}} (\theta_i - \theta_k) \tag{5.66}$$

and

$$P_{ik} = B_{ik}^{\text{bus}} (\theta_i - \theta_k) . \quad (5.67)$$

Given the type of equations that are used in this formulation; the state variables are defined to include only the voltage phase angles at all buses. The magnitude of voltage phasors is assumed to be 1 per unit at all buses; see the explanation regarding this assumption in Section 1.3.1 in Chapter 1. As for the measurements, they include only active power measurements. This is because reactive power does not appear in the linearized power flow equations in (5.66) and (5.67).

The vector of state variables and the vector of measurements are defined as

$$\mathbf{x} = \begin{bmatrix} \vdots \\ \theta_i \\ \theta_k \\ \vdots \end{bmatrix} \quad \text{and} \quad \mathbf{z} = \begin{bmatrix} \vdots \\ P_i \\ P_k \\ \vdots \\ P_{ik} \\ \vdots \end{bmatrix} . \quad (5.68)$$

The measurements are related to the state variables as follows:

$$\mathbf{z} = \mathbf{\Psi x} + \epsilon, \quad (5.69)$$

$$\min_{\mathbf{x}} \ \|\mathbf{z} - \mathbf{\Psi x}\|_2 . \quad (5.70)$$

The above is a standard LS optimization problem. It can be solved by using the command `lsqlin` in MATLAB [102]. As in (5.61), we assume that the voltage phase angle at the reference bus is zero. The voltage magnitude is assumed to be 1 at all buses, including at the reference bus.

Example 5.13 We can solve the state estimation problem in Example 5.12 also by using the approximate linearized formulation. The vector of state variables, the vector of measurements, and the measurements matrix are

$$\mathbf{x} = \begin{bmatrix} \theta_1 \\ \theta_2 \\ \theta_3 \\ \theta_4 \end{bmatrix}, \quad \mathbf{z} = \begin{bmatrix} -0.7786 \\ 1.2600 \\ -3.9927 \\ 1.3394 \\ 1.8858 \\ 1.5975 \\ -1.5850 \end{bmatrix}, \quad (5.71)$$

and

$$\mathbf{H} = \begin{bmatrix} -10 & 20 & 0 & -10 \\ -10 & 0 & 20 & -10 \\ -10 & -10 & -10 & 30 \\ 10 & -10 & 0 & 0 \\ 10 & 0 & 0 & -10 \\ 0 & 0 & 10 & -10 \\ 0 & 0 & -10 & 10 \end{bmatrix}, \tag{5.72}$$

respectively. The state estimation results are obtained as

$$\begin{aligned} \theta_1 &= 0^\circ, \\ \theta_2 &= -7.6439^\circ, \\ \theta_3 &= -1.7514^\circ, \\ \theta_4 &= -10.7790^\circ. \end{aligned} \tag{5.73}$$

By comparing the above results with those in (5.65), we can see that the obtained phase angles in the two methods are generally similar. If we rotate the phase angles in (5.73) by 38.8884° clockwise, then they can be presented equivalently as 38.8884°, 31.2445°, 37.1370°, and 28.1094°. These new values are comparable with the phase angles that are obtained in Example 3.20 in Chapter 3.

5.8.3 Convex Relaxation and Other Formulations

Besides the approximate linearized formulation that we used in Section 5.8.2, there are also other methods that use linearization or other approximation techniques in formulating and solving the power flow equations. An overview of some of these methods is provided in [362]. In general, any alternative formulation of the power flow equations can potentially be used also for state estimation.

Furthermore, there have been important advancements in recent years in the field of power flow analysis by using different methods for *convex relaxation*; e.g., see [363–365]. The idea is to start from the original power flow equations in complex domain, such as (1.54) and (1.55) in Chapter 1, and formulate the power flow analysis as a non-convex quadratic optimization problem in complex domain. Then use techniques for convex relaxation, such as relaxation of the original problem formulation to a *semi-definite program* (SDP), to solve the power flow equations. In general, the solutions that are obtained by applying convex relaxation techniques are approximate. However, under certain conditions, such as in a balanced symmetric radial distribution feeder, the solutions can be exact; e.g., see the summary discussions in [366, 367]. These new approaches have already been used in state estimation. Some examples include the studies in [368–371].

It should be noted that the state estimation problems that we discussed in this chapter and throughout this book are about *static* states, where the state variables at any instance depend only on the measurements at that same instance. However, state estimation problems can also be defined with respect to *dynamic* states of the system, such as generator rotor angles and speeds. Tools such as Kalman Filters are used to conduct dynamic state estimation; e.g., see [372, 373].

5.9 Three-Phase Power and Energy Measurements

The total power in a three-phase power system is the summation of the power that is measured separately at each of the three phases:

$$\begin{aligned} P &= P_A + P_B + P_C, \\ Q &= Q_A + Q_B + Q_C. \end{aligned} \tag{5.74}$$

Similarly, the total energy in a three-phase power system is the summation of the energy that is measured separately at each of the three phases:

$$E = E_A + E_B + E_C. \tag{5.75}$$

As for apparent power and power factor, they can be measured individually on each phase. In general, it is not common to express them as overall quantities in three-phase systems. We will discuss this subject in Section 5.9.4.

5.9.1 Two-Wattmeter Method

If the three-phase power system is *balanced*, then it is sufficient to use only one wattmeter to measure active power on one phase, and we can multiply the measurements by three to obtain the total active power for the overall three-phase system. However, if the three-phase power system is *unbalanced*, then we usually need three wattmeters to make *separate* measurements on each phase.

In some special cases, we may need fewer sensors. An example is shown in Figure 5.23. Here, the three-phase load has a three-wire star topology. In this topology, we can measure the total power usage of the three-phase load by using only *two* (instead of three) wattmeters. Wattmeter W_1 is connected to Phase A and wattmeter W_2 is connected to Phase B [374]. Unlike in Figure 5.1(a), where the potential coil of the wattmeter has a ground connection, the potential coils of wattmeters W_1 and W_2 in Figure 5.23 are connected to Phase C. Therefore, W_1 measures the *line-to-line* voltage between Phase A and Phase C; and W_2 measures the *line-to-line* voltage between Phase B and Phase C.

The instantaneous power usage that is measured by W_1 is

$$i_A(t)\, v_{AC}(t) = i_A(t)\, (v_A(t) - v_C(t))\,. \tag{5.76}$$

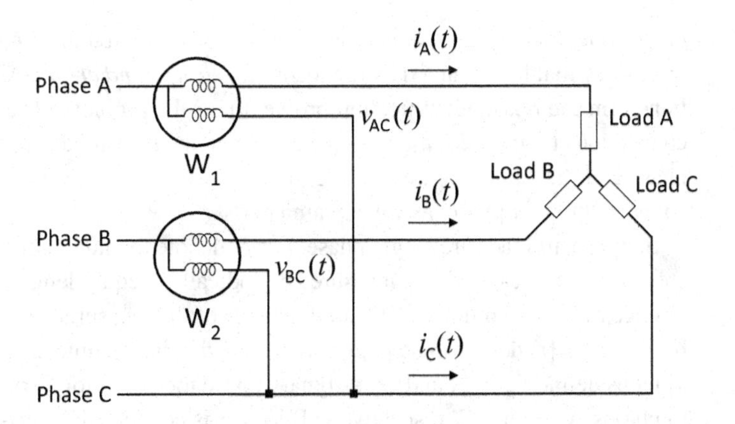

Figure 5.23 Measuring the total power usage of a three-phase load with only two wattmeters. The wattmeters are connected such that they measure line-to-line voltages.

Similarly, the instantaneous power usage that is measured by W_2 is

$$i_B(t)\, v_{BC}(t) = i_B(t)\,(v_B(t) - v_C(t))\,. \tag{5.77}$$

The sum of the instantaneous power that is measured by W_1 and W_2 is

$$i_A(t)\, v_A(t) + i_B(t)\, v_B(t) - (i_A(t) + i_B(t))\, v_C(t). \tag{5.78}$$

However, we know that in a three-wire star topology, we have:

$$i_A + i_B + i_C = 0 \quad \Rightarrow \quad i_C = -\,(i_A + i_B)\,. \tag{5.79}$$

By replacing (5.79) in (5.78), we can see that the sum of the instantaneous power usage that is measured by wattmeters W_1 and W_2 is equal to the total instantaneous power usage across all three phases:

$$i_A(t)\, v_A(t) + i_B(t)\, v_B(t) + i_C(t)\, v_C(t) = p_A(t) + p_B(t) + p_C(t). \tag{5.80}$$

It should be noted that the above two-wattmeter method can also be applied to a three-phase load with delta connections; see Exercise 5.21. Also, if we replace the two wattmeters with two varmeters or two watthour meters, then we can similarly measure the total reactive power usage or the total energy usage of this three-phase load, respectively, still using only two sensors in each case.

5.9.2 Phase Identification by Power and Energy Measurements

A fundamental problem in any three-phase power system is phase identification. We previously discussed solving the phase identification problem in power distribution systems by using the voltage or current measurements in Section 2.8.3 in Chapter 2 and the phase angle measurements in Section 3.6.4 in Chapter 3.

The phase identification problem can also be solved by using power and energy measurements. The basic idea is to use the principle of conservation of electric charge, i.e., the fact that energy that is supplied by a power distribution feeder must be equal to the energy that is consumed by the loads, *plus losses and errors*. Accordingly, one can transform the phase identification problem into the problem of identifying the phase connections to *minimize the mismatch* between the measured supplied energy on each phase and the summation of the measured consumed energy by all the loads that are connected to that phase, as we explain next.

Suppose all loads are single phase. Let N denote the number of loads. Suppose the energy usage of each load is measured periodically at equal-length time intervals, such as once every five minutes. The total number of the measurement intervals is denoted by T. Let $E_n[t]$ denote the energy usage of load n during time interval t. For each load n, let us define ζ_n^A, ζ_n^B, and ζ_n^C as binary phase identification variables corresponding to phases A, B, and C, respectively. If load n is connected to phase A, then $\zeta_n^A = 1$; otherwise $\zeta_n^A = 0$. Variables ζ_n^B and ζ_n^C are defined similarly. The following equality must hold because each single-phase load must be connected to exactly one phase:

$$\zeta_n^A + \zeta_n^B + \zeta_n^C = 1. \tag{5.81}$$

The total load that is identified on phase A at time interval t is calculated as

$$\sum_{n=1}^{N} \zeta_n^A E_n[t]. \tag{5.82}$$

Next, suppose $E^A[t]$, $E^B[t]$, and $E^C[t]$ denote the total energy usage of the *entire* power distribution feeder that is measured at the distribution substation during time interval t on phases A, B, and C, respectively. If phase identification is done correctly, i.e., if ζ_n^A, ζ_n^B, and ζ_n^C are set correctly for all loads $n = 1, \ldots, N$, then the amount in (5.82) matches $E^A[t]$ at all time intervals $t = 1, \ldots, T$. A similar statement is true for phases B and C. Therefore, in order to identify the unknown phase connections, we seek to select ζ_n^A, ζ_n^B, and ζ_n^C for all loads $n = 1, \ldots, N$ such that we minimize the following expression subject to the constraint in (5.81):

$$\begin{aligned} \sum_{t=1}^{T} \Bigg[& \left(\sum_{n=1}^{N} \zeta_n^A E_n[t] - E^A[t] \right)^2 + \\ & \left(\sum_{n=1}^{N} \zeta_n^B E_n[t] - E^B[t] \right)^2 + \\ & \left(\sum_{n=1}^{N} \zeta_n^C E_n[t] - E^C[t] \right)^2 \Bigg]. \end{aligned} \tag{5.83}$$

Hence, we can formulate the phase identification problem as a binary LS problem:

$$\begin{aligned} & \underset{\zeta}{\text{minimize}} \quad \|\mathbf{f} - \mathbf{E}\,\zeta\|_2, \\ & \text{subject to} \quad \zeta_n^A + \zeta_n^B + \zeta_n^C = 1, \quad n = 1, \ldots, N. \end{aligned} \tag{5.84}$$

Table 5.7 Energy usage measurements in Example 5.14

Time	Phase (kWh)			Loads (kWh)				
Interval	A	B	C	1	2	3	4	5
$t = 1$	5.5	3.2	7.1	2.1	3.3	3.0	2.6	4.3
$t = 2$	6.2	8.7	12.5	2.9	3.1	8.6	5.5	7.1

Here, **f** is the $3T \times 1$ vector of per-phase feeder load measurements, ζ is the $3N \times 1$ vector of per-load phase identification variables, **E** is the $3T \times 3N$ matrix of single-phase load measurements, and ϵ is the $3T \times 1$ vector of per-phase losses and errors. The constrained binary LS problem in (5.84) can be solved using techniques such as branch and bound or convex relaxation; cf. [375]. It can also be solved in MATLAB by using any convex optimization toolbox that supports binary or integer variables, such as CVX [178, 376] or CPLEX [377].

Example 5.14 Consider the energy measurements in Table 5.7 over $T = 2$ time intervals. We would like to identify how each of the $N = 5$ loads is connected to a phase. The vector of unknown phase identification variables is formulated as

$$\zeta = \begin{bmatrix} \zeta_1^A & \zeta_1^B & \zeta_1^C & \ldots & \zeta_5^A & \zeta_5^B & \zeta_5^C \end{bmatrix}^T. \tag{5.85}$$

The other parameters of the phase identification problem are

$$\mathrm{f} = \begin{bmatrix} 5.5 & 3.2 & 7.1 & 6.2 & 8.7 & 12.5 \end{bmatrix}^T, \tag{5.86}$$

$$\mathbf{E} = \begin{bmatrix} 2.1 & 0 & 0 & 3.3 & 0 & 0 & 3.0 & 0 & 0 & 2.6 & 0 & 0 & 4.3 & 0 & 0 \\ 0 & 2.1 & 0 & 0 & 3.3 & 0 & 0 & 3.0 & 0 & 0 & 2.6 & 0 & 0 & 4.3 & 0 \\ 0 & 0 & 2.1 & 0 & 0 & 3.3 & 0 & 0 & 3.0 & 0 & 0 & 2.6 & 0 & 0 & 4.3 \\ 2.9 & 0 & 0 & 3.1 & 0 & 0 & 8.6 & 0 & 0 & 5.5 & 0 & 0 & 7.1 & 0 & 0 \\ 0 & 2.9 & 0 & 0 & 3.1 & 0 & 0 & 8.6 & 0 & 0 & 5.5 & 0 & 0 & 7.1 & 0 \\ 0 & 0 & 2.9 & 0 & 0 & 3.1 & 0 & 0 & 8.6 & 0 & 0 & 5.5 & 0 & 0 & 7.1 \end{bmatrix}. \tag{5.87}$$

The solution is obtained as $\zeta_1^A = \zeta_2^A = \zeta_3^B = \zeta_4^C = \zeta_5^C = 1$ and $\zeta_3^A = \zeta_4^A = \zeta_5^A = \zeta_1^B = \zeta_2^B = \zeta_4^B = \zeta_5^B = \zeta_1^C = \zeta_2^C = \zeta_3^C = 0$. That is, loads 1 and 2 are connected to phase A, load 3 is connected to phase B, and loads 4 and 5 are connected to phase C. The minimum l_2 norm of the measurement residues is obtained as 0.3873 kWh, which corresponds to power loss and measurement errors.

As the number of loads increases, more time intervals need to be considered in order to collect enough information to accurately identify the phases.

In principle, the above phase identification problem can be formulated based on not only active power measurements but also reactive power measurements to utilize more information. The problem formulation can also be adjusted to include not only single-phase but also two-phase and three-phase loads. When data is available, the above phase identification problem can also be reinforced by power measurements at several load transformers across the distribution feeder.

5.9.3 Other Applications of Three-Phase Power and Energy Measurements

Many of the applications of power and energy measurements that we discussed in this chapter can be extended to three-phase measurements.

Load Modeling

Load modeling can be done at each of the three phases. All the load models that we discussed in Section 5.7 can also be used to model two-phase or three-phase loads. For example, for a three-phase load, we can extend the exponential load model in (5.44) to:

$$P = P_{A,0}\left(\frac{V_A}{V_{A,0}}\right)^{\gamma_{A,P}} + P_{B,0}\left(\frac{V_B}{V_{B,0}}\right)^{\gamma_{B,P}} + P_{C,0}\left(\frac{V_C}{V_{C,0}}\right)^{\gamma_{C,P}}, \qquad (5.88)$$

where the parameters are $\gamma_{A,P}$, $\gamma_{B,P}$, and $\gamma_{C,P}$.

Disaggregation

Load disaggregation can also be done at each of the three phases, or across a combination of single-phase, two-phase, and three-phase loads [323]. The methods that we learned in Section 5.6.1 can also be used in these cases. For example, switching of a three-phase load can be recognized based on its signature on all three phases, i.e., the changes that it causes in active power consumption and reactive power consumption, as in Figure 5.16, but on each phase.

Net load disaggregation, such as solar generation disaggregation, can also be done at each phase. One potentially helpful note is that, in practice, most three-phase PV and wind inverters are *balanced*, with equal power generation per phase across the three phases [378]; therefore, we can assume that the generation component of the net load is balanced across the three phases, while it is rather only the consumption component of the net load that can be unbalanced.

Per-Phase Smart Pricing

Pricing methods are usually designed based on *system-wide* considerations in the power system, such as with respect to the overall load profile of the utility and the price of power procurement in the wholesale electricity market. However, there is a growing interest in also developing pricing methods that reflect the operation challenges at the *power distribution level*. The prices in this new pricing paradigm can be designed based on objectives such as voltage regulation, balancing load across phases, or integrating distributed energy resources. Depending on the purpose of the pricing mechanism, some of these pricing methods are defined on each phase, i.e., they are *per-phase prices*; e.g., see the studies in [379–381].

Three-Phase DSSE

Three-phase state estimation is often not necessary at the transmission level; because the power systems at that level are mostly balanced. In contrast, the power system at the distribution level is usually unbalanced; therefore, there is a need in practice to develop efficient three-phase DSSE solutions. The convex relaxation methods that we briefly mentioned in Section 5.8.3 may not result in exact solutions when the three-phase power distribution system is unbalanced. One option is to approximately decompose the three-phase DSSE problem into three *separate* single-phase DSSE problems, e.g., by ignoring the mutual impedance across different phases of the distribution lines. In that case, each single-phase DSSE problem can be solved separately by using a convex relaxation method. Another option is to solve the original three-phase DSSE problem by either using the standard methods such as the Gauss–Newton method that we learned in Section 5.8.1 [382], or using some advanced optimization techniques such as those in [383–385].

5.9.4 Three-Phase Apparent Power and Power Factor

There are different ways to define apparent power in three-phase power systems [386]. The following definition is referred to as *arithmetical apparent power* [387]:

$$S = V_{A,\mathrm{rms}}\, I_{A,\mathrm{rms}} + V_{B,\mathrm{rms}}\, I_{B,\mathrm{rms}} + V_{C,\mathrm{rms}}\, I_{C,\mathrm{rms}}. \tag{5.89}$$

The following definition is referred to as *geometrical apparent power* [387]:

$$S = \sqrt{P^2 + Q^2}, \tag{5.90}$$

where P and Q are defined in (5.74). Here is another definition from [388]:

$$S = \sqrt{V_{A,\mathrm{rms}}^2 + V_{B,\mathrm{rms}}^2 + V_{C,\mathrm{rms}}^2}\,\sqrt{I_{A,\mathrm{rms}}^2 + I_{B,\mathrm{rms}}^2 + I_{C,\mathrm{rms}}^2}. \tag{5.91}$$

Note that these different definitions result in different values; see Exercise 5.22.

As for the power factor, the common approach is either to look at the power factor at each phase individually, or to simply use the *average* of the power factors across the three phases. Alternatively, the overall power factor may also be defined by dividing the total active power P as defined in (5.74) to S, where S can be any of the definitions of apparent power in (5.89), (5.90), or (5.91).

5.10 Accuracy in Power and Energy Measurements

5.10.1 Accuracy Classes

One of the standards that defines the accuracy classes for electricity metering is ANSI C12. The accuracy is expressed in terms of limits on measurement error; which is limited to 0.1%, 0.2%, 0.5%, and 1% for Accuracy Classes 0.1, 0.2, 0.5, and 1.0, respectively [389]. These accuracy levels are defined at the normal operating load

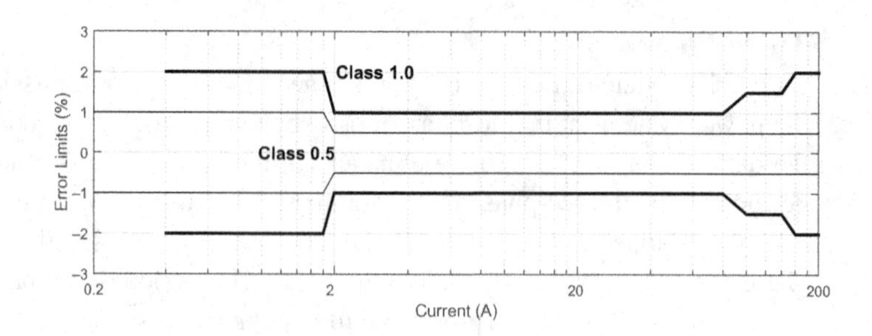

Figure 5.24 Comparison between accuracy classes 0.5 and 1.0 in electricity meters [389].

current, i.e., between 2 A and 100 A. The error limits are typically higher at both low load and high load current conditions.

Two ANSI C12 accuracy classes for power and energy metering are compared in Figure 5.24. The accuracy is expressed in terms of limits on measurement error, which are shown versus the operating load current, ranging from 0.2 A to 200 A on the logarithmic scale. Besides providing a better metering accuracy at normal loads, an Accuracy Class 0.5 meter is saturated at a higher load current, beyond 100 A which is the saturation level at Accuracy Class 1.0; it also continues to meter down to 0.1 A, whereas an Accuracy Class 1.0 meter may stop metering below 0.3 A.

It should be noted that the error limits shown in Figure 5.24 are under the assumption that the power factor is one, i.e., the load is purely resisitive. If the power factor reduces, then error limits may increase. For example, the error for an Accuracy Class 0.5 meter may increase up to ±2.5% if the power factor reduces to 0.25. This level of error is much higher than the maximum error limit of ±1% that we can see in Figure 5.24 using unity power factor.

5.10.2 Meter Accuracy versus System Accuracy

The accuracy of a power measurement system or an energy measurement system depends on the accuracy of all its components, which includes the meter itself and any instrument transformer that is being used; see Section 2.1 in Chapter 2.

Instrument transformers can affect the accuracy in measuring voltage and current. They can also affect the accuracy in measuring the phase shift between voltage and current, i.e., the accuracy in measuring power factor.

In order to obtain the total accuracy in the metering system, it is often assumed that the error that is caused by each component in the system has a Gaussian distribution. Therefore, the total accuracy of the system is obtained as [390, 391]:

$$\epsilon = \sqrt{\epsilon_{\mathrm{M}}^2 + \epsilon_{\mathrm{CT}}^2 + \epsilon_{\mathrm{PT}}^2 + \epsilon_{\mathrm{PF}}^2}, \tag{5.92}$$

where ϵ_{M}, ϵ_{CT}, ϵ_{PT}, and ϵ_{PF} denote the meter accuracy, the CT accuracy, the PT accuracy, and the power factor (phase shift) accuracy, respectively. Note that, ϵ, ϵ_{M}, ϵ_{CT}, ϵ_{PT}, and ϵ_{PF} are all expressed in percentage error.

Example 5.15 A Class 0.5 meter, i.e., with an accuracy level of $\epsilon_M = 0.5\%$, is used to measure power consumption of a load. A CT with an accuracy level of $\epsilon_{CT} = 0.75\%$ is used in this measurement system. No PT is used; because voltage level of the load is already within the operating range of the meter. Therefore, $\epsilon_{PT} = 0\%$. The power factor of the load is around 0.75. At this level of power factor, the CT can cause $\epsilon_{PF} = 1.16\%$ error in measuring power factor. From (5.92) the total accuracy level of this measurement system is obtained as

$$\epsilon = \sqrt{0.5^2 + 0.75^2 + 0^2 + 1.16^2} = 1.47\%. \tag{5.93}$$

Thus, if the meter indicates 4 kW load, the true load is 4 kW $\pm$ 59 W.

Phase Shift

The error in phase shift is sometimes expressed in minutes, where 60 minutes equal one degree and 30 minutes equal 0.5 degrees. This information is useful because it allows us to obtain the accuracy in measuring power factor at different power factor levels. Let ϵ_{Shift} denote the error level in phase shift. We have [392]:

$$\epsilon_{PF} = 100\% \times \left| 1 - \frac{\cos(\theta - \phi + \epsilon_{\text{Shift}})}{\cos(\theta - \phi)} \right|, \tag{5.94}$$

where $\theta - \phi$ denotes the difference between the phase angle of voltage, i.e., θ, and the phase angle of current, i.e., ϕ. For instance, suppose ϵ_{Shift} is 45 minutes, i.e., 0.75°. If power factor is 0.75, i.e., $\theta - \phi = 41.41°$, then from (5.94), we have:

$$\epsilon_{PF} = 100\% \times \left| 1 - \frac{\cos(41.41° + 0.75°)}{\cos(41.41°)} \right| = 1.16\%, \tag{5.95}$$

which is the same number that we used in Example 5.15. If power factor is 1, then $\epsilon_{PF} = 0.009\%$; and if power factor is 0.5, then $\epsilon_{PF} = 2.28\%$ [392].

Revenue Meters

Error in revenue metering is of concern because of its impact on billing and financial transactions. Any major error in the recording of energy usage or power usage can result in a loss to the utility, when *understating*, or to the customer, when *overstating*. Of course, higher accuracy metering does cost more in equipment and maintenance; however, the higher cost can often be justified when compared with the reduced level of uncertainty that it can offer in billing, especially for larger customers.

Example 5.16 Consider a commercial customer that is required to pay *peak-load charges* based on its peak power usage, measured in kW, during on-peak hours, mid-peak hours, and off-peak hours; see the paragraph on Other Pricing Methods in Section 5.4.1. The rates for calculating the peak-load charges are

- On-Peak Hours: 6.88 $/kW
- Mid-Peak Hours: 2.74 $/kW
- Off-Peak Hours: 1.31 $/kW.

The revenue meter indicates that the peak power usage is 11,760 kW during on-peak hours, 10,467 kW during mid-peak hours, and 8,732 kW during off-peak hours. First, assume that a *high accuracy* power measurement system is used:

$$\epsilon_{\mathrm{M}} = 0.5\%, \quad \epsilon_{\mathrm{CT}} = 0.3\%, \quad \epsilon_{\mathrm{PT}} = 0.3\%, \quad \epsilon_{\mathrm{PF}} = 0.77\%. \tag{5.96}$$

From (5.92), the overall accuracy of the metering system is 1.01%. This introduces $0.0101 \times 11760 = 119$ kW, $0.0101 \times 10467 = 106$ kW, and $0.0101 \times 11760 = 88$ kW uncertainty in measuring peak demand during on-peak hours, mid-peak hours, and off-peak hours, respectively. This results in a total of

$$119 \times 6.88 + 106 \times 2.74 + 88 \times 1.31 = \$1{,}224 \tag{5.97}$$

uncertainty in the monthly peak-load charges for this customer. Next, assume that a *low accuracy* power measurement system is used:

$$\epsilon_{\mathrm{M}} = 1\%, \quad \epsilon_{\mathrm{CT}} = 1.2\%, \quad \epsilon_{\mathrm{PT}} = 1.2\%, \quad \epsilon_{\mathrm{PF}} = 1\%. \tag{5.98}$$

Thus, the overall accuracy of the metering system is 2.29%. This introduces 269 kW, 240 kW, and 200 kW uncertainty in measuring peak demand during on-peak hours, mid-peak hours, and off-peak hours, respectively. This results in a total of

$$269 \times 6.88 + 240 \times 2.74200 \times 1.31 = \$2{,}770 \tag{5.99}$$

uncertainty in the monthly peak-load charges for this customer. The lower accuracy of the second power measurement system creates an additional $18,552 uncertainty in the peak-load charges for this customer over the course of a year.

5.10.3 Other Factors that Affect Accuracy

The most common mode of failure for the traditional electromechanical energy meters is *reduced registeration*. Anything that increases the drag on the rotating disk can cause a meter to run slow, resulting in reduced billing amounts. Failure modes also exist that could cause an electromechanical meter to run fast, but they are less common. Digital energy meters are also prone to error due to sampling. Furthermore, they are susceptible to line voltage transient events. Traditional elechtromechanical meters are generally more immune to standard surge events [393].

The accuracy in measuring energy depends also on clock accuracy. Due to cost considerations, the internal clock in customer meters has limited accuracy, certainly much less than the time accuracy in D-PMUs. For example, if a meter reports 56 Wh energy usage for the 6:00:00 PM to 6:15:00 PM time interval, and its internal clock lags the true clock by one second, then in reality, the reported 56 Wh was consumed from 6:00:01 PM to 6:15:01 PM. Such error in time may not cause a major issue for billing purposes. However, it may have impact in certain applications, such as state estimation and phase identification.

Ambient temperature can also affect measurement accuracy. The accuracy levels that are shown in Figure 5.24 are at 23°C as the *reference temperature*. At higher (or lower) temperatures, the error limits can be higher [394].

Exercises

5.1 File `E5-1.csv` contains the voltage and current waves at a single-phase load location. Both voltage and current are purely sinusoidal.
(a) Obtain the average power that is delivered to the load per AC cycle.
(b) What is the power factor?

5.2 Consider the minutely power consumption measurements in file `E5-2.csv`.
(a) Plot the power profile based on the original measurements.
(b) Plot the power profile based on the hourly average of the measurements.
(c) What is the load factor based on the power profile in Part (a)?
(c) What is the load factor based on the power profile in Part (b)?

5.3 Consider the minutely power generation measurements in file `E5-3.csv` that are obtained at a PV generation unit for a period of one day.
(a) At what time does PV generation start, and at what time does it stop?
(b) Obtain the *average* and *variance* of the generated power at *each* hour.

5.4 File `E5-4.csv` contains the active power consumption profile and the reactive power consumption profile for a power distribution feeder.
(a) Plot the power factor profile.
(b) Use the power factor profile in Part (a) to identify any potential capacitor bank switching event that may have occurred on this feeder.
(c) Does the distribution substation experience reverse active power flow or reverse reactive power flow at any time? Elaborate your answer.

5.5 Consider the definition of active power in (5.5) and the definition of reactive power in (5.18). Show that $S^2 \geq P^2 + Q^2$, where $S = V_{\text{rms}} I_{\text{rms}}$. One option is to apply the Cauchy-Schwarz inequality to the terms in P^2 and Q^2.

5.6 File `E5-6.csv` contains the voltage and current waves at a single-phase load location. Both voltage and current waves are distorted.
(a) How much active power is delivered to the load?
(b) Obtain the DPF for this load.
(c) Obtain the TPF for this load.

5.7 A pulse meter generates 20 pulses in a duration of five minutes as shown in Figure 5.25. Each pulse is equivalent to 0.5 kWh.
(a) How much is the minute-by-minute energy usage?
(b) How much is the total energy usage?

5.8 File `E5-8.csv` contains the hourly load profile of a residential customer over a period of one month during the summer. How much is the monthly electricity bill of this customer under the following two different scenarios?
(a) The price of the electricity is flat at 8.28 ¢/kWh.

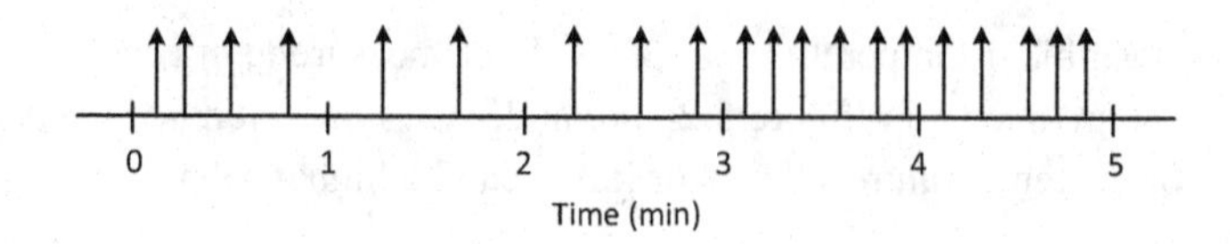

Figure 5.25 The train of pulses generated by a pulse watthour meter in Exercise 5.7.

(b) The price of electricity during the on-peak, mid-peak, and off-peak hours is 10.33 ¢/kWh, 8.28 ¢/kWh, and 7.27 ¢/kWh, respectively. The on-peak, mid-peak, and off-peak hours are previously shown in Figure 5.8.

5.9 Peak-load charges, also known as *demand charges*, are additional fees that are applied to the electric bills of commercial and industrial customers based upon the *highest* amount of their *power consumption* during a given time interval, typically 15 minutes, during the monthly billing period. Answer the following questions based on the power consumption measurements in file `E5-9.csv` that are obtained at a commercial building during one month:

(a) What is the monthly peak load during off-peak hours: 9 PM–8 AM?

(b) What is the monthly peak load during mid-peak hours: 8 AM–5 PM?

(c) What is the monthly peak load during on-peak hours: 5 PM–9 PM?

(d) Use the results in Parts (a), (b), and (c) to calculate the monthly demand charge for this commercial customer if the following peak-price rates are applied by the utility for demand charges: 6.88 $/kW during on-peak hours; 2.74 $/kW during mid-peak hours; and 1.31 $/kW during off-peak hours. Note that, these rates are in per kW, not in per kWh. This is because peak-load pricing is done based on peak power usage, not energy usage.

5.10 Recall the "High 5 of 10" method that we discussed for baseline calculation in Section 5.4.2. Other similar methods include the "High 3 of 10" method and the "High 4 of 5" method [395]. In the "High 3 of 10" method, baseline calculation is done based on the *three* days, out of the 10 previous (non-event) days in the baseline window, that have the highest average load for the corresponding duration of the DR event. In the "High 4 of 5" method, baseline calculation is done based on the *four* days, out of the five previous (non-event) days in the baseline window, that have the highest average load for the corresponding duration of the DR event. Answer the following questions based on the load scenario in Example 5.8 and Table 5.3:

(a) Use the "High 3 of 10" method to calculate the baseline.

(b) Use the "High 4 of 5" method to calculate the baseline.

5.11 Consider the load profiles in Figure 5.12 and their features in Table 5.4.

(a) Repeat the clustering but this time solely based on Feature 1.

(b) Repeat the clustering but this time solely based on Feature 2.

5.12 Consider the load profiles in Figure 5.12 and their features in Table 5.4. Repeat the clustering but this time assume that $k = 3$, i.e., consider three sets.

5.13 Consider the load profile in Table 5.8.

Table 5.8 Load profile in Exercise 5.13.

Hour	Load (kWh)	Hour	Load (kWh)	Hour	Load (kWh)
1	2.5545	9	0.8779	17	5.7395
2	2.0274	10	0.3607	18	5.6652
3	1.2027	11	0.6034	19	5.4435
4	0.9802	12	0.6359	20	4.2713
5	0.6986	13	0.6309	21	3.6523
6	1.0191	14	0.9193	22	2.3570
7	0.6980	15	3.3302	23	2.3303
8	1.1005	16	3.0359	24	2.6320

Table 5.9 Load parameters in Exercise 5.17.

Appliance	P_0	Q_0	α_Z	α_I	α_P	β_Z	β_I	β_P
Coffeemaker	1413	13	0.13	1.62	−0.75	3.89	−6	3.11
Microwave	1366	451	1.39	−1.96	1.57	50.1	−93.6	44.5
Vacuum	855	221	1.18	−0.38	0.2	4.1	−5.87	2.77
Air Compressor	1109	487	0.71	0.46	−0.17	−1.33	4.04	−1.7

(a) Extract its features based on Features 1 and 2 in Section 5.4.3.

(b) Cluster this load profile together with the load profiles in Table 5.4.

5.14 Consider the residential customer in Example 5.9. The total power usage of this customer is measured at 9.7 kW. What are the possible scenarios for the combinations of the major loads that can be on?

5.15 Consider a household which includes eight major appliances and equipment as we saw in Figure 5.16. Suppose we detect a *positive sharp edge* in both active power consumption, at 640 W, and reactive power consumption, at 95 VAR. Which appliance or equipment do you think has switched on?

5.16 Show that the optimization problem in (5.37) is the same as the following:

$$\min_{\alpha_Z,\alpha_I,\alpha_P} \sum_{i=1}^{n} \left[P_i - P_0 \left(\alpha_Z \left(\frac{V_i}{V_0} \right)^2 + \alpha_I \left(\frac{V}{V_0} \right) + \alpha_P \right) \right]^2 . \tag{5.100}$$

5.17 Consider the list of load parameters for multiple appliances and equipment in Table 5.9. The rated voltage for all these appliances and equipment is $V_0 = 120$ V. Plot the active power consumption profile and the reactive power consumption profile for each appliance or equipment assuming that the voltage varies based on the voltage measurements in file `E5-17.csv`.

5.18 Use the problem formulation in (5.47) to obtain γ_P and γ_Q for the load in Example 5.10. Recall that the measurements are provided in Table 5.5.

5.19 Obtain the formulation of the measurement Jacobian matrix $\mathbf{H}$ in Example 5.12. You need to calculate the partial derivative of active and reactive power injection

Table 5.10 Apparent power measurements in Exercise 5.20.

Line #	Apparent Power (p.u.)	Line #	Apparent Power (p.u.)
1,2	$0.8346 + j0.4537$	3,5	$-0.9700 - j0.1380$
1,4	$1.0885 + j0.6647$	4,1	$-1.0792 - j0.5355$
2,1	$-0.8392 - j0.3812$	4,2	$-0.2595 - j0.1800$
2,3	$0.7801 + j0.1985$	4,5	$-0.4559 - j0.1846$
2,4	$0.2590 + j0.1894$	5,2	$0.2022 + j0.0121$
2,5	$-0.2022 - j0.0085$	5,3	$0.9772 + j0.2233$
3,2	$-0.7770 - j0.1417$	5,4	$0.4558 + j0.2055$

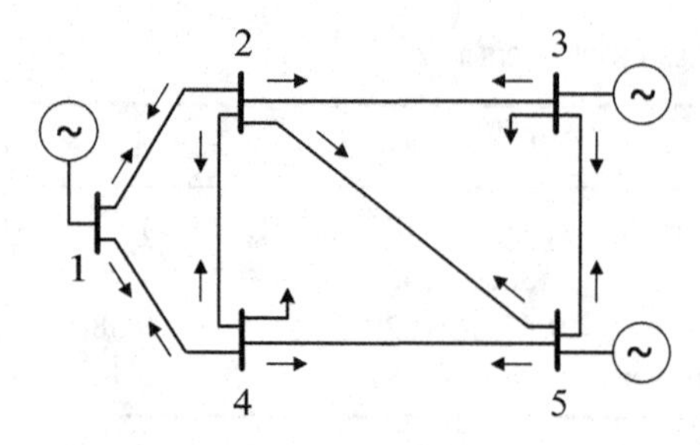

Figure 5.26 The power transmission network in Exercise 5.20.

at all buses and active and reactive power flow at all lines with respect to the voltage magnitude and voltage phase angle at all buses.

5.20 Consider the 5-bus transmission network in Figure 5.26. Suppose we measure apparent power flow at both ends of each transmission line. The measurements are given in Table 5.10. Estimate all the states of the system, i.e., the voltage magnitude and phase angle at all buses. Consider bus 1 as the reference bus. The voltage at bus 1 is measured at $1.077\angle 0^\circ$ per unit.

5.21 Consider the three-phase load in Figure 5.27. This load has a three-wire delta connection configuration topology. Explain why the sum of the instantaneous power usage that is measured by wattmeters W_1 and W_2 is equal to the total instantaneous power usage across all three phases.

5.22 File `E5-22.csv` contains the RMS voltage, RMS current, active power, and reactive power measurements at a three-phase load.

(a) Use the definition in (5.89) to plot the apparent power profile.

(b) Use the definition in (5.90) to plot the apparent power profile.

(c) Use the definition in (5.91) to plot the apparent power profile.

5.23 Consider the reactive power measurements in file `E5-23.csv` at an unbalanced three-phase distribution feeder. The feeder is equipped with a three-phase capacitor bank. The size of the capacitor bank is unknown. The reactive power measurements are made every minute. Estimate the size of the capacitor bank in

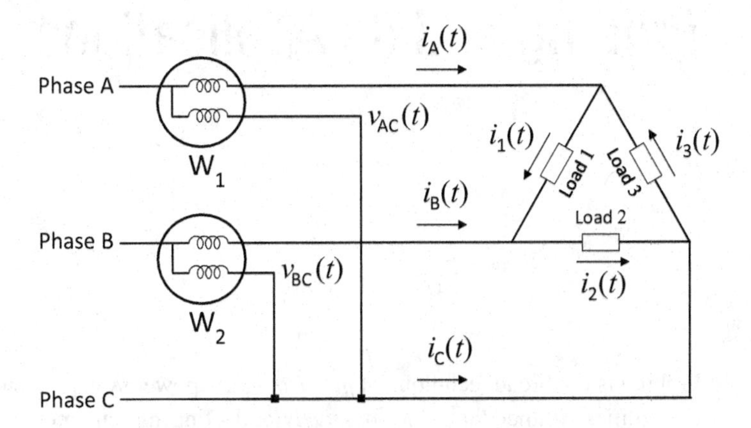

Figure 5.27 The three-phase three-wire delta topology in Exercise 5.21.

kVAR on each phase. Is the operation of the capacitor bank balanced across the three phases?

5.24 Consider the energy measurements in Table 5.11. The total energy usage is measured at each phase at feeder head. The individual energy usage is also measured at each load. All loads are single-phase with unknown phase connections. Identify the phase to which each load is connected.

5.25 Show that if ϵ_{Shift} is small and it is represented in radian, then we can approximate the error in power factor measurements in (5.94) as follows:

$$\epsilon_{\text{PF}} = 100\% \times \left| \tan(\theta - \phi)\, \epsilon_{\text{Shift}} \right| . \tag{5.101}$$

Table 5.11 Energy usage measurements in Exercise 5.24.

Energy Usage		Measurements in Each Time Interval (kWh)					
		$t = 1$	$t = 2$	$t = 3$	$t = 4$	$t = 5$	$t = 6$
Load	1	0.8598	0.8294	1.3148	3.7184	2.4262	2.9662
Load	2	0.4558	0.6994	0.5151	1.0585	1.4045	2.3076
Load	3	0.2922	0.2971	0.4493	0.7204	0.6462	0.2867
Load	4	0.9020	0.8699	1.0159	1.6155	1.5175	1.5944
Load	5	0.4612	0.6485	0.4823	3.7507	3.0228	2.0381
Load	6	0.6658	0.3821	0.4564	1.2185	1.5671	1.8570
Load	7	1.1095	0.6004	0.5992	0.4511	0.4685	3.5088
Load	8	0.6568	0.3187	0.3776	3.1684	4.5471	3.9740
Load	9	0.4605	0.4216	0.4070	2.0804	1.7104	1.4277
Phase	A	2.0261	1.9042	1.9145	5.9022	7.5384	7.9153
Phase	B	1.4311	1.1509	1.3320	4.0688	3.9687	3.5694
Phase	C	2.4473	2.0970	2.4514	8.0223	5.9182	8.5260

6 Probing and Its Applications

Probing is the broad technique of *perturbing* the power system to enhance monitoring capabilities. Rather than only *passively* collecting measurements, probing methods make use of various grid components in order to *actively* create opportunities to learn more about the power system and its unknowns.

The perturbation that is needed in order to do probing can be created in different ways, such as by load switching, harmonic current injection, and modulating power flow at power electronics devices. Probing could be a *one-time action*, or it could be a sequence of actions in order to create a *probing signal*.

In this chapter, we will discuss different types of probing actions and their various applications. Sections 6.1–6.3 are concerned with probing actions that can help with steady-state measurements. Section 6.4 is concerned with probing actions that can help with transient measurements. Sections 6.5–6.7 are concerned with the probing actions that can be achieved through a particular type of smart grid communications technologies, namely power line communications.

6.1 State and Parameter Estimation Using Probing

State and parameter estimation is an important tool for power system operation. We covered this subject in multiple chapters, such as in Section 3.8 in Chapter 3 and Section 5.8 in Chapter 5. So far, we have made the implicit assumption that the power system is *observable*; i.e., the available measurements are sufficient to allow obtaining the unknown states and/or unknown parameters. In fact, it is often assumed that the available measurements are more than sufficient, i.e., they provide *redundancy*, such that we can even address error in measurements.

Contrary to the above, in this section, we consider the cases where the available measurements cannot provide observability or redundancy. We will discuss how a probing action, such as switching a load, may help in such circumstances in order to estimate the unknown states and/or unknown parameters in the system.

6.1.1 Enhanced Observability

Consider the power distribution system in Figure 6.1(a). Suppose two meters are available. The first meter measures the total load of the feeder, as denoted by complex

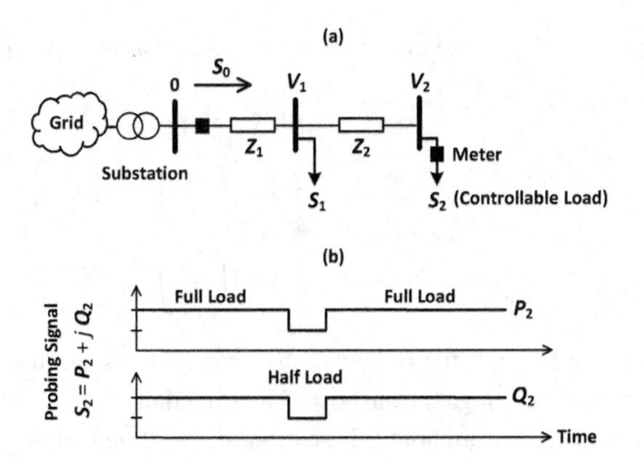

Figure 6.1 The setup for the state and parameter estimation problem in Example 6.1: (a) a power distribution system with two meters; (b) probing signal at bus 2.

power S_0. The second meter measures the load at bus 2, as denoted by complex power S_2. There is no other sensor on this network. The load at bus 1 is *not* metered. The parameters of the load at bus 1 are *not* known either.

Based on the exponential load model in Section 5.7.1 in Chapter 5, the unknown load at bus 1 is modeled as follows, where $V_{1,\text{base}}$ and α are *not* known:

$$S_1 = \left(\frac{|V_1|}{V_{1,\text{base}}}\right)^{\text{Re}\{\alpha\}} + j\left(\frac{|V_1|}{V_{1,\text{base}}}\right)^{\text{Im}\{\alpha\}}. \tag{6.1}$$

In addition to the above unknown load parameters, the voltage phasors at bus 1 and bus 2 are also not known. The voltage at the substation is fixed at $V_0 = 1\angle 0^\circ$ p.u. Overall, the *unknowns* in the system are as follows:

$$V_{1,\text{base}}, \alpha, V_1, V_2. \tag{6.2}$$

Therefore, the unknowns include *both* load parameters and state variables.

In order to estimate the unknowns in this power system, we need to first formulate and then solve a system of equations based on the available measurements. From the Kirchhoff's Voltage Law (KVL) at each bus, we have:

$$V_1 = V_0 - Z_1\left[\left(\frac{S_1}{V_1}\right)^* + \left(\frac{S_2}{V_2}\right)^*\right], \tag{6.3}$$

$$V_2 = V_0 - Z_1\left[\left(\frac{S_1}{V_1}\right)^* + \left(\frac{S_2}{V_2}\right)^*\right] - Z_2\left[\left(\frac{S_2}{V_2}\right)^*\right]. \tag{6.4}$$

Also, from the law of complex power conservation, we have:

$$\begin{aligned} S_0 &= S_1 + S_2 \\ &+ Z_1 \left| \left(\frac{S_1}{V_1}\right)^* + \left(\frac{S_2}{V_2}\right)^* \right|^2 \\ &+ Z_2 \left| \left(\frac{S_2}{V_2}\right)^* \right|^2, \end{aligned} \tag{6.5}$$

where the second line indicates the power losses on distribution lines. The operator $\{\cdot\}^*$ denotes conjugate transpose. The impedance on each distribution line is assumed to be known. Throughout this section, we assume that:

$$\begin{aligned} Z_1 &= 0.000059292 + j0.000030225, \\ Z_2 &= 0.000317040 + j0.000161478. \end{aligned} \tag{6.6}$$

All the above values and also all the quantities in this section are in *per unit*.

Once we replace S_1 with its expression in (6.1), the system of equations in (6.3)–(6.5) would provide us with *three* equations that are independent. Thus, they are sufficient to obtain *three* independent unknowns. However, the number of independent unknowns in the system is *four*, as we previously listed in (6.2). There is no other independent equation that one can consider in this power system. Therefore, this power system is *under-determined* and *unobservable*.

Achieving Observability with Probing Action

Probing can help achieve observability in the above system. Suppose we can *control* the load at bus 2. Suppose we *momentarily* cut the load *by half*, such as for a duration of only one second. This action can create a *probing signal* at bus 2, as shown in Figure 6.1(b). The measurements that are collected in the system during this *momentary probing action* provide us with additional information about the understudy power system that can help us make the system observable.

During full load operation, we collect the following measurements:

$$S_0^{\text{full}}, S_2^{\text{full}}. \tag{6.7}$$

The unknowns in the system during full load operation are:

$$V_{1,\text{base}}, \alpha, V_1^{\text{full}}, V_2^{\text{full}}. \tag{6.8}$$

Similarly, during half load operation, we collect the following measurements:

$$S_0^{\text{half}}, S_2^{\text{half}}. \tag{6.9}$$

The unknowns in the system during half load operation are:

$$V_{1,\text{base}}, \alpha, V_1^{\text{half}}, V_2^{\text{half}}. \tag{6.10}$$

The system of equations in (6.3)–(6.5) holds during *both* full load and half load operation. That is, it holds over (6.7) and (6.8) as well as over (6.9) and (6.10).

Given the fact that the probing action is done over a short period of time, the two unknown parameters of the load at bus 1, i.e., $V_{1,\text{base}}$ and α, can be reasonably assumed to remain unchanged during the probing action.

We now have an *equal number* of unknowns and equations. We have *six* unknowns, as listed in (6.8) and (6.10); and *six* independent equations, which include equations (6.3)–(6.5) during full load operation and equations (6.3)–(6.5) during half load operation. The system is now observable.

Example 6.1 Consider the power system in Figure 6.1(a). Suppose a momentary probing action is done at bus 2. The following measurements are collected:

$$\begin{aligned} S_0^{\text{full}} &= 3.539551 + j1.629843 \\ S_2^{\text{full}} &= 1 + j0.25 \\ S_0^{\text{half}} &= 2.287646 + j1.328913 \\ S_2^{\text{half}} &= 0.5 + j0.125. \end{aligned} \tag{6.11}$$

The unknowns in this power system are obtained as

$$\begin{aligned} V_{1,\text{base}} &= 0.980 \\ \alpha &= 1.808 + j1.401 \end{aligned} \tag{6.12}$$

and

$$\begin{aligned} V_1^{\text{full}} &= 0.999741\angle -0.000593^\circ \\ V_2^{\text{full}} &= 0.998850\angle -0.012841^\circ \\ V_1^{\text{half}} &= 0.999824\angle 0.000553^\circ \\ V_2^{\text{half}} &= 0.999379\angle -0.005568^\circ. \end{aligned} \tag{6.13}$$

Thus, with the assistance of probing, we are able to estimate not only the unknown load parameters but also the unknown state variables of the power systems under both full loading and half loading conditions of the controllable load.

Note that probing may *not* always resolve the lack of observability. For example, suppose we extend the distribution feeder in Figure 6.1(a) to include one more bus; *without* adding any additional meter. In that case, the parameters and the states at the new bus, i.e., bus 3, would remain unobservable, no matter what probing signal we use at bus 2. The conditions under which probing can enhance observability in power distribution systems are discussed in [176, 396, 397].

Probing in Practice

Probing action is meant to be done over a short period of time, because we want all the parameters in the system to remain constant, except for the probing parameter.

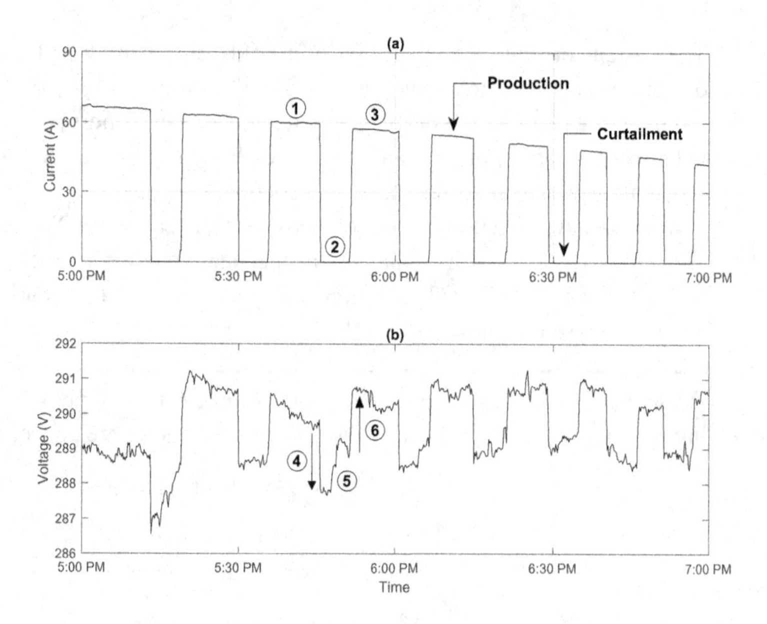

Figure 6.2 Several probing actions that are taken at a PV inverter: (a) the current that is injected by the inverter to the grid; (b) the voltage at the point of interconnection.

However, in practice, what we measure during a probing action includes also the impact of the ongoing changes in various aspects of the system. Thus, the measurements should be examined to remove the impact of such unintended changes in the system. It is also helpful if we can repeat the probing actions to better capture the changes in the system that are directly the result of the probing actions.

Example 6.2 Figure 6.2 shows the results of several back-to-back probing actions that are done in a 90 kW PV inverter. Only one phase is shown in this figure. The probing actions are done on a summer day over a period of two hours from 5 PM till 7 PM; i.e., during a period when the production is gradually declining in the evening. As shown in Figure 6.2(a), the probing parameter is the generation output of the PV inverter, which is switched between full production, as in ①, and full curtailment, as in ②, and then we go back to full production, as in ③. The corresponding impact of the probing actions on the voltage at the point of the inverter's interconnection with the grid is shown in Figure 6.2(b). We can see that the sudden curtailment from ① to ② results in a sudden decrease in voltage, as marked by ④. The subsequent changes in voltage over the next few minutes, which are marked by ⑤, are *not* due to the probing action. The next impact of probing is in the sudden increase in voltage, as marked by ⑥, which is due to the sudden return to full production from ① to ②. Therefore, we must focus only on the changes at ④ and ⑥ as the direct impact of the probing action in our analysis. Of course, since the probing actions are repeated in this example, we can better estimate how much the voltage is impacted per each 1 A change in the current that is injected by the PV unit to the power grid.

6.1.2 Enhanced Redundancy

Even when the power system is already observable, probing may help by creating additional *redundancy* in measurements to enhance accuracy in the presence of measurement error. For instance, consider again the network in Example 6.1. Suppose we *do* know $V_{1,\text{base}}$. That is, suppose the only unknown load parameter is α. In that case, the power system under the scenario in Example 6.1 becomes observable even without probing. The reason is that, since $V_{1,\text{base}}$ is no longer unknown, the number of independent unknowns reduces from four to three. Thus, the number of unknowns becomes *equal* to the number of independent equations.

Nevertheless, if we use probing, we can enhance the accuracy in estimating the remaining unknown load parameter α. Note that, if we use probing, then the number of independent unknowns would be five, as listed below:

$$\alpha, V_1^{\text{full}}, V_2^{\text{full}}, V_1^{\text{half}}, V_2^{\text{half}}. \tag{6.14}$$

Yet, the number of equations remains at six. This creates redundancy in our analysis. An example is given in Exercise 6.2; also see [176, 396].

6.2 Topology and Phase Identification Using Probing

There are at least two different ways that probing can help with topology identification. First, recall from Section 3.8.3 in Chapter 3 that topology identification can be formulated as a parameter estimation problem, where the unknown parameters are the status (open or closed) of the switches. Accordingly, probing can help by enhancing observability or redundancy in the topology identification problem, based on the same principles that we discussed in Section 6.1.

Second, probing can help with topology identification also by using a group of *signal generators* and *signal discriminators*. For the rest of this section, our focus will be on this second possibility. We will show that signal generators and signal discriminators can also be used to help with phase identification.

6.2.1 Topology Identification

Consider a power distribution system with two switches, denoted by ① and ②, as shown in Figure 6.3(a). One signal generator and two signal discriminators are installed on this network. The signal generator is a current source. It is the *probing device*. It generates a current signal that can be distinguished from the fundamental 60 Hz (or 50 Hz) power signal as well as any other existing source in the network. For example, the signal generator can be a *harmonic current source* that generates a harmonic current at a frequency that is unique in the network. As for the two signal discriminators that are denoted by letters A and B, they can be any current sensor that is capable of detecting the signal that is generated by the signal generator. For example,

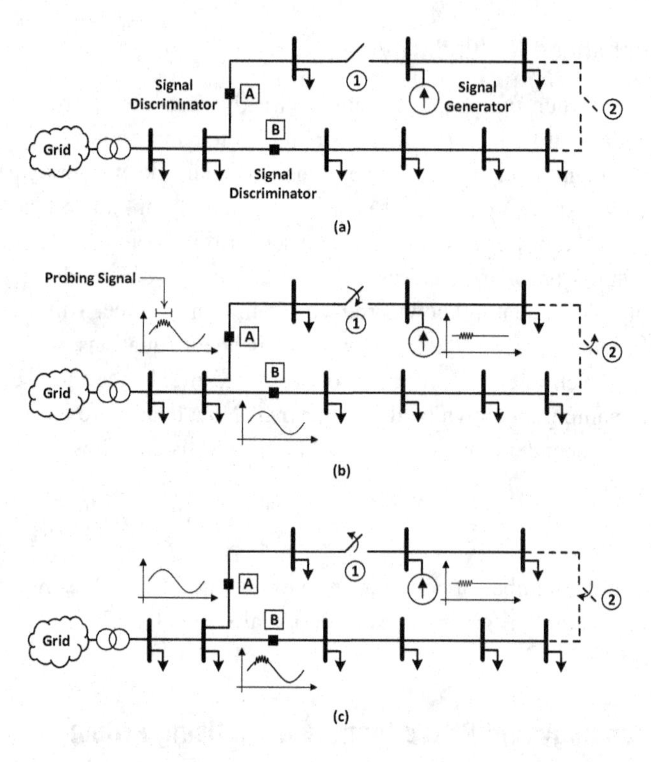

Figure 6.3 Application of probing in topology identification: (a) a power distribution system with two switches; (b) signal discriminator A detects the probing signal; (c) signal discriminator B detects the probing signal.

depending on the choice of the probing signal, the signal discriminators can be power quality meters, waveform sensors, H-PMUs, etc.; see Chapter 4.

Recall from the analysis in Section 4.5.2 in Chapter 4 that the harmonic current flows almost entirely through the substation and not through the loads. Therefore, one can identify the network topology by checking whether any of the signal discriminators can detect the probing signal that is generated by the signal generator. Two examples are illustrated in Figures 6.3(b) and (c).

In Figure 6.3(b), switch ① is closed and switch ② is open. Since the harmonic current flows through the substation, the probing signal that is generated by the signal generator flows through switch ①. As a result, the probing signal is detected by signal discriminator A and not by signal discriminator B.

In Figure 6.3(c), switch ② is closed and switch ① is open. Since the harmonic current flows through the substation, the probing signal that is generated by the signal generator flows through switch ②. As a result, the probing signal is detected by signal discriminator B and not by signal discriminator A.

By comparing the results in Figures 6.3(b) and (c), we can conclude that the network topology can be identified based on whether each of the signal discriminators can detect the probing signal that is generated by the signal generator.

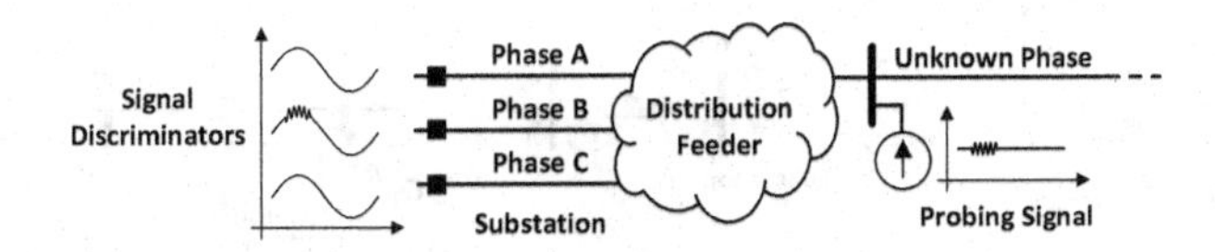

Figure 6.4 Application of probing to phase identification.

Note that the above analysis is applicable only to a radial network, where there is only one path from the signal generator to the substation.

For a larger radial network with multiple laterals and several switches, one may need to install multiple signal generators in order to fully identify the network topology. The signal generators should generate signals that can be distinguished from each other. For example, they may have to inject harmonic currents at frequencies that are different from each other; see [252].

It should be also noted that the above analysis is very similar to the topology identification method in Section 4.5.2 in Chapter 4. The main difference is whether the harmonic current is injected *intentionally* for the purpose of topology identification. If the harmonic current is injected into the power system for the purpose of topology identification, then it is indeed a probing signal.

6.2.2 Phase Identification

The probing method in Section 6.2.1 can also be used for phase identification. Consider the phase identification problem in Figure 6.4. On the right-hand side, there is a single-phase bus with an unknown phase. A signal generator is installed at this bus in order to generate a probing signal. On the left-hand side, there are three phases, labeled as Phases A, B, and C. A signal discriminator is installed at each phase. The probing signal is detected by the signal discriminator at bus B. Therefore, the unknown phase is identified as Phase B.

Interestingly, phase labeling could be done even with only two signal detectors; such as one at Phase A and one at Phase B. If neither of the two signal detectors detects the probing signal, then we can conclude that the probing signal is generated on Phase C. However, by using three signal detectors, we can benefit from redundancy and achieve more reliable phase identification results. Additional details on using probing for phase identification is available in [398].

6.3 Model-Free Control Using Probing

One of the reasons to obtain power systems measurements is to support various smart grid control applications where the objective is to control certain quantities in the power system, such as to regulate voltage or to control power flow. It is common for smart grid control applications to use the measurements together with a *model* of the power system. The model may include the topology of the power system, the

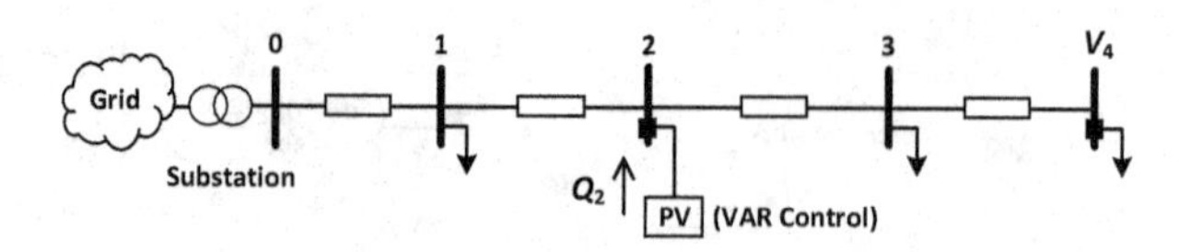

Figure 6.5 The PV inverter can use probing to obtain the relationship between reactive power injection at bus 2 and voltage at bus 4 in order to provide Volt/VAR control.

parameters, etc. However, what if such a model is *not* available? In this section, we explain how a probing action may help in such scenarios.

Consider a basic Volt/VAR Control (VVC) problem in a power distribution system, as shown in Figure 6.5. Here, the objective is to control reactive power injection at the PV inverter at bus 2 in order to regulate voltage at the end of the feeder at bus 4. Suppose the model of the system is *not* available. However, we are able to measure *reactive power injection* at bus 2 and also *voltage magnitude* at bus 4. These quantities are denoted by Q_2 and V_4, respectively.

In order to conduct VVC, we need to know the relationship between Q_2 and V_4. That is, we need to know function $f(\cdot)$ in the following expression:

$$V_4 = f(Q_2). \tag{6.15}$$

The above function can be obtained if the system model is available and all the loads also are measured and known. However, even if such information is not available, then we can still estimate $f(\cdot)$ using a probing action.

First, consider the normal operation of the PV inverter. The following relationship must hold between the two quantities that are being measured:

$$V_4^{\text{normal}} = f(Q_2^{\text{normal}}). \tag{6.16}$$

Next, suppose the PV inverter *perturbs* its reactive power injection as follows:

$$Q_2^{\text{perturbed}} = Q_2^{\text{normal}} + \Delta Q, \tag{6.17}$$

where ΔQ denotes the amount of perturbation. The following relationship must hold between the available measurements during the perturbation action:

$$V_4^{\text{perturbed}} = f(Q_2^{\text{perturbed}}). \tag{6.18}$$

Assuming that the perturbation is small, we can obtain a *first-order approximation* for the unknown relationship between the quantities, as follows:

$$f(Q_2) \approx f(Q_2^{\text{normal}}) + \frac{f(Q_2^{\text{perturbed}}) - f(Q_2^{\text{normal}})}{Q_2^{\text{perturbed}} - Q_2^{\text{normal}}}(Q_2 - Q_2^{\text{normal}}). \tag{6.19}$$

Therefore, we can approximate the relationship in (6.15) as

$$V_4 \approx V_2^{\text{normal}} + \frac{V_2^{\text{perturbed}} - V_2^{\text{normal}}}{Q_2^{\text{perturbed}} - Q_2^{\text{normal}}}(Q_2 - Q_2^{\text{normal}}). \tag{6.20}$$

Here, the perturbation in reactive power injection serves as a probing action to help us address the lack of access to a system model.

Example 6.3 Again, consider the distribution system in Figure 6.5. Even though the network topology is known; the line impedances and the loads are unknown. Thus, the system model overall is unknown. Suppose the measurements are

$$Q_2^{\text{normal}} = 10, \qquad V_4^{\text{normal}} = 0.989849,$$
$$Q_2^{\text{perturbed}} = 30, \qquad V_4^{\text{perturbed}} = 0.989888.$$

All the quantities are in per unit. From (6.20), the relationship between reactive power injection at bus 2 and the voltage at bus 4 is obtained as

$$V_4 \approx 0.989849 + 1.95 \times 10^{-6}\,(Q_2 - 10). \tag{6.21}$$

We can use (6.21) to choose the amount of reactive power injection by the PV inverter in order to regulate voltage V_4. For instance, suppose we seek to increase V_4 to 0.99 p.u. The PV inverter must set its reactive power injection at

$$Q_4 = 10 + (0.99 - 0.989849)\,/\,1.95 \times 10^{-6} \approx 87.4 \text{ p.u.} \tag{6.22}$$

Probing can be used similarly to support model-free control in other smart grid applications, such as power flow control in transmission systems; see Exercise 6.8.

Important advancements have been reported in recent years to use probing in *model-free optimal control* in smart grid control applications. For example, in [399], probing is used to learn the *gradient* of the objective function in a model-free power loss minimization problem in a power distribution system.

6.4 Modal Analysis with Probing

Recall from Section 2.6.3 in Chapter 2 that the electromechanical modes of the power system can be estimated by applying modal analysis, such as the Prony method, to various power system measurements during *transient oscillations*. However, our ability to conduct modal analysis depends on how significant the magnitudes of the oscillations are compared to the ambient noise in the system, and also how often a significant transient oscillatory event occurs.

Accuracy and repeatability in estimating the electromechanical modes of the power system can be improved by using probing signals. In this approach, instead of relying on random natural events to occasionally excite the electromechanical modes of the system, these modes are excited by injecting a probing signal.

A probing signal may be injected into the bulk power system using a number of different actuators, such as by momentarily switching on and off a *resistive brake*, or by modulating power flow on a *high voltage direct current* (HVDC) intertie using power electronics. Both of these options are currently available in the Western Interconnection in the United States, as shown in Figure 6.6.

Figure 6.6 The resistive brake at Chief Joseph substation and the HVDC intertie between Celilo and Sylmar converter stations can be used to generate probing signals on the Western Interconnection. Measurements are done at Malin substation.

6.4.1 Probing with Resistive Brake

A resistive brake is used in power systems to quickly dissipate a large amount of energy. Brakes are often primarily designed to enhance transient stability in power systems during a fault contingency. However, they can also be used to excite the power system when needed in order to conduct modal analysis.

The resistor brake at Chief Joseph substation (see Figure 6.6) is designed to dissipate 1400 MW of power when energized at 240 kV. It is capable of withstanding operation for up to three seconds before it is de-energized for cooling [400]. Brake insertion is done by switching on the brake for a short period of time and then switching it off, thus creating a momentary excitation to the system.

Example 6.4 Figure 6.7 shows power system measurements at Malin substation on the Western Interconnection during two brake insertion events. The second brake insertion occurred only five minutes after the first brake insertion. During the first brake insertion, the change in the power flow on the transmission line between the Malin substation and the Round Mountain substation is as shown in Figure 6.7(a). During the second brake insertion, the change in the power flow on the transmission line between the Malin substation and the Round Mountain substation is as shown in Figure 6.7(b). We can see that the system excitation was *repeatable*, and the measurements of the inter-area oscillations are similar in these two experiments.

The fact that probing is repeatable allows creating redundancy in estimating the electromechanical modes of the power system. For instance, by applying the modal analysis to the transient power flow oscillations in Figure 6.7(a), we obtain the corresponding dominant modes during the first brake insertion as:

$$0.241 \text{ Hz with } 12.4\% \text{ Damping}, \quad 0.379 \text{ Hz with } 8.5\% \text{ Damping}. \tag{6.23}$$

Similarly, by applying the modal analysis to the power flow oscillations in Figure 6.7(b), we obtain the dominant modes during the second brake insertion as:

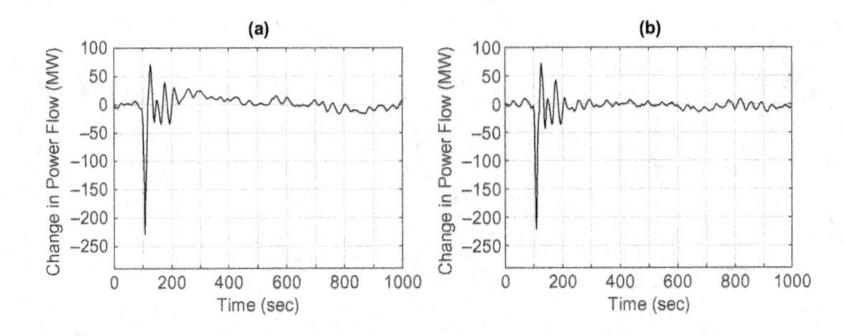

Figure 6.7 Response of the power system to two brake insertions at Cheif Joseph substation [401]: (a) the first brake insertion; (b) the second brake insertion.

$$0.244 \text{ Hz with } 10.7\% \text{ Damping}, \quad 0.360 \text{ Hz with } 9.1\% \text{ Damping}. \tag{6.24}$$

Having multiple estimations such as the above can help improve the accuracy in estimating the electromechanical modes of the system. For instance, we can take the average of the results or apply other statistical analysis to the results.

6.4.2 Probing with Intermittent Wave Modulation

Probing in modal analysis can also be done by modulating a *probing waveform* on an HVDC intertie using power electronics. Recall from Section 1.2.9 in Chapter 1 that each HVDC intertie has two converter stations, one on each side, in order to convert the three-phase AC power to and from DC power. For example, for Pacific HVDC Intertie on the Western Interconnection, Celilo Converter Station in the state of Oregon is at the northern terminus of the intertie, and Sylmar Converter Station in Southern California is at the southern terminus of the intertie; see Figure 6.6.

Example 6.5 Figure 6.8(a) shows a probing wave during a modal analysis experiment at the Pacific HVDC Intertie. The power electronics converters modulated two cycles of a ± 125 MW square wave on the DC transmission line. The experiment was done twice. The measured voltages at the Malin substation during the two experiments are shown in Figures 6.8(b) and (c), respectively. The dominant modes of the system during the first experiment are estimated as [402]:

$$0.303 \text{ Hz with } 10.2\% \text{ Damping}, \quad 0.416 \text{ Hz with } 9.0\% \text{ Damping}. \tag{6.25}$$

The dominant modes during the second experiment are estimated as

$$0.296 \text{ Hz with } 10.0\% \text{ Damping}, \quad 0.416 \text{ Hz with } 8.0\% \text{ Damping}. \tag{6.26}$$

Again, we can see that the probing experiment was *repeatable* and provided redundancy in estimating the electromechanical modes of the power system. Figure 6.8 is courtesy of the Pacific Northwest National Laboratory, operated by Battelle for the U.S. Department of Energy.

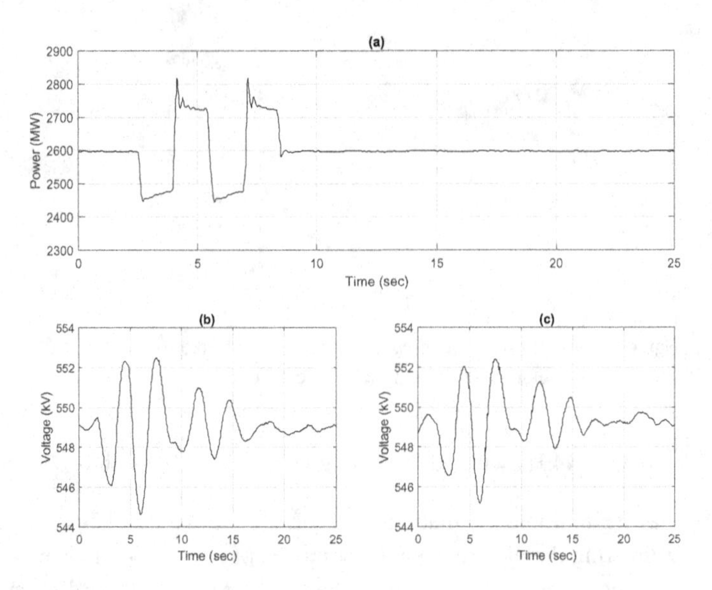

Figure 6.8 Two probing experiments at Pacific HVDC Intertie [402]: (a) probing signal; (b) voltage at the Malin substation during the first test; (c) voltage at the Malin substation during the second test.

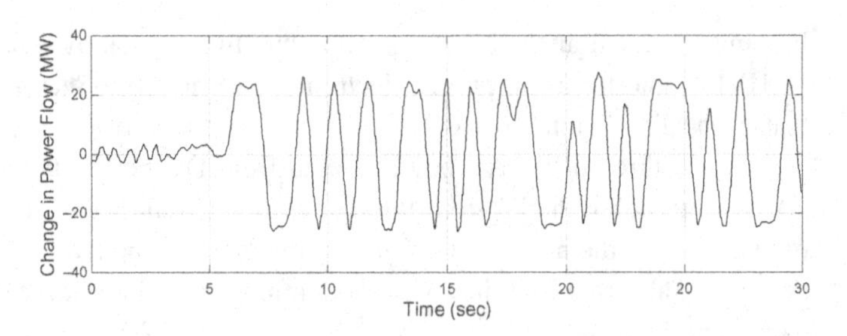

Figure 6.9 A probing signal with ±25 MW noise modulation at the Pacific HVDC Intertie [403]. The modulation is done on the power flow that goes into the intertie.

Longer complex probing signals may also be used for the purpose of modal analysis. For example, Figure 6.9 shows a longer ±25 MW noise wave that has been modulated on the Pacific HVDC Intertie. The goal of this type of noise modulation probing signals is to excite the power system across a range of local oscillation modes. For example, the probing signal in Figure 6.9 can excite the power system at a frequency spectrum from 0.1 Hz to 1.1 Hz; see [403].

6.4.3 Input–Output System Identification

With *known* input signals, one can go beyond modal analysis and seek to also obtain an *input–output model* for the power system. Here, the input to the system is the probing signal, and the output from the system is the measurement. For instance, the input-output relationship in Example 6.5 is shown in Figure 6.10.

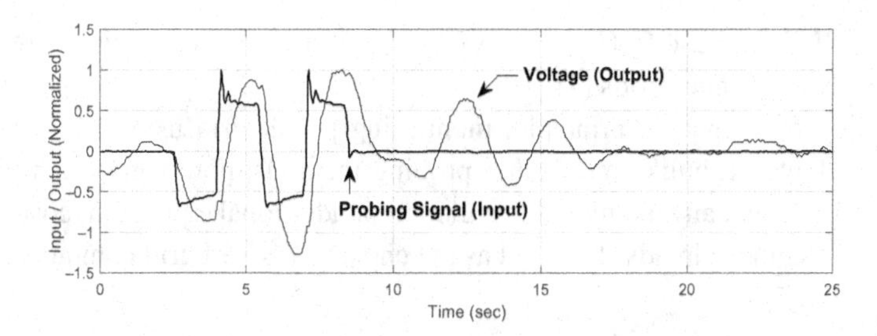

Figure 6.10 Input–output relationship during the first probing experience in Example 6.5. Both signals are normalized so that they can be compared with each other.

A simple input–output model that can be considered is the discrete linear *AutoRegressive eXogenous* (ARX) model. It can be formulated as follows:

$$\begin{aligned} x(m) = {} & a_1 x(m-1) + a_2 x(m-2) + \cdots + a_m x(0) \\ & + b_1 u(m-1) + b_2 u(m-2) + \cdots + b_p u(m-p), \end{aligned} \tag{6.27}$$

where $p \leq m$. Here, power system measurement $x(t)$ at time sample m is approximated based on its previous samples and also the previous samples of the probing signal $u(t)$, which is an input to the power system. In (6.27), it is assumed that the measurements do *not* carry error; however, measurement error can also be added to this model. By comparing (6.27) with (2.10) in Chapter 2, we see that the ARX model is an extension of the AR model. The difference is that the ARX model also includes the exogenous variables associated with the input to the system.

Input–output system identification is beyond the scope of this book. However, there is already a tremendous amount of literature on system identification that readers can refer to. A good classic textbook on this subject is [404].

6.5 Power Line Communications as a Probing Tool

Power line communications (PLC) is one of the communication methods that is currently used by many electric utilities around the world to support smart grid applications [405, 406]. PLC is a *wired* (as opposed to *wireless*) communication technology that reuses power lines as the media for the purpose of data transmission.

PLC superimposes a high-frequency communication signal over the 60 Hz (or 50 Hz) electrical signal in an existing power cable. Some of the common applications of PLC in the field of smart grids is to provide communications for automated meter reading and demand side management [407–409].

The fact that PLC signals travel through power lines makes reliable communication challenging. However, it also provides us with a distinct advantage that is of interest in this book: By observing the PLC signal that is sent through the power grid, we can learn about the status of the power grid itself, such as to diagnose incipient faults in power line cables or to identify the power grid topology. This is a unique property of

PLC, because PLC is the only "through the grid" communications technology in the field of smart grids [410].

Note that, in principle, monitoring power grid using PLC is a *probing* method. However, unlike most other probing methods, power grid monitoring via PLC does *not* entail any additional installation or additional equipment costs, as long as the PLC system is already deployed as the choice for smart grid communications.

6.5.1 Basics of Power Line Communications

The basic architecture for power line communications is shown in Figure 6.11(a). The power cable serves both as the conductor to transfer power and as the media to transfer communications data. The PLC transmitter is located at one end of the power line. It comprises a coupling device and a modem (modulator-demodulator). The PLC receiver is located at the other end of the power line. It too comprises a coupling device and a modem. At the transmitter, data is first modulated into a *communication carrier signal* by the modem. The communication carrier signal is then injected into the power line by the coupler. At the receiver, the communication carrier signal is extracted from the power line by the coupler. It is then demodulated by the modem to obtain the original data.

Coupling devices are used to inject or extract communication signals into or from power lines. Coupling can be done in different ways. For example, in inductive coupling, a coil is wound around and clamped over the power cable, as shown in Figure 6.11(b). When current flows through the winding coil, it yields an electromagnetic field which inductively loads the communication signal into the conductor. A survey of coupling technologies for PLC is available in [411].

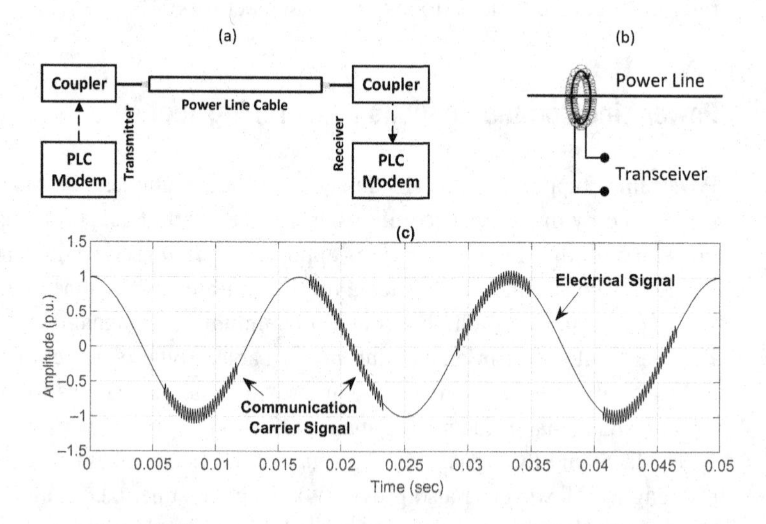

Figure 6.11 Basics of PLC: (a) system setup with a PLC transmitter and a PLC receiver at two ends of a power line; (b) inductive coupling; (c) high-frequency communication carrier signals that are superimposed on an electrical signal.

PLC modems are used to modulate data into a high-frequency communication carrier signal at the PLC transmitter, and to demodulate the high-frequency communication carrier signal to data at the PLC receiver. Once the modulated high-frequency communication carrier signal is *superimposed* to the power cable, the resulting voltage or current waveform will be as shown in Figure 6.11(c). The frequency of the communication carrier signal depends on the PLC technology being used and the communication standard being enforced. For example, in the United States, the Federal Communications Commission (FCC) allows frequencies of 10–490 kHz for narrow-band PLC (NB-PLC) [410]. Most existing NB-PLC technologies use Orthogonal Frequency-Division Multiplexing (OFDM) as the modulation method; cf. [412].

To support smart grid applications, we often need a *network* of PLC devices. They communicate with each other directly, when they are within the communication range of each other. Otherwise, they communicate via PLC relays. PLC relays may also be needed at a location with step-change transformers [413].

6.5.2 Signal Attenuation and Channel Estimation

As in any other communications system, transmitted PLC signals experience *attenuation*, i.e., they lose strength, as they travel through the power line. Attenuation is affected by several factors, such as the length and the type of the conductor. Importantly, it is affected also by the frequency of the signal.

The extent of attenuation at each frequency can be captured by obtaining the transfer function of the channel. This is shown in Figure 6.12(a), where $H(f)$ denotes the *channel transfer function* (CTF). It is the Fourier transformation of the *channel impulse response*. As for $X(f)$ and $Y(f)$, they denote the transmitted communication signal and the received communication signal, both in frequency domain. From the definition of CTF, at any frequency f, we have

$$Y(f) = H(f)X(f). \tag{6.28}$$

The magnitude of $H(f)$ for a real-world NB-PLC channel is shown in Figure 6.12(b). At $f = 70$ kHz, the magnitude of $H(f)$ is -43 dB, i.e., the communication signal has

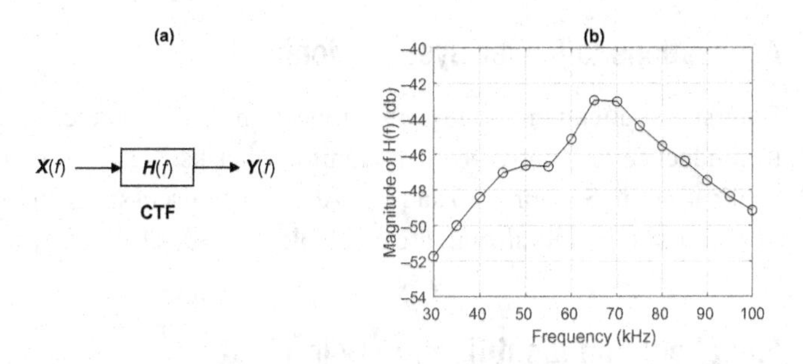

Figure 6.12 Channel transfer function in power line communications: (a) block diagram of the system; (b) experimental measurements [414].

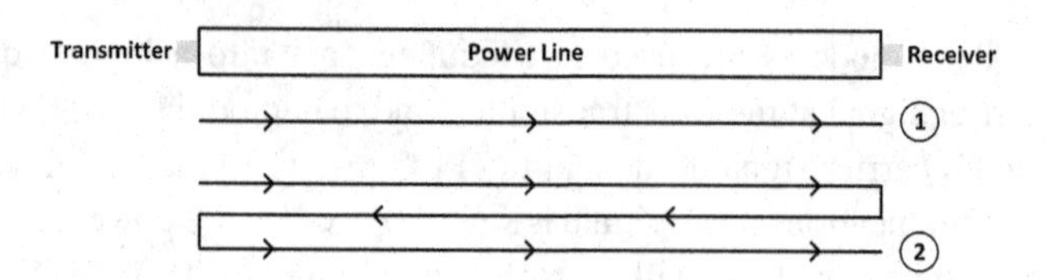

Figure 6.13 Reflection of the PLC signals in a healthy power cable.

only about 0.7% of its original strength. We can see that the attenuation level varies significantly across the examined frequency band.

CTF often fluctuates over day and night due to various reasons. Therefore, most PLC protocols perform channel estimation when they need to transmit new data. For example, they may perform the following steps:

- Idle Channel: to measure noise levels;
- Transmit Test Sequence: to estimate the channel transfer function;
- Transmit Data: to transfer the actual data.

The results from channel estimation are used by digital signal processing algorithms in PLC devices to improve the accuracy of data transfer.

6.5.3 Signal Reflection at Impedance Discontinuities

Another physical phenomenon in PLC systems is the *reflection* of the communication signal at locations with *impedance mismatch*. That is, there is an "echo" of the transmitted communication signal at any point where there is a change in impedance. At the very least, reflection occurs at the receiver and also subsequently at the transmitter, as shown in Figure 6.13. Here, signal ① is the communication signal that is received at the receiver through the direct path. Signal ② is a reflected signal that was first echoed at the receiver and then subsequently also echoed at the transmitter. Additional reflected signals arrive at the receiver, after four echos, six echos, and so on, although they are not shown in this figure. Reflected signals are attenuated more than the original signal because they travel longer through the power cable. Nevertheless, their impact must be addressed by using proper signal processing methods; cf. [412].

6.5.4 Applications to Power System Monitoring

Both signal attenuation and signal reflection can be studied as underlying principles to conduct certain power system monitoring tasks. Next, we will discuss two common applications in Sections 6.6 and 6.7. Additional discussions on the applications of PLC in power system monitoring are available in [410, 415–426].

6.6 Fault Location Identification Using PLC

PLC can be used to detect and identify the location of a fault in a power line. One idea is to utilize the fact that damage in a cable causes a *change in impedance* of the cable at the location of the damage. From Section 6.5.3, we know that a change in impedance causes *reflection* in the communication signal. In a healthy cable, reflections occur only at the receiver and the transmitter. In a damaged cable, *reflections occur also at the point of damage*. Thus, the location of the fault can be estimated by investigating the characteristics of the reflected signal.

Example 6.6 Consider a PLC system over a power cable. The length of the cable is 1000 m. The cable is partially damaged at a location that is 150 m away from the PLC transmitter. Due to the mismatch in impedance, the communication waves experience reflection when they reach the location of the cable that is damaged. This is illustrated in Figure 6.15. Signal ① is the communication signal that is received at the receiver through the direct path. Signal ② is a reflected signal that is first echoed at the damaged location and then subsequently also echoed at the transmitter. Signal ③ is a reflected signal that is first echoed at the receiver and then subsequently also echoed at the damaged location. Signals ④ and ⑤ can be explained similarly. Other reflected signals are also created and ultimately arrived at the receiver, which are not shown here; cf. [427].

By comparing the scenarios in Figure 6.14 and those in Figure 6.13, we can see that there are major differences between the reflected signals that the PLC modems receive in a healthy cable versus in a damaged cable.

Recall from Section 6.5.1 that the PLC modems frequently estimate the CTF, i.e., $H(f)$. The CTF in a cable that has a damage at an unknown location can be written as a product of the channel transfer function $H_{\text{healthy}}(f)$ of the healthy cable, and the channel transfer function $H_{\text{dmg}}(f)$ of the damaged cable:

$$H(f) = H_{\text{healthy}}(f)\, H_{\text{dmg}}(f). \tag{6.29}$$

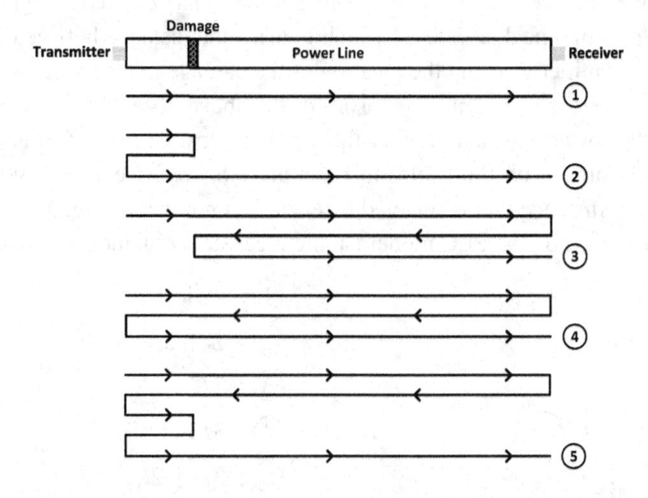

Figure 6.14 Reflection of the PLC signals in a damaged power cable.

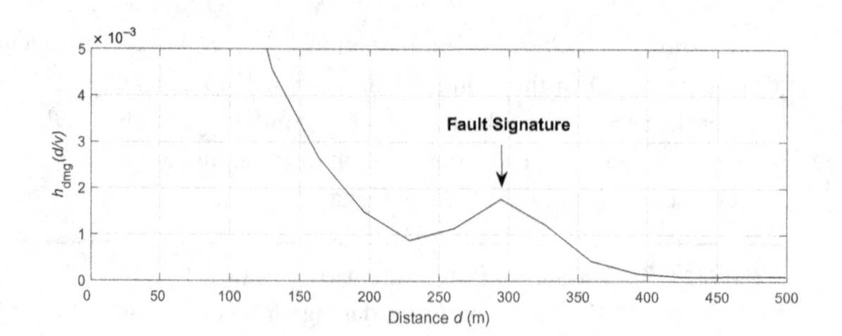

Figure 6.15 Fault signature in the damage impulse response of the cable [427].

In practice, one can assume that $H_{\text{healthy}}(f)$ is equal to the $H(f)$ at the time of installing the PLC modems. In other words, one can assume that the cable was *initially healthy* at the time that the PLC modems were first installed. From (6.29), and based on the historical estimation of $H_{\text{healthy}}(f)$, we can obtain:

$$H_{\text{dmg}}(f) = \frac{H(f)}{H_{\text{healthy}}(f)}. \tag{6.30}$$

By applying the Inverse Discrete Fourier Transform (IDFT), we can obtain the impulse response of the damaged part of the cable, denoted by $h_{\text{dmg}}(t)$. This impulse response is sometimes referred to as the *damage impulse response*.

Figure 6.15 shows the *damage impulse response* of the damaged cable in Example 6.6 that was estimated by the PLC modem on the receiver side. Here, the damage impulse response is plotted over *distance* instead of over *time*. Note that time can be expressed as the distance that communication wave travels, denoted by d, divided by the propagation speed of communication wave, denoted by v:

$$t = \frac{d}{v}. \tag{6.31}$$

We can see that the damage in the cable has created a signature in $h_{\text{dmg}}(d/v)$, in form of a peak at 300 m, which is *twice* the distance between the PLC modem at the transmitter side and the location of the damage in the cable.

Next, we explain the reason for the above observation. Let L denote the length of the power line, and l denote the distance between the PLC transmitter and the location of damage on the cable. In Example 6.6, we have $L = 1000$ m and $l = 150$ m. The distance that is traveled by each communication signal in Figure 6.14 before it is received by the PLC modem at the receiver is obtained as follows:

$$\begin{aligned}
&①: \; L, \\
&②: \; L + 2l, \\
&③: \; 3L - 2l, \\
&④: \; 3L, \\
&⑤: \; 3L + 2l,
\end{aligned} \tag{6.32}$$

respectively. In fact, for each reflected signal, the distance traveled by the signal wave is a multiple of L plus a positive or a negative multiple of $2l$. As a result, a peak is created in the damage impulse response $h_{\text{dmg}}(d/v)$ at the following distance, as we previously saw in Figure 6.15:

$$d = 2l = 300\ m. \tag{6.33}$$

Based on the above analysis, we can estimate the location of the damage in a cable by taking the following steps. First, the PLC devices continuously estimate the CTF and watch for major changes in $H(f)$ compared to the historical estimation of the CTF. Second, if major changes are detected in $H(f)$, then the relationship in (6.30) is used to estimate $H_{\text{dmg}}(f)$. Third, by applying IDFT, the damage impulse response $h_{\text{dmg}}(t)$ is obtained. Fourth, by using the relationship in (6.31), the damage impulse response is analyzed to identify the fault signature at distance d, similar to the analysis in Figure 6.15. Finally, by using (6.33), the location of the damage is estimated at distance $l = d/2$.

Other studies on cable monitoring using PLC include [415–417].

6.7 Topology and Phase Identification Using PLC

6.7.1 Topology Identification Using PLC

When PLC devices communicate with each other, they can measure the time that it takes for them to send and receive communication messages with each other. Such information can be used to identify the topology of the underlying power grid that is the physical media for their communication.

The *time-of-flight* (TOF) between two PLC devices is the time that it takes for the communication signal to travel from one PLC device to the other PLC device. If PLC devices *are* equipped with GPS, then TOF can be measured by using GPS signals to precisely time stamp data transmissions, similar to how time synchronization is done in PMUs and WMUs. If PLC devices are *not* equipped with GPS, then PLC devices can still measure TOF via a simple *two-way handshake* by using the Network Time Protocol (NTP) [142]. For example, consider PLC device A and PLC device B. The handshake consists of the following steps [418]:

- Device A sends at time 0 a message to device B;
- Device B receives the message at time $t_1 = \tau_{\text{AB}}$, with τ_{AB} being the TOF;
- Device B sends back a message to device A after a known delay t_2;
- Device A receives the message at a known time $t_3 = 2t_1 + t_2$;
- Device A estimates the TOF as

$$\tau_{\text{AB}} = \frac{t_3 - t_2}{2}. \tag{6.34}$$

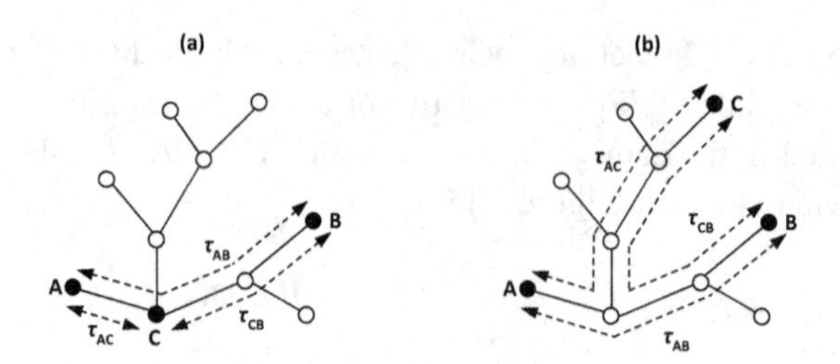

Figure 6.16 Application of PLC TOF estimation in topology identification [418]: (a) PLC device C is on the path between PLC device A and PLC device B; (b) PLC device C is not on the path between PLC device A and PLC device B.

Note that, in general, if there are multiple communication paths available between two PLC devices, then TOF is the duration of the *shortest time path* between the two PLC devices. However, when it comes to a radial network topology, such as in most power distribution systems, there is only one path to communicate between any two PLC devices.

Next, consider three PLC devices A, B, and C, as shown in Figure 6.16. Suppose TOF is measured among these three PLC devices and the results are denoted by τ_{AB}, τ_{AC}, and τ_{CB}. If PLC device C *is* on the path between PLC device A and PLC device B, as in Figure 6.16(a), then we would have:

$$\tau_{AB} = \tau_{AC} + \tau_{CB}. \tag{6.35}$$

If PLC device C is *not* on the path between PLC device A and PLC device B, as in Figure 6.16(b), then we would have:

$$\tau_{AB} < \tau_{AC} + \tau_{CB}. \tag{6.36}$$

Similarly, we can determine whether PLC device A is on the path between PLC device B and PLC device C; and whether PLC device B is on the path between PLC device A and PLC device C. This analysis results in fully identifying the *relative locations* of PLC devices A, B, and C with respect to each other.

Suppose there is a PLC device at each bus in a radial power distribution system. Also suppose TOF is measured between any two PLC devices. By repeating the above analysis on any group of three PLC devices, we can identify the relative location of *all* PLC devices on the network with respect to each other. For a radial topology, such information is sufficient to identify the network topology.

Example 6.7 Consider a power distribution system with six buses. Bus 1 is the substation. The network topology is radial but unknown. There is one PLC device at each bus. Therefore, we can refer to the PLC devices and buses interchangeably. The network is fully connected and all PLC devices can talk to each other. The symmetric matrix of TOF measurements is obtained as

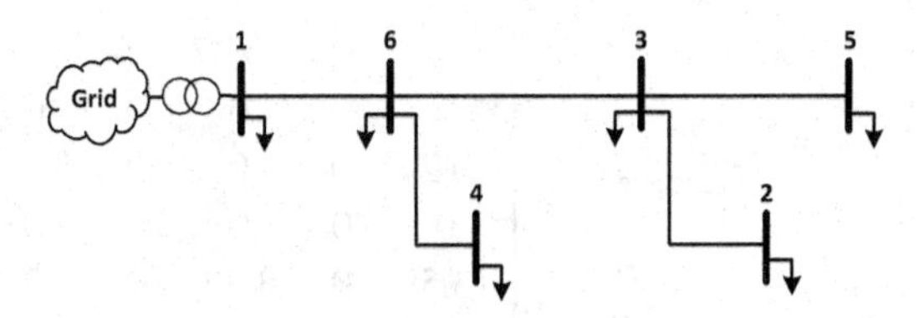

Figure 6.17 The result of network topology identification in Example 6.7.

$$\tau = \begin{bmatrix} 0 & 12.64 & 10.27 & 8.04 & 13.59 & 4.47 \\ 12.64 & 0 & 2.38 & 11.74 & 5.70 & 8.17 \\ 10.27 & 2.38 & 0 & 9.37 & 3.32 & 5.80 \\ 8.04 & 11.74 & 9.37 & 0 & 12.68 & 3.57 \\ 13.59 & 5.70 & 3.32 & 12.68 & 0 & 9.12 \\ 4.47 & 8.17 & 5.80 & 3.57 & 9.12 & 0 \end{bmatrix}. \tag{6.37}$$

All the entries in (6.37) are in micro-seconds. The diagonal entries are zero because they denote the TOF between a PLC device and itself. By examining the conditions in (6.35) and (6.36), we can conclude that:

- Bus 2 is not on the path between bus 1 and any other bus;
- Bus 3 is on the path between bus 1 and buses 2 and 5;
- Bus 4 is not on the path between bus 1 and any other bus;
- Bus 5 is not on the path between bus 1 and any other bus;
- Bus 6 is on the path between bus 1 and buses 2, 3, 4, and 5.

Therefore, the network topology is identified as shown in Figure 6.17.

6.7.2 Phase Identification Using PLC

PLC can be used also to conduct phase identification. One option is to extend the topology identification method in Section 6.7.1. Suppose each PLC device is able to send and receive handshake messages on all the phases that are connected to the bus where it is installed. Accordingly, we can estimate the TOF on *each phase*. As a result, we can obtain matrix τ separately for each phase. We can then conduct phase identification by examining the three obtained matrices.

For instance, consider Example 6.7. Suppose it is a three-phase network. However, suppose some buses are connected only to one or two phases. Suppose the TOF measurements are obtained separately on each phase as follows:

$$\tau^{\text{Phase A}} = \begin{bmatrix} 0 & 0 & 0 & 8.04 & 0 & 4.47 \\ 0 & 0 & 0 & 0 & 0 & 0 \\ 0 & 0 & 0 & 0 & 0 & 0 \\ 8.04 & 0 & 0 & 0 & 0 & 3.57 \\ 0 & 0 & 0 & 0 & 0 & 0 \\ 4.47 & 0 & 0 & 3.57 & 0 & 0 \end{bmatrix}, \tag{6.38}$$

$$\tau^{\text{Phase B}} = \begin{bmatrix} 0 & 0 & 10.27 & 0 & 13.59 & 4.47 \\ 0 & 0 & 0 & 0 & 0 & 0 \\ 10.27 & 0 & 0 & 0 & 3.32 & 5.80 \\ 0 & 0 & 0 & 0 & 0 & 0 \\ 13.59 & 0 & 3.32 & 0 & 0 & 9.11 \\ 4.47 & 0 & 5.80 & 0 & 9.11 & 0 \end{bmatrix}, \tag{6.39}$$

$$\tau^{\text{Phase C}} = \begin{bmatrix} 0 & 12.64 & 10.27 & 0 & 0 & 4.47 \\ 12.64 & 0 & 2.38 & 0 & 0 & 8.17 \\ 10.27 & 2.38 & 0 & 0 & 0 & 5.80 \\ 0 & 0 & 0 & 0 & 0 & 0 \\ 0 & 0 & 0 & 0 & 0 & 0 \\ 4.47 & 8.17 & 5.80 & 0 & 0 & 0 \end{bmatrix}. \tag{6.40}$$

Note that there are several zero entries in these matrices. This is because several buses are not connected to all three phases, and they do not send and receive handshake messages on the phases that they are not connected to.

By examining the matrices in (6.38)–(6.40), we can conclude that

- Bus 2 is connected to Phase C;
- Bus 3 is connected to Phases B and C;
- Bus 4 is connected to Phase A;
- Bus 5 is connected to Phase B;
- Bus 6 is connected to all three phases.

Extending the TOF measurement process is not the only option to use PLC for phase identification. Other methods may also be used; e.g., see [423, 424].

Exercises

6.1 Consider the analysis in Example 6.1.
(a) Write all of the six independent equations that are used in this analysis.
(b) Use the measurements in (6.11) and the fact that $V_0 = 1\angle 0°$ to solve the equations in Part (a). You may use `fsolve` in MATLAB [34].

6.2 Again, consider the analysis in Example 6.1. Suppose, we know that $V_{1,\text{base}} = 1$ per unit. Therefore, the only unknown parameter in this system is α.
(a) Suppose the load at bus 2 operates at full load. Estimate the unknown parameter α at bus 1 based on the following per unit measurements:

$$S_0 = 3.502596 + j1.601236, \quad S_2 = 2.5 + j0.6. \tag{6.41}$$

(b) Suppose the load at bus 2 operates at half load. Estimate the unknown parameter α at bus 1 based on the following per unit measurements:

$$S_0 = 2.250666 + j1.300281, \quad S_2 = 1.25 + j0.3. \tag{6.42}$$

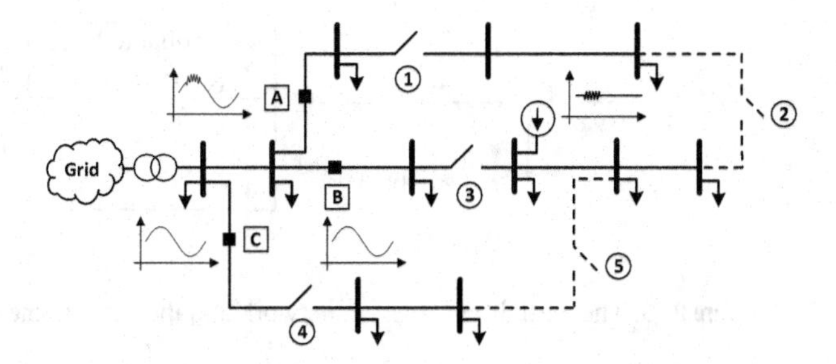

Figure 6.18 The power distribution network that is considered in Exercise 6.5.

(c) Explain how the combination of the results in Parts (a) and (b) can provide redundancy in estimating parameter α.

6.3 File `E6-3.csv` contains the current and voltage measurements corresponding to the probing experience in Figure 6.2.

(a) Calculate how much the current decreases and how much the voltage decreases every time that the generation is curtailed at the PV unit.

(b) Calculate how much the current increases and how much the voltage increases every time that the generation is resumed at the PV unit.

(c) Present the results in Parts (a) and (b) in a scatter plot where the x-axis is the change in current and the y-axis is the change in voltage.

6.4 File `E6-4.csv` contains the current waveform measurements at four signal discriminators on the same distribution feeder. A signal generator generates a probing signal at a frequency of 1.2 kHz for a short period of time.

(a) Which one of the signal discriminators detects the probing signal?

(b) What is the time duration of the probing signal?

6.5 Consider the power distribution network in Figure 6.18. The network has five switches. Switches ①, ③, and ④ are the normally closed switches. Switches ② and ⑤ are the normally open switches. One probing signal generator device is installed on this network to generate high frequency probing signals, as shown with a current source in the figure. The probing signal *is* detected by signal discriminator A. The probing signal is *not* detected by signal discriminators B and C. What are the possible radial topology configurations in this network based on the probing results?

6.6 Again, consider the probing experiment in Exercise 6.3.

(a) Use the available voltage and current measurements to obtain a first-order approximation for the relationship between voltage and current. In other words, express voltage as a linear function of current at the PV unit:

$$V = a + bI, \tag{6.43}$$

where a and b are unknown parameters that need to be estimated. Parameters a and b can be estimated by using a least square optimization based on the collection of the points that we plotted in Part (c) in Exercise 6.3.

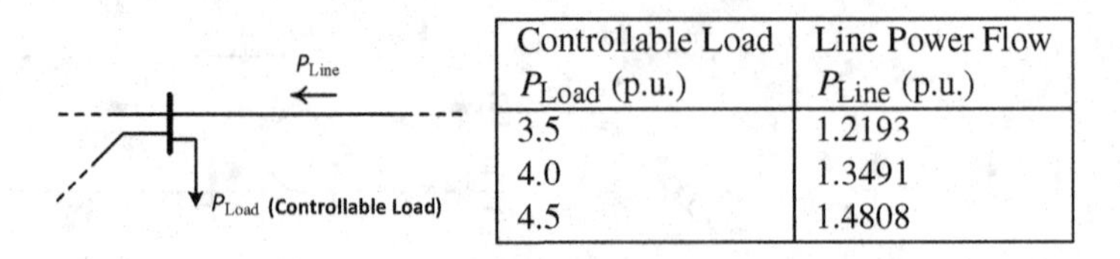

Controllable Load P_{Load} (p.u.)	Line Power Flow P_{Line} (p.u.)
3.5	1.2193
4.0	1.3491
4.5	1.4808

Figure 6.19 The partial transmission network and the measurements in Exercise 6.8.

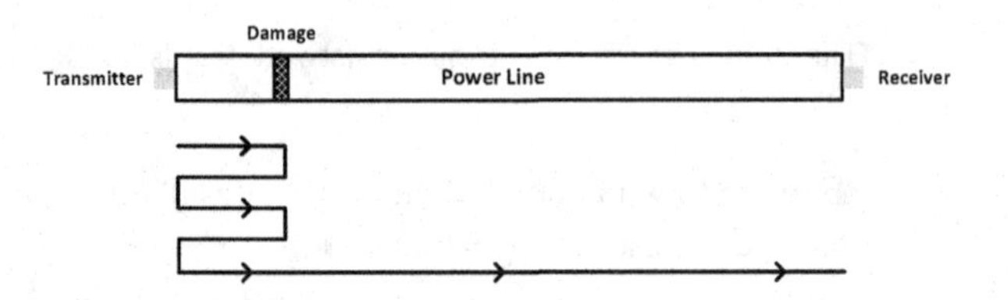

Figure 6.20 Reflections of the PLC signal in a damaged power cable in Exercise 6.9.

(b) Use the results in Part (a) to indicate the current level at the PV unit that is required in order to regulate the voltage at 290 V.

6.7 File `E6-7.csv` contains the damping oscillations in voltage measurements at a transmission line that result from two subsequent probing experiments. Obtain the dominant modes of oscillation in each case. You can use Prony analysis that we learned in Section 2.6.3 in Chapter 2. Set $m = 101$.

6.8 Figure 6.19 shows a portion of a power transmission network. It contains a bus and three transmission lines. The load at this bus is controllable; thus, it can serve as a probing device. The controllable load is set to three different levels; and at each level the power flow on one of the transmission lines is measured accordingly. The measurements are shown in the figure.

(a) Express P_{Line} as a linear function of P_{Load}.

(b) Use the results in Part (a) to set P_{Load} such that $P_{\text{Line}} = 1.3$ per unit.

6.9 Consider the damaged power line cable in Figure 6.20.

(a) Explain the reflections of the PLC signal that are shown in this figure.

(b) What is the distance traveled by the PLC signal? Is it a multiple of L plus a positive or a negative multiple of $2l$? Elaborate your answer.

6.10 Consider the damaged power line cable in Figure 6.21. The length of the cable is 1000 m. The location of the damage is 150 m from the PLC modem on the *receiver* side. Thus, the damage is closer to the *receiver* side.

(a) Repeat the analysis in Example 6.6 and draw the signal path for five different examples of the reflected PLC signals, in a way similar to Figure 6.14.

(b) Indicate the distance that the communication wave travels in each case in Part (a). Compare the results with the distances that are listed in (6.32).

(c) What conclusions can you make about the possible location of the fault? Express your conclusions in comparison with the results in Example 6.6.

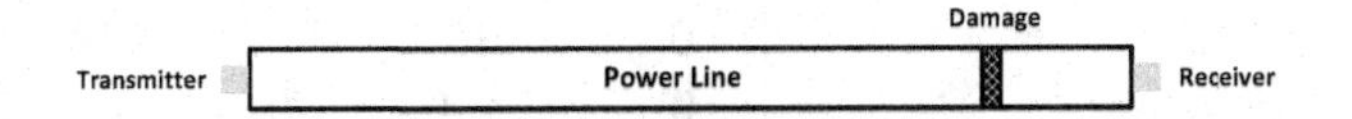

Figure 6.21 The damaged power cable in Exercise 6.10.

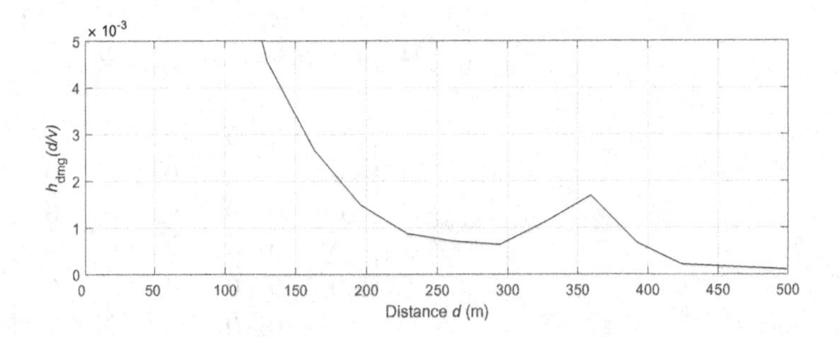

Figure 6.22 The damage impulse response of the power line cable in Exercise 6.11.

6.11 Figure 6.22 shows the damage impulse response of a damaged power line cable that is estimated by the PLC modem on the receiver side. The length of the cable is 1000 m. What is the distance of the location of the damage on the cable from the PLC modem on the *transmitter* side?

6.12 Consider a power distribution system with six buses. Bus 1 is the substation. The network topology is unknown but radial. There is one PLC device at each bus. The network is fully connected, and all PLC devices can talk to each other. The symmetric matrix of the TOF measurements is obtained as

$$\tau = \begin{bmatrix} 0 & 13.30 & 13.24 & 4.55 & 10.19 & 7.94 \\ 13.30 & 0 & 10.66 & 8.75 & 7.61 & 5.36 \\ 13.24 & 10.66 & 0 & 8.69 & 3.05 & 5.30 \\ 4.55 & 8.75 & 8.69 & 0 & 5.64 & 3.39 \\ 10.19 & 7.61 & 3.05 & 5.64 & 0 & 2.25 \\ 7.94 & 5.36 & 5.30 & 3.39 & 2.25 & 0 \end{bmatrix}. \tag{6.44}$$

Identify the topology of this power distribution network.

6.13 Repeat Exercise 6.12 for the matrix of TOF measurements that is provided in file `E6-13.csv`. The network has 10 buses. Bus 1 is the substation.

6.14 Suppose the TOF measurements are obtained separately on each phase of a three-phase power distribution system as follows:

$$\tau^{\text{Phase A}} = \begin{bmatrix} 0 & 13.30 & 0 & 4.55 & 0 & 7.94 \\ 13.30 & 0 & 0 & 8.76 & 0 & 5.36 \\ 0 & 0 & 0 & 0 & 0 & 0 \\ 4.55 & 8.76 & 0 & 0 & 0 & 3.39 \\ 0 & 0 & 0 & 0 & 0 & 0 \\ 7.94 & 5.36 & 0 & 3.39 & 0 & 0 \end{bmatrix}, \tag{6.45}$$

$$\tau^{\text{Phase B}} = \begin{bmatrix} 0 & 0 & 13.24 & 4.55 & 10.19 & 7.94 \\ 0 & 0 & 0 & 0 & 0 & 0 \\ 13.24 & 0 & 0 & 8.69 & 3.05 & 5.30 \\ 4.55 & 0 & 8.69 & 0 & 5.64 & 3.39 \\ 10.19 & 0 & 3.05 & 5.64 & 0 & 2.24 \\ 7.94 & 0 & 5.30 & 3.39 & 2.24 & 0 \end{bmatrix}, \qquad (6.46)$$

$$\tau^{\text{Phase C}} = \begin{bmatrix} 0 & 0 & 0 & 4.55 & 0 & 7.94 \\ 0 & 0 & 0 & 0 & 0 & 0 \\ 0 & 0 & 0 & 0 & 0 & 0 \\ 4.55 & 0 & 0 & 0 & 0 & 3.39 \\ 0 & 0 & 0 & 0 & 0 & 0 \\ 7.94 & 0 & 0 & 3.39 & 0 & 0 \end{bmatrix}. \qquad (6.47)$$

Identify which buses are on Phase A, Phase B, and Phase C.

7 Other Sensors and Off-Domain Measurements and Their Applications

The fundamentals of smart grid sensors and measurement-based applications were covered in Chapters 2–6. However, depending on the application, other sensors and measurements may also be used in the field of smart grids. Some of these sensors and measurements are discussed in this chapter. They range from electrical to mechanical and chemical measurements, to images and financial data. These additional data and measurements can be used as stand-alone data, or in cross examination with some of the measurements that we discussed in the previous chapters.

7.1 Device and Asset Sensors

A power grid is equipped with millions of transformers, capacitor banks, fuses, relays, switches, regulating devices, etc. The proliferation of renewable and distributed energy resources is also adding millions of solar panels, wind turbines, batteries, etc., that are interconnected to the power grid, whether they belong to the utility or to the customers. Keeping track of the *operation status* and *state of health* of these various pieces of equipment and assets is necessary in order to maintain grid efficiency, identify potential malfunctions, and forecast future issues or failures.

Many of the grid equipment and assets may have dedicated power system sensors to measure their voltage, current, and power. Such measurements can be studied using the same fundamental methods that we discussed in Chapters 2–6. Furthermore, some of these equipment and assets may have sensors that are specific to them and their unique characteristics and issues. In this section, we are particularly concerned with the sensors and measurements that may reveal *additional information* about specific types of grid equipment and assets, beyond what can be achieved by measuring only voltage, current, and power.

7.1.1 Transformers

Transformers are critical assets in power systems. Failure modes for transformers are related to the degradation of components such as the tap changer, bushings, windings, core, tank, and dielectric fluid, or to thermal aging of the insulating materials [428]. Advanced monitoring systems can be installed on transformers to measure voltage and current, oil temperature, ambient temperature, hot spot temperature, tap changer position, moisture, and the level of dissolved gas in the transformer's oil. These measurements are often processed on site, and the summary of results are reported to grid operation centers [429].

A common method to monitor the health of a transformer is *dissolved gas analysis* (DGA). In this method, the levels of dissolved gas contents of the transformer's oil are measured using *chemical sensors*, and if they exceed certain thresholds, then the transformer is flagged for maintenance.

The dissolved gas levels that are typically monitored include hydrogen, ethylene, methane, acetylene, ethane, and carbon monoxide. These chemical measurements can indicate not only the presence of a fault, but also the likely type of the fault. For instance, if the hydrogen and methane exceed their threshold, then the likely associated fault is arcing. If hydrogen and acetylene exceed their threshold, then the likely associated fault is partial discharge [430].

Example 7.1 The dissolved gas trends in parts per million (ppm) are shown in Figure 7.1 for a 700 MVA transformer over several years. Three events are marked. At Event ①, two earth faults occurred relatively close to this transformer. There was a step increase in the gas levels. It was observed that the production of the dissolved gases responded to increases and decreases in the loading. Therefore, the transformer loads were subsequently adjusted to allow it to be kept for operation at stable conditions for several years. Later, at around the point that is marked as ②, the gas levels started to elevate again. This time, the condition progressed quite steadily; which ultimately led to the transformer failure in Event ③; cf. [431].

Traditionally, dissolved gas levels are measured once every few months, and often manually. However, the more advanced transformer monitoring systems provide fast, continuous, and real-time monitoring of the dissolved gas levels and other on-site measurements, with communication capabilities to transfer the continuous stream of the measurements to the grid control centers. The DGA results can be reported frequently, such as once every hour. The reporting rates can be set to increase automatically when the gas concentrations exceed their thresholds; cf. [430]. Such real-time monitoring of

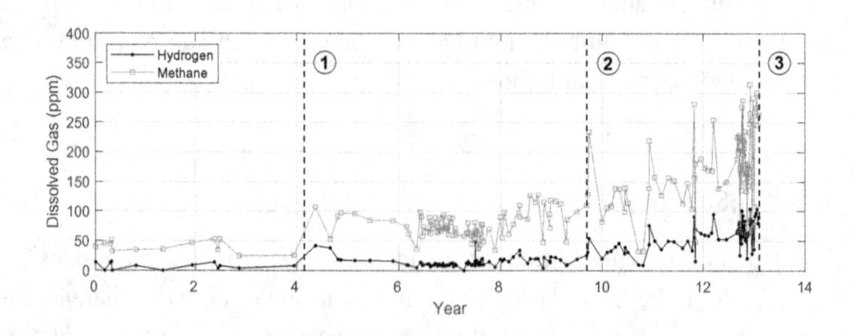

Figure 7.1 The dissolved gas trends for the transformer in Example 7.1 [431].

dissolved gas levels can help identify incipient issues before they cause catastrophic transformer failures.

The results from DGA can be used together with the power system measurements that we discussed in Chapters 2–5. For example, recall from Section 4.3.3 in Chapter 4 that a *zero-current event* that is captured by a waveform sensor may potentially indicate an incipient fault in a transformer tap changer. This issue can be *cross-examined* with the results from DGA in order to draw a more reliable conclusion with respect to the state of the health of the transformer.

7.1.2 Capacitor Banks

Capacitor banks are used in power systems for reactive power management and voltage support. Most capacitor banks in practice currently do not have dedicated asset sensors; however, it is becoming increasingly common to include built-in monitoring and communications capabilities in the new capacitor bank controller devices. Advanced capacitor bank monitoring systems measure voltage, current, power, frequency, power factor, harmonics, temperature, daily switching counts, etc. They can be set to trigger alarms in case of switching failures such as when the close counter is reached, blown fuses, reaching minimum or maximum voltage limits, reverse power, or when the target power factor is not achieved [432, 433].

However, smaller or older capacitor banks are usually *not* equipped with built-in monitoring systems; even though monitoring them could be of value. For example, the medium-voltage overhead capacitors that are widely used in power distribution systems are notorious for blowing fuses, which causes an imbalanced level of voltage and reactive power support; yet most of them do not have a built-in monitoring system. Therefore, utilities often need to do manual patrols and inspections to verify that their capacitors operate in proper working conditions. An alternative cost-effective solution is discussed in Example 7.2.

Example 7.2 Recall from Section 2.2 in Chapter 2 that most non-contact overhead line sensors *harvest power* from the conductor. Now, suppose a non-contact overhead line sensor is installed externally at a capacitor bank location to measure the neutral current i_n at the neutral terminal of the capacitor; see Figure 7.2. Initially, and during normal operation, the neutral current is zero (or very small) and the installed non-contact overhead line sensor is in *sleep mode*. It will remain in sleep mode until a blown fuse or multiple blown fuses cause considerable phase unbalance. At that point, the neutral current will start to harvest power. Shortly after, once the sensor is powered on, the sensor begins to report the RMS neutral current values at a pre-determined time interval. This alerts the utility that there is a recent issue at the capacitor and an investigation is initiated [434].

Note that, in Example 7.2, a generic non-contact line sensor was used *exclusively* to monitor the balanced operation of the capacitor.

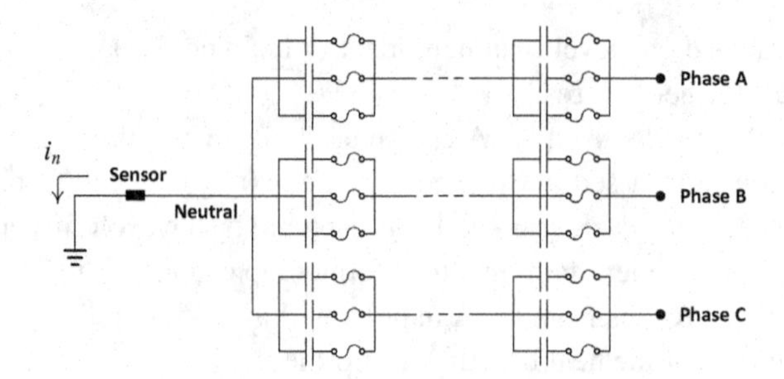

Figure 7.2 A capacitor bank that comprises several internal capacitor units. A power harvesting sensor is used at the neutral terminal to alert the utility about a major unbalanced operation.

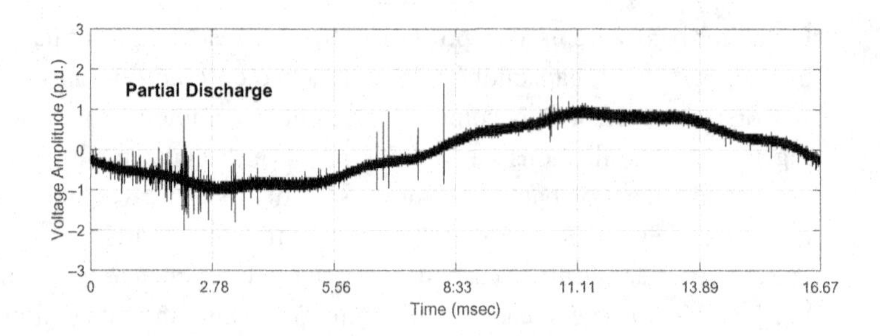

Figure 7.3 Voltage waveform that is captured on one phase by an asset sensor at a transmission line during a partial discharge event on the line conductor [260].

7.1.3 Line Conductors

Line conductors may also be monitored directly using dedicated asset sensors. For example, dedicated waveform sensors can be used in long transmission lines in order to monitor the health of the conductor and identify incipient failures such as *partial discharge*; see Figure 7.3. This type of incipient failure cannot be captured unless a dedicated sensor is used on site directly at the conductor.

Dynamic Line Rating

In recent years, overhead line sensors have been equipped to monitor not only voltage and current, but also some mechanical, thermal, and weather parameters such as conductor temperature, ambient temperature, wind speed, wind direction, solar irradiation, and line clearance. An example is shown in Figure 7.4. This additional information can be used in *dynamic line rating* (DLR).

Power flow on overhead transmission lines must be limited in order to keep the conductor temperature below the line's Maximum Allowable Conductor Temperature (MACT). This is necessary in order to maintain acceptable electrical clearances along the line and avoid excessive aging of the conductor system. Static line ratings (SLRs) equal the maximum line current for which the line conductor temperature is less than the MACT under *conservative weather assumptions*, which often overestimate the conductor temperature. For example, according to the International Council on

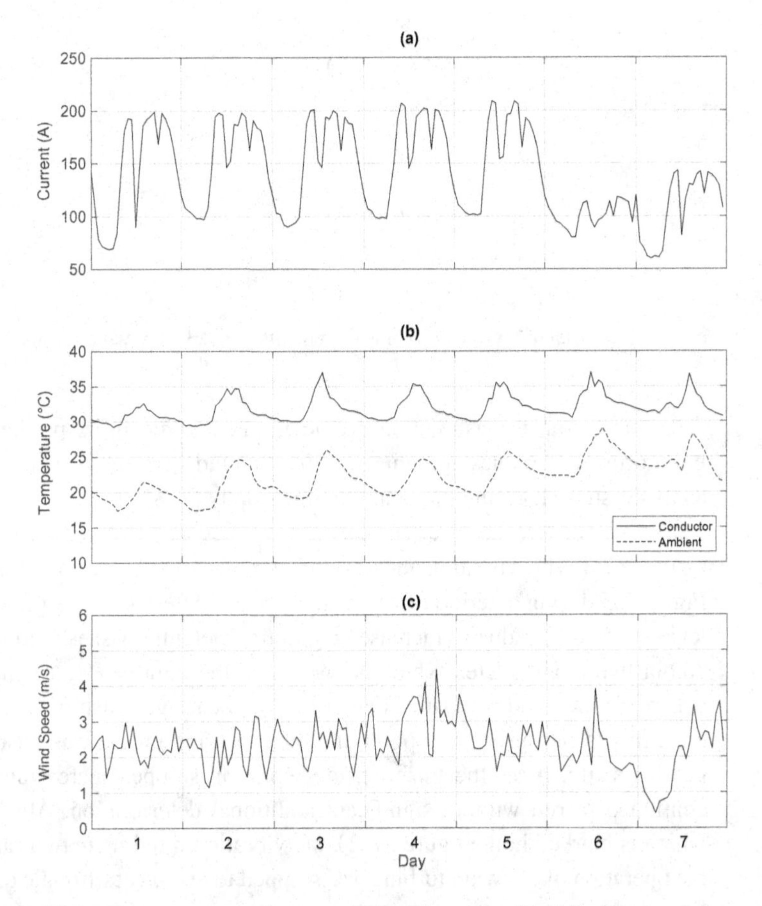

Figure 7.4 Various measurements that are provided by an overhead line sensor: (a) current; (b) ambient and conductor temperature; (c) wind speed [435].

Large Electric Systems (CIGRE), SLRs are calculated based on low-speed wind (e.g., 0.6 m/s) and therefore little cooling conditions, a seasonally high ambient temperature (e.g., 40°C), and full solar heating (e.g., 1000 W/m^2) [436].

DLRs too are equal to the line current for which the conductor temperature is less than the MACT; however, DLRs are calculated based on the *actual weather conditions*. Of course, since the weather conditions vary in time, DLRs are valid for only a limited time into the future, called the *thermal rating period*, such as the next one hour. DLRs are calculated by taking into account the actual ambient temperature, the actual wind speed, and the actual solar heating. Importantly, under most conditions, the DLR is higher than the SLR of a monitored line. Therefore, it can result in a *better utilization* of the existing power lines [437].

7.1.4 Wind Turbines

Condition monitoring and advanced prognostic systems are used in wind turbines to continuously monitor the health of various turbine components. The objective is to

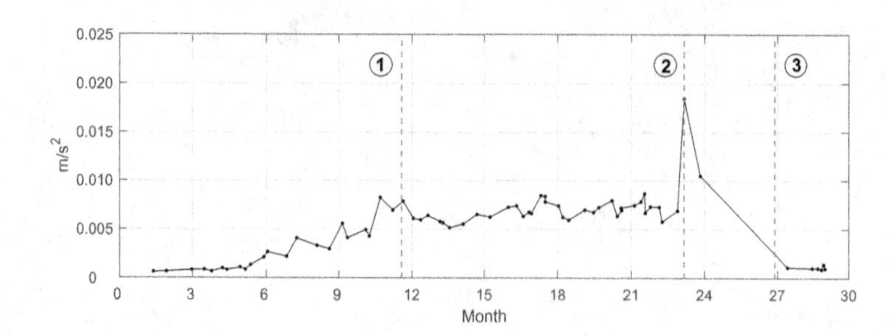

Figure 7.5 Vibration measurements at a 2.3 MW wind turbine [439].

detect faults early so as to minimize downtime and maximize productivity. Condition monitoring techniques for wind turbines include *vibration analysis*, acoustics, oil analysis, strain measurement, and thermography [438].

Example 7.3 The vibration measurements at a 2.3 MW wind turbine are shown in Figure 7.5 during a period of about three years. After about six months, the vibration level started to gradually increase, suggesting potential issues with the main bearing. About five months later, which is marked in the figure as ①, a physical inspection was performed and confirmed damage to the bearing, which in this case was a moderate macro-pitting. At this point, the main bearing grease was flushed to extend the bearing's life. After the flushing, the vibration stopped increasing and the bearing continued to run without significant additional deterioration. About one year later, which is marked in the figure as ②, the vibration trend increased rapidly. As a result the operation of the wind turbine was stopped to avoid catastrophic damage. The main bearing was replaced and the turbine restarted operation after about three months, which is marked in the figure as ③. Subsequently, the vibration trends went back to normal [439].

Condition monitoring can help conducting repairs and replacements only when needed in order to avoid unnecessary and costly up-tower jobs at wind turbines.

7.1.5 Solar Panels

Dust accumulation is one of the factors which negatively impact the PV module output in solar panels; because it obstructs solar radiation to the surface of the solar panels; thus reducing the overall performance of the solar power generation system [440]. Most solar PV systems are cleaned by rain or at scheduled cleaning services. Some solar PV systems are also equipped with an *automated cleaning system*, which activates the cleaning mechanism, for example, when the output power drops 20% below the average normal production; cf. [441]. All these factors have a major impact on the generation output of the solar panels.

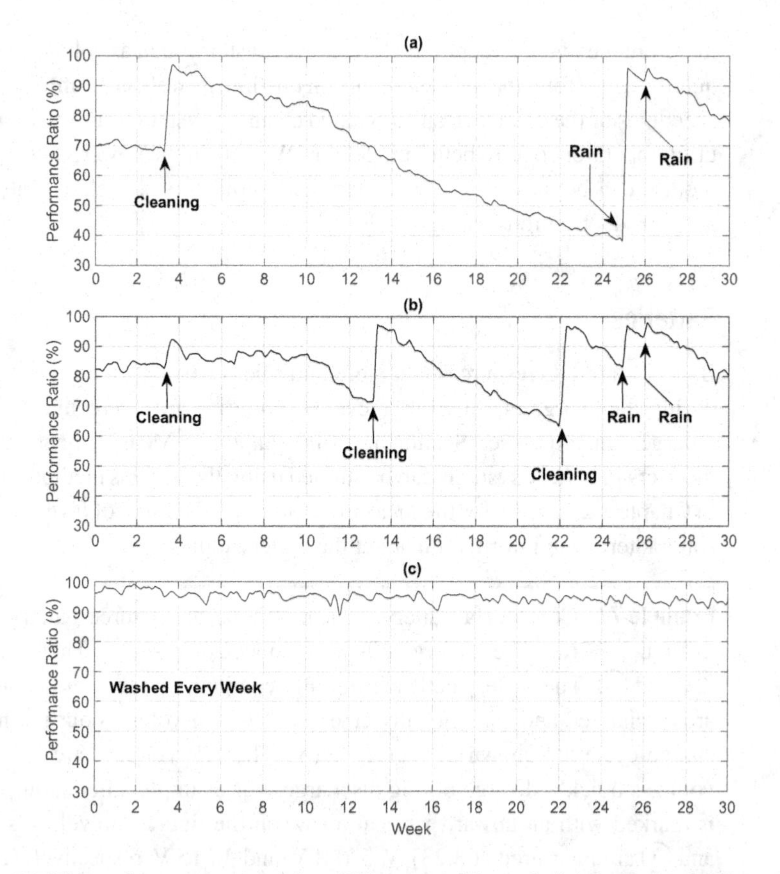

Figure 7.6 The performance ratio of three PV arrays in Example 7.4 under three different cleaning frequencies: (a) low wash; (b) medium wash; (c) high wash [442].

Example 7.4 The effect of dust and weather conditions on PV performance was investigated in [442]. The experiments were done in Doha, Qatar. Due to the desert environment, the PV panels are often affected by ambient dust. Three PV arrays were considered with different cleaning frequencies: *low wash*, *medium wash*, and *high wash*. The performance ratio was measured at the three PV panels over 30 weeks; see the measurements in Figure 7.6. For the case in Figure 7.6(a), the PV array was washed once every six months. For the case in Figure 7.6(b), the PV array was washed once every two months. For the case in Figure 7.6(c), the PV array was washed once every week. We can see that the profiles for the performance ratio vary drastically across these three PV arrays. In particular, the power production may drop to as low as only 40% of the normal production rate under low wash conditions; while the power production is almost always close to the normal production rate under high wash conditions.

Depending on the size and the penetration rate of the PV units, the impact of dusting and cleaning schedules can be significant on the production output of PV panels and therefore the overall operation of the power system, such as voltage profiles and power quality. Therefore, the measurements on dusting, as well as the information of cleaning schedules or the details on an automated cleaning system, could be of great value to the utility operator so as to better model the PV units in the power system. Such improved models can be used in various smart grid applications, such as state estimation and solar production forecasting.

7.1.6 Batteries

Batteries are DC resources that store and release energy in electrochemical reactions during the charge and discharge cycles. Batteries are interconnected to the grid via (charger) inverters, see Section 1.2.9 in Chapter 1. Measurements at the AC side of inverters in battery systems can be studied using the various methods that we discussed in Chapters 2 to 5. As for the measurements at the DC side of inverters, they can reveal some interesting information about the battery units.

Example 7.5 Consider a battery pack that consists of three battery cells, as shown in Figure 7.7(a). The battery cells are connected in series. They are labeled as ①, ②, and ③. The battery cells are initially charged, and they are balanced. Then they are discharged and subsequently charged. The cell voltages during the discharge and charge cycle are shown in Figure 7.7(b). The cell voltages are DC. Notice that the voltages quickly diverge during discharge. For example, after about 1.5 hours, which is marked with an upward vertical arrow on the figure, the voltages at cells ①, ②, and ③ are measured at 3.231 V, 3.198 V, and 3.215 V, respectively. At the end of the discharge and charge cycle, the voltages across the three cells are clearly unbalanced. Importantly, these battery cells belong to a *used* battery pack. They have aged differently over time; thus, their effective capacities are different, and hence they quickly lose balance in their state of charge after only one discharge and charge cycle. This condition results in an ineffective use of the battery cells. This battery pack needs to be reconfigured to increase the usable capacity of the battery cells; see [443].

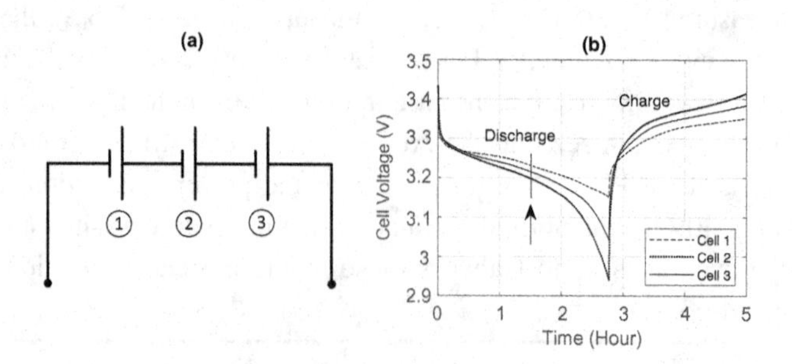

Figure 7.7 Measurements at a battery pack with three battery cells: (a) the connection of the battery cells; (b) cell voltage during a full discharge and charge cycle.

7.2 Building Sensors

Energy usage in buildings currently accounts for about 40% of the total energy usage in the United States [444]. Electricity is the largest source of energy in buildings, currently accounting for nearly half of the total energy usage in buildings. Given the enormous amount of electric power that is consumed in buildings, smart building technologies and building energy management systems play an important role in developing a smart grid [445–447].

Recall from Section 5.6.3 in Chapter 5 that sub-metering can help with monitoring individual appliances or other individual load components in buildings. In this section, we will discuss examples of sensors and measurements that are not directly related to power system measurements, but may reveal *additional information* about different load types in a building, beyond what can be achieved by sub-metering. Such additional information may help us better characterize different building loads and also help better control different building loads in order to improve building energy consumption efficiency.

7.2.1 Occupancy

Occupancy sensors are widely used in building energy management, e.g., to control lighting or air conditioning. Different occupancy sensors work based on different principles. Some sensors detect *movement*. Some sensors detect heat, i.e., they measure *infrared radiation* from the human body. Other types of occupancy sensors include *ultrasonic* sensors that send high frequency sound waves into the area and check for their reflected patterns; see the survey in [448].

The basic output of an occupancy sensor is a binary number; one means occupied and zero means not occupied. By recording the output of an occupancy sensor during the day, one can obtain the daily *occupancy pattern* in the area.

Example 7.6 Figure 7.8 shows occupancy measurements in different rooms in a house during a weekday. We can see that the occupancy patterns are very different across different rooms. The bedroom is occupied during the night and in early afternoon. The living room is occupied occasionally. The kitchen is occupied in the morning for breakfast and also in the evening around dinner time.

Occupancy measurements can be used to control lighting. The lights may *turn on* as soon as the room is detected as occupied; and the lights may *turn off* after a period of time that the room is detected as unoccupied. This simple mechanism can help reduce power consumption without affecting the building's occupants.

Occupancy measurements can be used also to control air conditioning. For example, during the summer, the cooling set point can be set to 76°F (24°C) when the room is occupied and 78°F (25°C) when the room is not occupied. This too can help reduce power consumption without affecting the occupants.

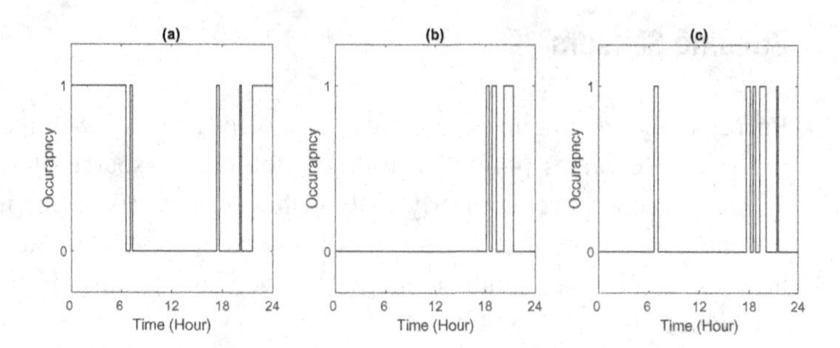

Figure 7.8 Readings from three occupancy sensors: (a) bedroom; (b) living room; (c) kitchen. One means occupancy is detected, while zero means no occupant.

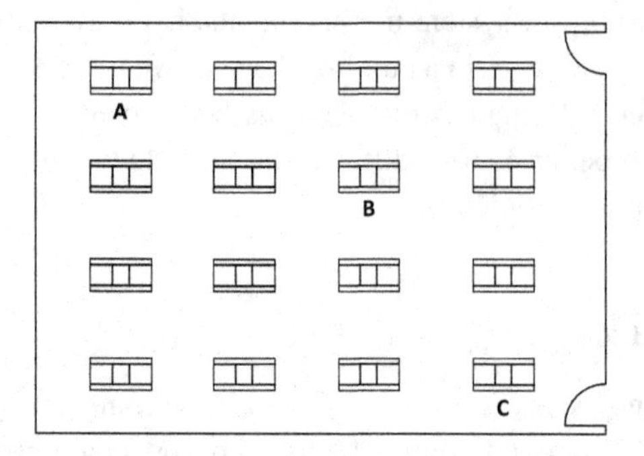

Figure 7.9 Layout of a classroom with 16 occupancy sensors in Example 7.7.

Zonal Occupancy Sensors

A single occupancy sensor might be sufficient for a small room, such as in a single-family house. However, there are advantages to use multiple occupancy sensors in large rooms, such as in offices and commercial buildings, including large classrooms, libraries, and warehouses. Using multiple occupancy sensors can help increase *spatial granularity* in sensing occupancy. This may help with fine-tuning *zonal control* in a single room; for both lighting and air conditioning. It can also help with understanding *zonal occupancy patterns*; because occupancy patterns may not necessarily be the same across different areas in the same room.

Example 7.7 Consider a classroom with the capacity of about 100 college students. A total of 16 light fixtures are installed in this classroom, as shown in Figure 7.9. Each lighting fixture has a built-in occupancy sensor of the type that is discussed in [449, 450]. Occupancy is reported once every five minutes. The daily occupancy patterns of three representative sensors, denoted by A, B, and C, are shown in Figure 7.10. We can see that, although the three sensors are in the same room, they detect considerably different occupancy patterns.

Figure 7.11 shows a monthly summary of the occupancy measurements in the classroom in Example 7.7. At least one sensor is activated, i.e., it detects occupancy,

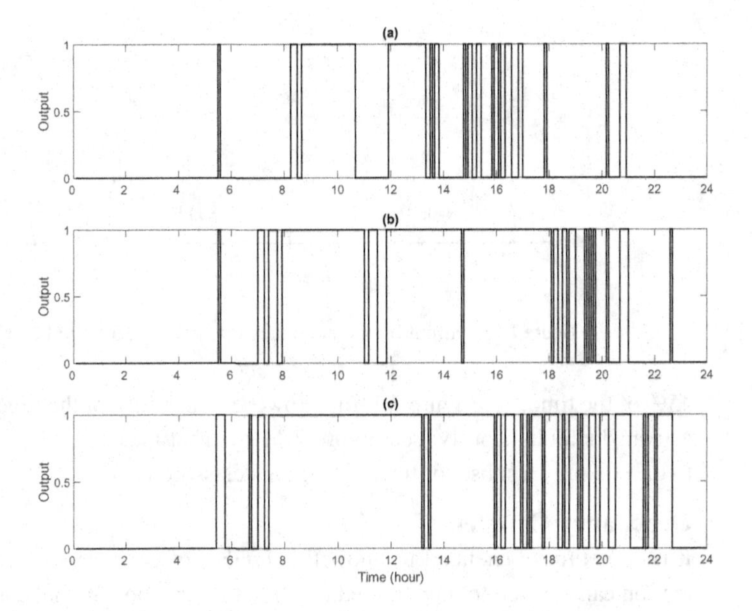

Figure 7.10 The daily occupancy patterns that are recorded by three representative sensors for the classroom in Example 7.7: (a) Sensor A; (b) Sensor B; (c) Sensor C.

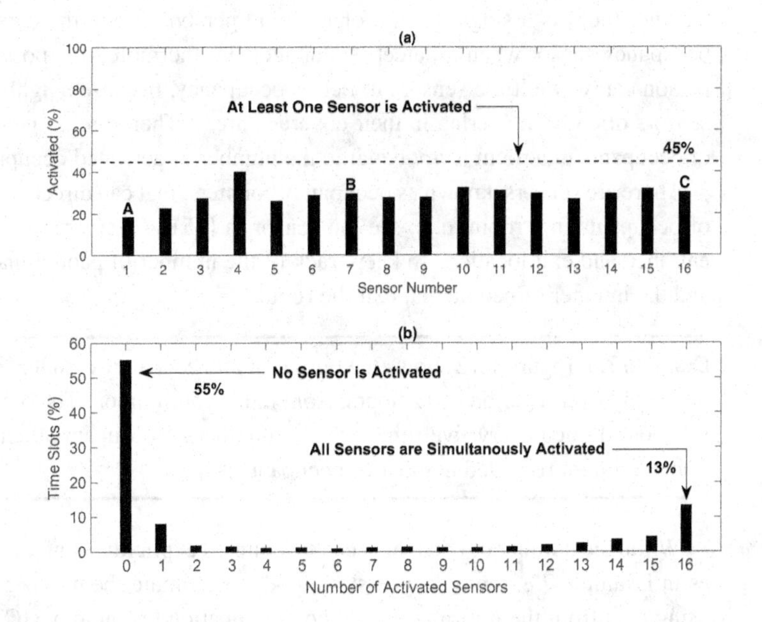

Figure 7.11 Analysis of one month of the occupancy measurements recorded by the 16 occupancy sensors in Example 7.7: (a) the percentage of the time that each sensor is activated; (b) the percentage of the time that a number of sensors are activated simultaneously.

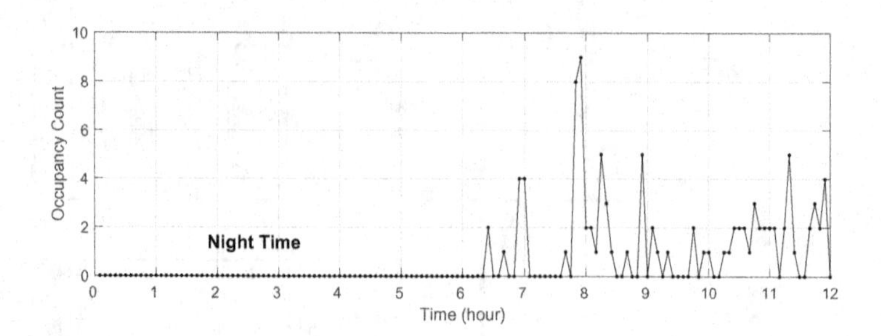

Figure 7.12 Output of an occupancy counter over a period of 12 hours.

45% of the time; see Figure 7.11(a). However, only 13% of the time are all 16 sensors activated simultaneously; see Figure 7.11(b). That means, 32% of the time, we need to turn on only a subset of the lights in this classroom.

Occupancy Counter

It is sometimes beneficial to know the *number* of occupants in a room. Such information can be used to model, predict, and minimize operational energy consumption, such as for air-conditioning systems. However, the number of occupants cannot be directly obtained from standard occupancy sensors. In fact, even if there are multiple occupancy sensors in a room, such as in Example 7.7, we still do not know whether there is a single person or multiple persons inside the coverage area of an occupancy sensor when it detects occupancy. Furthermore, it is possible that a single person causes multiple sensors to detect occupancy, because neighboring occupancy sensors often have overlap in their coverage areas. Therefore, in general, the number of occupants in a room is *not* equal to the number of activated occupancy sensors.

There are sensors, known as occupancy counters, that can directly count the number of occupants in a room; e.g., see the sensor in [451]. They are often installed at the entrance and exit locations to keep track of the number of people that enter the room and the number of people that exit the room.

Example 7.8 Figure 7.12 shows the output of an occupancy counter in a lab space in a university over a period of 12 hours, from midnight till noon. The output of the sensor is reported once every five minutes. The number of occupants fluctuates over time. The maximum recorded number of occupants is nine.

While the number of occupants can be counted explicitly using occupancy counters as in Example 7.8, there also exist methods to estimate the number of occupants by using data from the information and communication technology (ICT) systems, such as the number of WiFi connections; see [452].

7.2.2 Temperature and Illuminance

Measuring indoor temperature is widely used in building energy management systems to control *heating, ventilation, and air conditioning* (HVAC). The common practice is

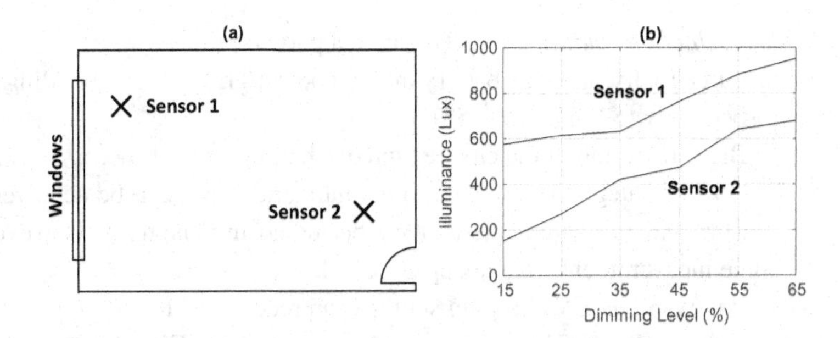

Figure 7.13 Illuminance measurements: (a) layout of the room; (b) illuminance at each sensor location versus the dimming level of the lights in the room [455].

to have one or only a few temperature sensors in each room. However, there are new technologies that can support spot temperate measurements at several locations in a room. For example, some lighting fixtures can measure not only spot occupancy (see Section 7.2.1), but also spot temperature [453]. Spot temperate measurements can be used to achieve *zonal control* in the HVAC system in order to improve building energy efficiency [338].

Illuminance can be measured in order to improve *daylight harvesting* in buildings. By measuring illuminance at different locations in a room, we can do *locational and time-of-day dimming control* at different lighting fixtures in each room [454]. The lighting fixtures at the locations that are closer to ambient light sources, such as windows and skylights, can be set to reduce their light and therefore reduce their power consumption, as we can see in Example 7.9.

Example 7.9 Consider a room as shown in Figure 7.13(a). The windows are in the back of the room. The room does not have skylight. Illuminance is measured in two locations, as marked on the figure. The illuminance measurements are done under different dimming levels of the ceiling lights in this room, as shown in Figure 7.13(b). The two locations in the room receive different illuminance flux per unit area, as measured in lux. At all dimming levels, the illuminance at Location 1, which is close to the windows, is significantly higher than the illuminance at Location 2, which is far from the windows. Therefore, even if the entire room is occupied, one can improve building energy efficiency by lowering the dimming level at the lighting fixtures that are closer to the windows.

7.2.3 Electric Vehicles

Supporting electric vehicles (EVs) is one of the objectives of developing a smart grid [8, 9]. On one hand, the growing number of EVs can significantly change the load on power distribution systems. This can require making upgrades in grid equipment and in the way that the power grid is operated [456]. On the other hand, the energy storage capacity of EVs can support new smart grid concepts such as *vehicle-to-grid* (V2G) and *vehicle-to-building* (V2B); where a parked plugged-in EV is treated as an energy resource to discharge its battery into the power grid or into a building's electric circuit, respectively [58–60, 457, 458].

Basic metering of the charge (and discharge) power is necessary in order to support the proper integration of EVs to a smart grid. This can be achieved by the kind of sensors and measurements that we discussed in Chapter 5. However, there are also some measurements that are specific to EVs.

For example, EV charging stations can record and report *plug-in time* and *plug-out time* for each EV. The former is the time when the EV is plugged in to the charging station; the latter is the time when the EV is plugged out, i.e., unplugged. Note that, plug-in time and plug-out time may not be obtained from power metering because an EV may not necessarily be charged (or discharged) for the entire time duration that it is parked and plugged in to the charging station.

Example 7.10 Consider a charging station at the parking lot of an office building. The charging station has four charging ports, numbered from 1 to 4. Table 7.1 shows the plug-in time, the plug-out time, and the total charged energy for all the EVs that are plugged into this charging station on a weekday. The first EV is plugged in at 7:51 AM, and the last EV is plugged out at 7:12 PM.

The measurements in Table 7.1 can be used to characterize and even forecast the load of the charging station; cf. [459, 460]. These measurements can be used also to identify which EVs might be able to participate in a demand response program as a *time-shiftable* load; see Section 5.4.1 in Chapter 5.

The EV measurements in Table 7.1 can be better visualized in the form of the diagram in Figure 7.14. Here, each EV is represented by a two-sided arrow. The arrow

Table 7.1 Measurements at an EV charging station with four charging ports.

Port	Plug-in Time	Plug-out Time	Energy (kW)
1	7 : 51	9 : 51	7.17
4	7 : 53	9 : 11	8.01
2	8 : 02	8 : 57	5.58
3	8 : 06	10 : 20	7.29
2	8 : 59	11 : 02	12.45
4	9 : 22	11 : 12	11.34
1	9 : 57	12 : 12	6.15
3	10 : 21	11 : 43	4.46
4	11 : 22	13 : 20	12.09
2	11 : 55	16 : 09	13.78
1	12 : 37	14 : 49	2.58
3	12 : 49	16 : 42	9.90
4	14 : 22	16 : 40	6.73
1	14 : 55	17 : 18	3.53
3	17 : 16	19 : 12	6.44

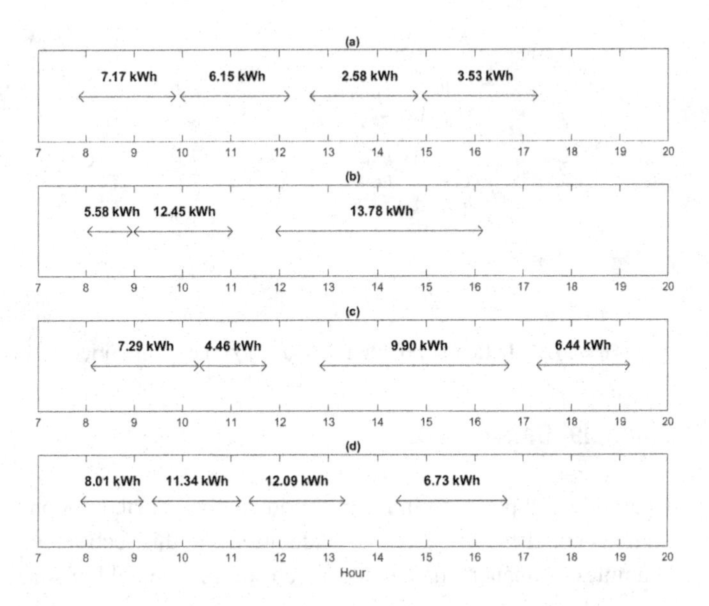

Figure 7.14 The visualization of the EV measurements in Table 7.1 at each port of the charging station: (a) Port 1; (b) Port 2; (c) Port 3; (d) Port 4. Each arrow indicates the time period that each EV was plugged in to the charging station. The number above the arrow indicates the total energy that was charged to the EV [88].

starts at the plug-in time and ends at the plug-out time. The number on each arrow is the charged energy. Each sub-figure is associated with one of the four charging ports at the charging station. Let us now compare the second arrow in Figure 7.14(b) with the third arrow in Figure 7.14(c). The former represents an EV that charges *more* energy (12.45 kWh) but stays at the charging station for a *shorter* period of time (123 minutes). The latter represents an EV that charges *less* energy (9.90 kWh) but stays at the charging station for a *longer* period of time (233 minutes). Accordingly, the latter arrow represents an EV that is more *flexible* in its charging load. Such flexibility can be used in order to shift the actual charging task to the best point in time within the time frame during which the EV is plugged in to the charging station.

Example 7.11 At the charging station in Example 7.10, consider the EV that was plugged in to Port 3 from 12:49 till 16:42. Suppose this EV is charged at 7.6 kW. Therefore, it takes 9.90 kWh / 7.6 kW = 1.3 hours, i.e., 1 hour and 18 minutes, to finish charging for this EV. Suppose the rate of electricity is 10.6 ¢/kWh from 10 AM till 3 PM; and 8.83 ¢/kWh from 3 PM till 6 PM. If the EV is charged right after it is plugged in, i.e., from 12:49 till 14:08, then the cost of charging it becomes $9.90 \times 10.6/100 =$ \$1.05. However, if the EV *shifts* its charging load to later in the afternoon, from 15:00 till 16:18, then the cost of charging becomes $9.90 \times 8.83/100 =$ \$0.87. Note that, in both cases, the charging of this EV is completed by the time that the EV departs the charging station.

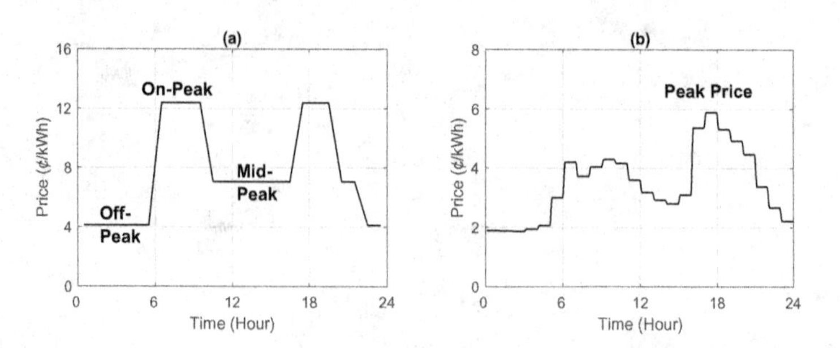

Figure 7.15 Different pricing methods: (a) time-of-use pricing; (b) real-time pricing.

7.3 Financial Data

Operation and planning in a smart grid can be affected not only by technical considerations but also by financial considerations. In this section, we overview some of the examples of financial data that can be useful in the field of smart grids.

7.3.1 Pricing and Billing

Pricing data are important for electricity customers so that they can take actions to minimize their electricity cost. Moreover, pricing is also a mechanism for utilities to influence the behavior of electricity customers, such as encouraging them to shift some of their major load to off-peak hours.

A wide range of pricing methods are currently used in practice. For example, we previously discussed *time-of-use* pricing in Section 5.4.1 in Chapter 5. Under ToU pricing, electricity is most expensive during on-peak hours and least expensive during off-peak hours; e.g., see Figure 7.15(a) [461]. The on-peak and off-peak hours are usually *fixed* for a whole season. Another example is *real-time pricing* (RTP). Under RTP, the price of electricity is more volatile, as it reflects the time-varying cost of generation more directly; see Figure 7.15(b) [281]. In fact, the prices under RTP are influenced by the prices in the wholesale electricity market. We will discuss wholesale electricity market prices in Section 7.3.2.

Other financial data include billing information. Traditionally, at each month, the electricity bill shows the customer's electricity usage and the breakdown of the cost based on the pricing method being used by the utility.

Example 7.12 Table 7.2 shows the monthly electricity bill for a commercial building. ToU pricing is used by the utility based on pre-determined on-peak hours, mid-peak hours, and off-peak hours. The customer is charged not only for its energy usage in kWh, but also for its peak power usage in kW.

In addition to serving the primary purpose of billing the customers, billing data could be used also for power system operation; such as in form of pseudo-measurements in distribution system state estimation; cf. [462, 463].

Table 7.2 Monthly electricity bill for a commercial building.

	On-Peak Hours	Mid-Peak Hours	Off-Peak Hours
Energy (kWh)	5952	13056	25632
Peak Power (kW)	76.8	84.5	73.9

If *smart meters* are installed, then they provide even more details about the energy usage of the consumers. See Section 5.4 in Chapter 5 for further discussion.

7.3.2 Electricity Market

Many regions in the United States and across the world operate competitive wholesale markets for electricity. The markets are often operated by Independent System Operators (ISOs), such as the California ISO [464].

In a wholesale electricity market, the prices for electricity change periodically, such as once every hour or even once every five minutes, according to factors such as demand, generation cost, and more recently the wind and solar generation levels.

Many ISOs manage electricity markets in two settlements, such as a day-ahead market (DAM) and a real-time market (RTM). In each settlement of the market, the ISO processes the *demand bids* and the *supply bids* that are submitted by the market participants, such as utilities and power plants, respectively; accordingly, the ISO determines the prices of the electricity. The price of electricity may vary depending on the location and the grid operation conditions.

Example 7.13 Figure 7.16 shows the changes in the price of electricity during a day in early May in the wholesale electricity market that is operated by California ISO. The DAM prices are determined once every hour, as shown in Figure 7.16(a). The prices are lower after midnight when the demand is low and higher in the evening when the demand is high. The RTM prices are determined once every five minutes, as shown in Figure 7.16(b). The RTM prices are much more volatile; since they directly reflect the minutely changes in the load and generation conditions. Occasionally, RTM prices become negative, as marked on the figure.

Both DAM prices and RTM prices are often determined by ISOs in the form of *locational marginal prices* (LMPs), which can be different at different buses across the power transmission system. For example, consider the three different buses that are marked on the map of California in Figure 7.17(a), namely Davis in the north, Huron in the center, and Chino in the south. The LMPs in the DAM during a day in July are shown in Figure 7.17(b). We can see that the prices are similar during the night and in early morning; however, they start diverging from each other during the afternoon because of the different conditions in the market at different locations, before they converge again to similar amounts later in the evening.

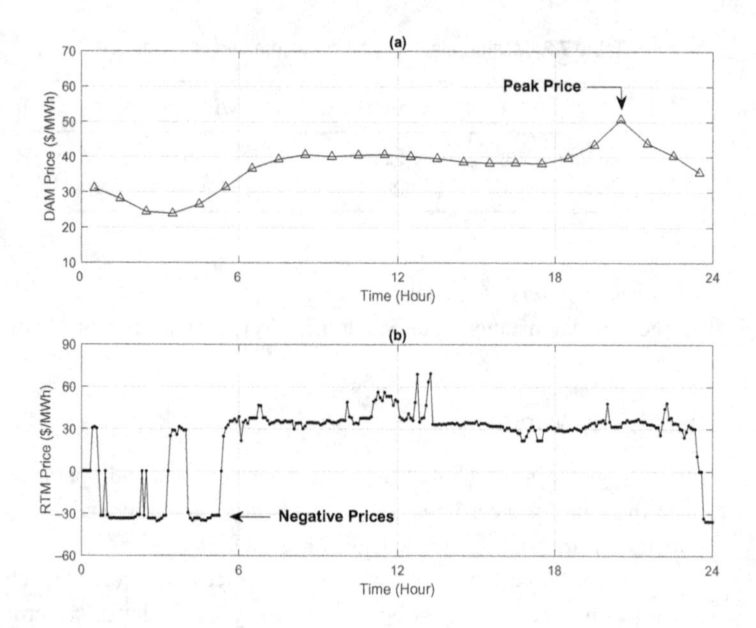

Figure 7.16 An example for the electricity prices in the California ISO two-settlement wholesale electricity market: (a) DAM prices; (b) RTM prices [465].

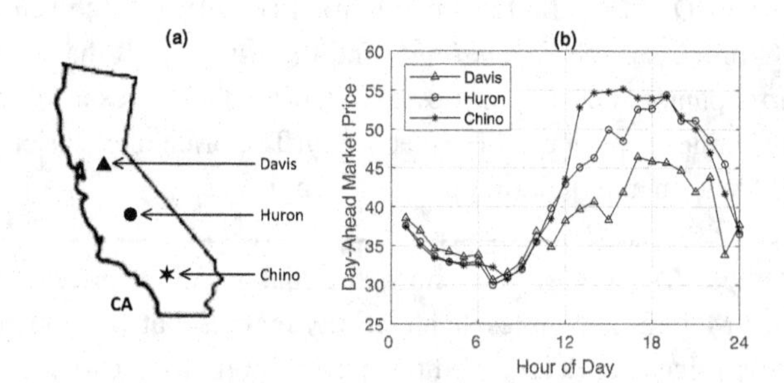

Figure 7.17 LMPs at different locations: (a) three market locations that are marked on the map of California; (b) LMPs in the DAM at those three locations [466].

7.4 Images, LIDARs, Drones, and Robots

The field of smart grid sensors is broad and growing rapidly. There are sensor technologies that are borrowed from other fields, but they have proved to be useful in the field of smart grids. In this section we briefly overview some of these technologies and their existing applications in smart grid monitoring.

7.4.1 Images

Different types of images may have applications in the field of smart grids. For example, satellite images can be used to monitor *clouds*, as shown in Figure 7.18(a). Cloud

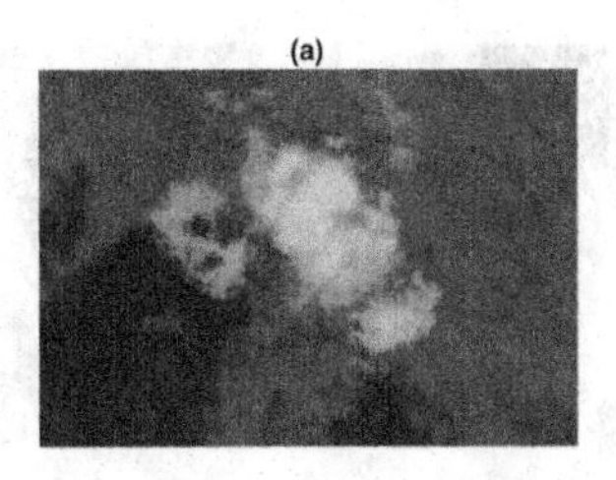

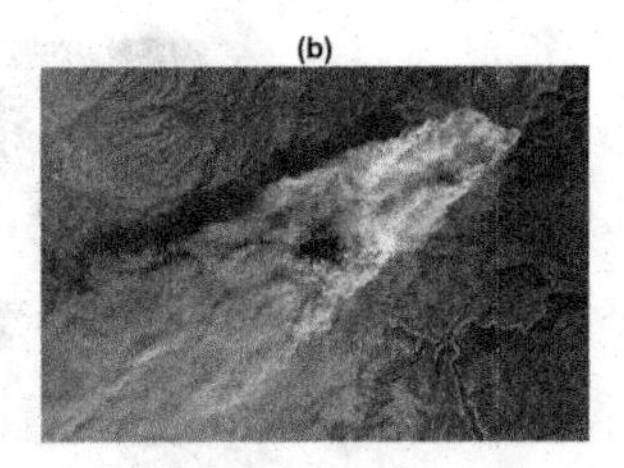

Figure 7.18 Satellite images: (a) cloud coverage [468]; (b) smoke from wildfire [469].

images can be used to enhance solar production forecasting [467]. By analyzing cloud movements in cloud images, one can significantly enhance *short-term forecasting* of the drop in solar production level that is caused at each PV location due to temporary cloud cover. Such analysis can help the grid operator to cope better with the intermittency in solar power generation.

Satellite images can also be used to detect and monitor wildfire, as shown in Figure 7.18(b). Recall from Example 4.15 in Chapter 4 that a wildfire near transmission lines and other grid assets can cause major disruptions with cascading impacts across the power system. By analyzing which transmission lines are likely to be impacted by within the next few minutes, the grid operator will have enough time to conduct proper grid reconfiguration, prevent unintended automated protection system triggering, or conduct other remedial actions.

Even if satellite images are not available, one can still use ground-based images as an alternative in certain applications. For example, *ground-based sky imaging* can also be used to analyze cloud coverage to enhance short-term forecasting at solar PV sites. An example for ground-based sky imaging is given in Exercise 7.14.

Given the low cost of ground-based sky imaging, there have been several studies in this area in recent years, including one on developing a network of cameras at different locations and facing different directions; see [470].

Other existing applications of images in the field of smart grids include asset health monitoring [471] and physical security monitoring [472].

7.4.2 LIDARs

Light Detection and Ranging (LiDAR), also known as LIDAR, is a remote sensing technology that uses laser beams to measure distance to various objects. LIDAR-derived images can be used to understand the surrounding environment; cf. [473].

LIDAR-derived imaging has been used in recent years to monitor transmission line corridors, to obtain images such as the one shown in Figure 7.19. LIDAR images can be used to detect different types of damage in poles and towers and line conductors, not only during routine inspections but also particularly after a storm or an earthquake or other natural disasters to quickly identify the damaged components.

LIDAR images can show not only the power line components but also the vegetation around the power lines. Therefore, LIDAR imaging can be used also in vegetation

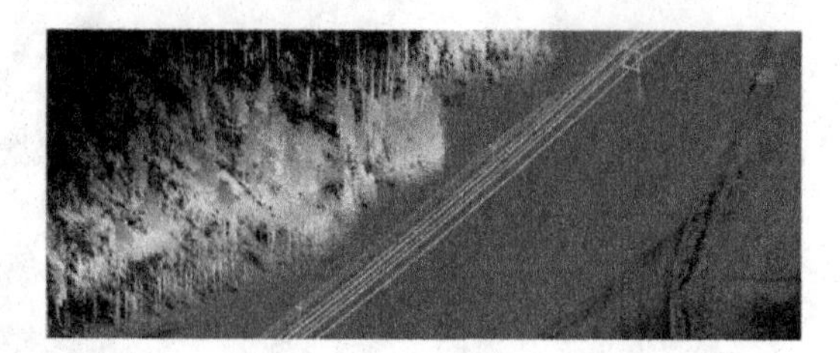

Figure 7.19 Airborne LIDAR-derived image of a transmission line corridor [475].

management. Recall from Section 4.3.2 in Chapter 4 that tree-contact is among the major causes of incipient faults in an overhead power line. LIDAR images can be used to determine the location of *alert trees* which require vegetation clearance, as explained in detail in [474].

7.4.3 Drones and Robots

The LIDAR-derived image that we saw in Figure 7.19 was obtained by an Unmanned Aerial Vehicle (UAV), also known as a *drone*. Drones can be used to monitor electrical infrastructure, such as power lines, by capturing regular aerial images [476], LIDAR-derived images [477], or even infrared thermal images [478].

While drones are used for aerial monitoring, other robotics technologies can be used, for example, to monitor underground cables; cf. [479]. Autonomous robotic technologies can also be used for maintenance of power system components [480, 481]. A summary of the recent advances in robotics applications in power systems is available in [482–484].

7.5 Other Off-Domain Measurements

Many smart grid technologies rely on different forms of off-domain measurements, i.e., the measurements that are *not* primarily intended for the power sector. For instance, weather data, data from the National Lightning Detection Network, and Geographic Information System (GIS) data are used to enhance power system operations at different levels and time scales; cf. [485, 486].

Ambient temperature data are a prominent example of the weather data that are used for electric load forecasting, due to its impact on cooling and heating loads [487]. Other types of weather data that are commonly used in smart grid applications include wind speed and solar irradiation, which are used for wind power integration and solar power integration, respectively.

Traffic data also has applications in the field of smart grids, such as for EV load forecasting or location selection and planning of EV charging stations [488–490]. Another example of off-domain measurements include data from *social media* that can be used to learn customer behavior or identify and engage customers in energy efficiency programs; see some examples in [491, 492].

There could be other forms of off-domain data that are yet to be explored to identify their relevance and usefulness in smart grid applications. There is virtually no limit on the possibilities of the intelligence that can be brought to operation and planning of power systems from all sources of data.

Exercises

7.1 Under what conditions can the sensor in Example 7.2 fail to identify blown fuses in capacitor units? How does this affect grid operation?

7.2 File `E7-2.csv` contains the current measurements and the conductor temperature measurements at a power line cable. It also contains the ambient temperature. Generate a scatter plot to show the *difference* between the conductor temperature and the ambient temperature versus the current.

7.3 File `E7-3.csv` contains the cell voltage measurements at a battery pack with three cells. Let V_1, V_2, and V_3 denote the voltage at the first, the second, and the third battery cells, respectively. Calculate and plot the following index to evaluate *cell imbalance* in this battery pack:

$$\frac{1}{\Gamma} \max_i |V_i - \Gamma| \times 100\%, \tag{7.1}$$

where Γ denotes the average voltage across the three battery cells.

7.4 File `E7-4.csv` contains the hourly wind speed measurements over a period of three weeks [493]. Answer the following questions:

(a) What is the minimum wind speed during this period?

(b) What is the maximum wind speed during this period?

(c) Plot the histogram for the wind speed measurements during this period.

(d) What is the probability that wind speed is between 6 m/s and 9 m/s?

(e) What is the probability that wind speed is between 3 m/s and 6 m/s?

(f) If wind speed in the present hour is between 6 m/s and 9 m/s, then what is the probability that wind speed in the next hour will be between 3 m/s and 6 m/s? Note that this is a *conditional* probability.

7.5 File `E7-5.csv` contains occupancy measurements in a room.

(a) In what percentage of the time is the room occupied?

(b) In what percentage of the time during off-peak hours (9:00 PM–8:00 AM), during mid-peak hours (8:00 AM–5:00 PM), and during on-peak hours (5:00 PM–9:00 PM) is the room occupied?

7.6 Consider the illuminance measurements that we previously discussed in Figure 7.13. What should be the dimming level of the lighting fixtures so that we have at least 400 lx of illuminance at each sensor location?

7.7 Suppose you park your EV at a charging station at work at 8:00 AM with an initial charge level at 1 kWh. The full charge capacity of your EV is 16 kWh.

Table 7.3 Price of electricity at the charging station in Exercise 7.7.

Hour	Price (¢/kWh)
8:00–9:00	13.0
9:00–10:00	13.9
10:00–11:00	12.5
11:00–12:00	14.1
12:00–13:00	13.3
13:00–14:00	14.8
14:00–15:00	15.1
15:00–16:00	15.3

You can charge your EV at 5 kW power rating at any time while you park. There are a 60% probability and a 40% probability that you will depart at 12:00 PM and 4:00 PM, respectively. You will know your departure time at 11:00 AM. The price of electricity is given in Table 7.3. Plan your charge schedule to minimize the expected value of the cost of fully charging your EV.

7.8 One month of driving and parking data is given for three EVs in file `E7-8.csv`. Zero means driving and one means parking [494].
(a) Calculate at what percentage of the time each EV is parked.
(b) Plot the histogram for the *duration* of parking events for each EV.

7.9 Consider an EV charging station with two ports. File `E7-9.csv` contains the record of all EVs that are charged at this charging station over a period of one month. Calculate the *blocking probability* at this charging station, i.e., the probability that both ports of the charging station are occupied.

7.10 In Exercise 7.9, calculate the total charging load of each port of the charging station on each day. Accordingly, plot their monthly charging load profile.

7.11 How much is the total cost of electricity for the monthly bill in Table 7.2 if the energy usage is charged at 10.33 ¢/kWh (on-peak), 8.82 ¢/kWh (mid-peak), and 7.27 ¢/kWh (off-peak); and the peak demand is charged at 6.88 $/kW (on-peak), 2.74 $/kW (mid-peak), and 1.31 $/kW (off-peak)?

7.12 File `E7-12.csv` contains the hourly average of real-time prices at a market location in a wholesale electricity market [465].
(a) At what percentage of the time is the real-time price negative?
(b) At what time of the day do the negative real-time prices usually occur?

7.13 File `E7-13.csv` contains the hourly day-ahead market prices at the same market location and during the same period of time as in Exercise 7.12. At what percentage of the time is the hourly day-ahead market price higher than the hourly average real-time market price?

7.14 Files `E7-14a.jpg` and `E7-14b.jpg` contain two sky images that are taken at a PV site. Either visually or by using an image processing tool, identify what percentage of each image is covered by clouds. Briefly explain how this type of analysis can be useful in smart grid applications.

References

[1] J. Momoh, *Smart Grid: Fundamentals of Design and Analysis*. Hoboken, NJ: John Wiley and Sons, 2012.

[2] S. F. Bush, *Smart Grid: Communication-Enabled Intelligence for the Electric Power Grid*. Hoboken, NJ: Wiley-IEEE, 2014.

[3] G. M. Masters, *Renewable and Efficient Electric Power Systems*, 2nd ed. Hoboken, NJ: John Wiley and Sons, 2013.

[4] B. M. Buchholz and Z. Styczynski, *Smart Grids–Fundamentals and Technologies in Electricity Networks*. Berlin, Germany: Springer, 2014.

[5] F. P. Sioshansi, *Smart Grid: Integrating Renewable, Distributed and Efficient Energy*. Waltham, MA: Elsevier, 2012.

[6] A. Keyhani, *Design of Smart Power Grid Renewable Energy Systems*, 2nd ed. Hoboken, NJ: John Wiley and Sons, 2017.

[7] K. C. Budka, J. G. Deshpande, and M. Thottan, *Communication Networks for Smart Grids*. London, UK: Springer, 2014.

[8] H. Farhangi, "The Path of the Smart Grid," *IEEE Power and Energy Society Magazine*, vol. 8, no. 1, pp. 18–28, Jan. 2010.

[9] A. Ipakchi and F. Albuyeh, "Grid of the Future," *IEEE Power and Energy Society Magazine*, vol. 7, no. 2, pp. 52–62, Mar. 2009.

[10] S. M. Amin and B. F. Wollenberg, "Toward a Smart Grid: Power Delivery for the 21st Century," *IEEE Power and Energy Society Magazine*, vol. 3, no. 5, pp. 34–41, Sep. 2005.

[11] H. Farhangi, "A Road Map to Integration: Perspectives on Smart Grid Development," *IEEE Power and Energy Society Magazine*, vol. 12, no. 3, pp. 52–66, May 2014.

[12] A. R. Bergen and V. Vittal, *Power System Analysis*, 2nd ed. Upper Saddle River, NJ: Prentice Hall, 2000.

[13] J. D. Glover, T. Overbye, and M. S. Sarma, *Power System Analysis and Design*, 6th ed. Boston, MA: Cengage Learning, 2016.

[14] J. Momoh, *Electric Power Distribution, Automation, Protection, and Control*. Boca Raton, FL: CRC Press, 2008.

[15] A. J. Wood, B. F. Wollenberg, and G. B. Sheblé, *Power Generation, Operation, and Control*, 3rd ed. New York, NY: John Wiley and Sons, 2013.

[16] U.S. Department of Energy, "What Is the Smart Grid?" www.smartgrid.gov/the_smart_grid/.

[17] U.S. Department of Energy, "The Smart Grid: An Introduction," Sep. 2008. www.energy.gov/oe/downloads/smart-grid-introduction-0.

[18] European Commission, "Smart Grids," Feb. 2020. https://s3platform.jrc.ec.europa.eu/smart-grids.

[19] European Commission, "Smart Grids and Meters," Apr. 2021. https://ec.europa.eu/energy/topics/markets-and-consumers/smart-grids-and-meters/overview_en.

[20] The Chinese Central Government's Official Web Portal, "State Grid: Complete a Unified Strong Smart Grid in 2020," May 2009. www.gov.cn/jrzg/2009-05/21/content_1321530.htm.

[21] IEEE Smart Grid, "What Is the Smart Grid?" 2015. https://smartgrid.ieee.org/about-ieee-smart-grid.

[22] National Institute of Standards and Technology, "Smart Grid," 2021. https://www.nist.gov/el/smart-grid.

[23] Electric Power Research Institute, "Smart Grid Resource Center," 2020. https://smartgrid.epri.com/.

[24] The Institution of Engineering and Technology, "What is a Smart Grid?" 2013. www.theiet.org/media/1251/smart-grids.pdf.

[25] United States Congress, "H.R. 6 Energy Independence and Security Act of 2007," Jan. 2017.www.congress.gov/bill/110th-congress/house-bill/6.

[26] U.S. Department of Energy, "Large Power Transformers and the U.S. Electric Grid," Apr. 2014. www.energy.gov/sites/prod/files/2014/04/f15/LPTStudyUpdate-040914.pdf.

[27] L. Lamarre, "Problems with Power Quality," *EPRI Journal*, pp. 14–23, Jul. 1991.

[28] American National Standards Institute, "ANSI C84.1-2016: American National Standard for Electric Power Systems and Equipment–Voltage Ratings (60 Hz)." New York, NY: American National Standards Institute, 2016.

[29] Y. Zhang, N. Rahbari-Asr, J. Duan, and M.-Y. Chow, "Day-Ahead Smart Grid Cooperative Distributed Energy Scheduling with Renewable and Storage Integration," *IEEE Transactions on Sustainable Energy*, vol. 7, no. 4, pp. 1739–1748, Oct. 2016.

[30] A. Giani, E. Bitar, M. Garcia, M. McQueen, P. Khargonekar, and K. Poolla, "Smart Grid Data Integrity Attacks," *IEEE Transactions on Smart Grid*, vol. 4, no. 3, pp. 1244–1253, Sep. 2013.

[31] H. Mohsenian-Rad, "Coordinated Price-Maker Operation of Large Energy Storage Systems in Nodal Energy Markets," *IEEE Transactions on Power Systems*, vol. 31, no. 1, pp. 786–797, Jan. 2016.

[32] A. Garces, "A Linear Three-Phase Load Flow for Power Distribution Systems," *IEEE Transactions on Power Systems*, vol. 31, no. 1, pp. 827–828, Jan. 2016.

[33] M. E. Baran and F. F. Wu, "Network Reconfiguration in Distribution Systems for Loss Reduction and Load Balancing," *IEEE Transactions on Power Delivery*, vol. 4, no. 2, pp. 1401–1407, Apr. 1989.

[34] MathWorks, "Help Center–solve." www.mathworks.com/help/optim/ug/fsolve.html.

[35] International Energy Agency, "Electricity Statistics," Dec. 2019. www.iea.org/data-and-statistics/data-product/monthly-electricity-statistics.

[36] California Energy Commission, "2018 Total System Electric Generation in Gigawatt Hours," www.energy.ca.gov/data-reports/energy-almanac/california-electricity-data/2019-total-system-electric-generation/2018.

[37] J. Twidell and T. Weir, *Renewable Energy Resources*. New York, NY: E & FN Spon, 2015.

[38] S. Bhattacharya, *Design of Foundations for Offshore Wind Turbines*. Hoboken, NJ: Wiley, Apr. 2019.

[39] Power Technology, "Alta Wind Energy Center (AWEC), California." www.power-technology.com/projects/alta-wind-energy-center-awec-california/.

[40] Sun Power Inc., "Solar Star Projects Fact Sheet," https://us.sunpower.com/sites/default/files/cs-solar-star-projects-fact-sheet_0.pdf.

[41] J. Driesen and F. Katiraei, "Design for Distributed Energy Resources," *IEEE Power and Energy Magazine*, vol. 6, no. 3, pp. 30–40, May 2008.

[42] X. Han, K. Heussen, O. Gehrke, H. W. Bindner, and B. Kroposki, "Taxonomy for Evaluation of Distributed Control Strategies for Distributed Energy Resources," *IEEE Transactions on Smart Grid*, vol. 9, no. 5, pp. 5185–5195, Sep. 2018.

[43] P. Basak, S. Chowdhury, S. Nee, and D. S. Chowdhury, "A Literature Review on Integration of Distributed Energy Resources in the Perspective of Control, Protection and Stability of Microgrid," *Renewable and Sustainable Energy Reviews*, vol. 16, no. 8, pp. 5545–5556, Oct. 2012.

[44] N. Hatziargyriou, H. Asano, R. Iravani, and C. Marnay, "Microgrids," *IEEE Power and Energy Magazine*, vol. 5, no. 4, pp. 78–94, Jul. 2007.

[45] F. Katiraei, R. Iravani, N. Hatziargyrious, and S. Amires, "Microgrids: Control and Management," *IEEE Power and Energy Magazine*, vol. 6, no. 3, pp. 74–98, May 2008.

[46] B. Kroposki, R. Lasseter, T. Ise, S. Morozumi, S. Papathanassiou, and N. Hatziargyriou, "Making Microgrids Work," *IEEE Power and Energy Magazine*, vol. 6, no. 3, pp. 40–53, May 2008.

[47] N. Hatziargyriou, *Microgrids: Architectures and Control.* Chichester, UK: Wiley-IEEE Press, 2014.

[48] U. A. Bakshi and A. V. Bakshi, *Generation, Transmission, and Distribution.* Pune, India: Technical Publications, 2009.

[49] Riverside Public Utilities, "Schedule A: General Service and Schedule D: Domestic Service," www.riversideca.gov/utilities/businesses/rates-electric.asp.

[50] Y. Liu, B. Qiu, X. Fan, H. Zhu, and B. Han, "Review of Smart Home Energy Management Systems," *Energy Procedia*, vol. 104, pp. 504–508, Dec. 2016.

[51] S. Althaher, P. Mancarella, and J. Mutale, "Automated Demand Response from Home Energy Management System under Dynamic Pricing and Power and Comfort Constraints," *IEEE Transactions on Smart Grid*, vol. 6, no. 4, pp. 1874–1883, Jul. 2015.

[52] W. Tushar, N. Wijerathne, W.-T. Li, C. Yuen, H. V. Poor, T. K. Saha, and K. L. Wood, "Internet of Things for Green Building Management: Disruptive Innovations through Low-Cost Sensor Technology and Artificial Intelligence," *IEEE Signal Processing Magazine*, vol. 35, no. 5, pp. 100–110, Sep. 2018.

[53] M. Yu, S. H. Hong, and J. B. Kim, "Incentive-Based Demand Response Approach for Aggregated Demand Side Participation," in *Proceedings of the IEEE International Conference on Smart Grid Communications*, Sydney, NSW, Australia, Nov. 2016.

[54] H. Zhong, L. Xie, and Q. Xia, "Coupon Incentive-Based Demand Response: Theory and Case Study," *IEEE Transactions on Power Systems*, vol. 28, no. 2, pp. 1266–1276, May 2013.

[55] M. R. Sarker, M. A. Ortega-Vazquez, and D. S. Kirschen, "Optimal Coordination and Scheduling of Demand Response via Monetary Incentives," *IEEE Transactions on Smart Grid*, vol. 6, no. 3, pp. 1341–1352, May 2015.

[56] H. Mohsenian-Rad, V. Wong, J. Jatskevich, R. Schober, and A. Leon-Garcia, "Autonomous Demand-Side Management Based on Game Theoretic Energy Consumption

Scheduling for the Future Smart Grid," *IEEE Transactions on Smart Grid*, vol. 1, no. 3, pp. 320–331, Nov. 2010.

[57] J. Gorzelany, "Electric Vehicle Battery Basics." https://www.myev.com/research/ev-101/electric-vehicle-battery-basics.

[58] Y. Ota, H. Taniguchi, T. Nakajima, K. M. Liyanage, J. Baba, and A. Yokoyama, "Autonomous Distributed V2G (Vehicle-to-Grid) Considering Charging Request and Battery Condition," in *Proceedings of the IEEE PES Innovative Smart Grid Technologies Conference*, Gothenberg, Sweden, Oct. 2009.

[59] Y. Ota, H. Taniguchi, T. Nakajima, K. M. Liyanage, J. Baba, and A. Yokoyama, "Autonomous Distributed V2G (Vehicle-to-Grid) Satisfying Scheduled Charging," *IEEE Transactions on Smart Grid*, vol. 3, no. 1, pp. 559–564, Mar. 2012.

[60] K. Tanguy, M. R. Dubois, K. L. Lopez, and C. Gagné, "Optimization Model and Economic Assessment of Collaborative Charging Using Vehicle-to-Building," *Sustainable Cities and Society*, vol. 26, pp. 496–506, Oct. 2016.

[61] California Energy Commission, "Energy Storage," Aug. 2018. www.energy.ca.gov/sites/default/files/2019-12/energy_storage_ada.pdf.

[62] S. Rehmana, L. M. Al-Hadhramia, and M. M. Alam, "Pumped Hydro Energy Storage System: A Technological Review," *Renewable and Sustainable Energy Reviews*, vol. 44, pp. 586–598, Apr. 2015.

[63] G. Venkataramani, P. Parankusam, V. Ramalingam, and J. Wang, "A Review on Compressed Air Energy Storage – A Pathway for Smart Grid and Polygeneration," *Renewable and Sustainable Energy Reviews*, vol. 62, pp. 895–907, Sep. 2016.

[64] S. M. Mousavi, F. Faraji, A. Majazi, and K. Al-Haddad, "A Comprehensive Review of Flywheel Energy Storage System Technology," *Renewable and Sustainable Energy Reviews*, vol. 67, pp. 477–490, Jan. 2017.

[65] A. Arteconi, N. J. Hewitt, and F. Polonara, "Domestic Demand-Side Management (DSM): Role of Heat Pumps and Thermal Energy Storage (TES) Systems," *Applied Thermal Engineering*, vol. 51, no. 1, pp. 155–165, Mar. 2013.

[66] F. Sehar, S. Rahman, and M. Pipattanasomporn, "Impacts of Ice Storage on Electrical Energy Consumptions in Office Buildings," *Energy and Buildings*, vol. 51, pp. 255–262, Aug. 2012.

[67] A. R. Sparacino, G. F. Reed, R. J. Kerestes, B. M. Grainger, and Z. T. Smith, "Survey of Battery Energy Storage Systems and Modeling Techniques," in *Proceedings of the IEEE Power and Energy Society General Meeting*, San Diego, CA, Jul. 2012.

[68] M. Abdel-Monem, O. Hegazy, N. Omar, K. Trad, P. V. den Bossche, and J. V. Mierlo, "Lithium-Ion Batteries: Comprehensive Technical Analysis of Second-Life Batteries for Smart Grid Applications," in *Proceedings of the European Conference on Power Electronics and Applications*, Warsaw, Poland, sep 2017.

[69] Z. Taylor, H. Akhavan-Hejazi, E. Cortez, L. Alvarez, S. Ula, M. Barth, and H. Mohsenian-Rad, "Customer-side SCADA-Assisted Large Battery Operation Optimization for Distribution Feeder Peak Load Shaving," *IEEE Transactions on Smart Grid*, vol. 10, no. 1, pp. 992–1004, Jan. 2019.

[70] W. Jing, C. H. Lai, S. H. W. Wong, and M. L. D. Wong, "Supercapacitor Hybrid Energy Storage System in Standalone DC Microgrids: A Review," *IET Renewable Power Generation*, vol. 11, no. 4, pp. 461–469, Mar. 2017.

[71] Z. Taylor, H. Akhavan-Hejazi, and H. Mohsenian-Rad, "Optimal Operation of Grid-Tied Energy Storage Systems Considering Detailed Device-Level Battery Models," *IEEE Transactions on Industrial Informatics*, vol. 16, no. 6, pp. 3928–3941, Jun. 2020.

[72] H. Akhavan-Hejazi, Z. Taylor, and H. Mohsenian-Rad, "Optimal Cell Removal to Enhance Operation of Aged Grid-Tied Battery Storage Systems," *IEEE Transactions on Sustainable Energy*, vol. 12, no. 1, pp. 739–742, Jan. 2021.

[73] A. Sadeghi-Mobarakeh and H. Mohsenian-Rad, "Performance Accuracy Scores in CAISO and MISO Regulation Markets: A Comparison based on Real Data and Mathematical Analysis," *IEEE Transactions on Power Systems*, vol. 33, no. 3, pp. 3196–3198, May 2018.

[74] U. A. Bakshi and A. V. Bakshi, *Electronic and Electrical Measuring Instruments and Machines*. Pune, India: Technical Publications, 2009.

[75] V. Lackovic, "Voltage Transformers," Woodcliff Lake, NJ. www.cedengineering.com/userfiles/Voltage%20Transformers-R1.pdf.

[76] Lindsey Real Time Transmission Conductor Monitor, http://lindsey-usa.com,Mar. 2017.

[77] W. Houschild and E. Lemke, *High Voltage Test and Measuring Techniques*. Springer, 2013.

[78] W. R. Smith-Vaniz and R. L. Sieron, "Apparatus for Measuring the Potential of a Transmission Line Conductor," U.S. Patent Number 4,714,893.

[79] T. W. Cease and P. Johnston, "A Magneto-Optic Current Transducer," *IEEE Transactions on Power Delivery*, vol. 5, pp. 548–555, Apr. 1990.

[80] K. Bohnert, P. Gabus, and H. Brandle, "Fiber-Optic Current and Voltage Sensors for High-Voltage Substations," in *Proceedings of the IEEE International Conference on Optical Fiber Sensors*, Oct. 2003.

[81] Sentient Energy, "Product Brochures: MM3 Line Monitor," www.sentient-energy.com/products/mm3-intelligent-sensor.

[82] R. J. Marks, *Introduction to Shannon Sampling and Interpolation Theory*. Springer, 1991.

[83] S. Ramet, "A Low-Distortion Anti-Aliasing/Smoothing Filter for Sampled Data Integrated Circuits," *IEEE Journal of Solid-State Circuits*, vol. 23, no. 5, pp. 1267–1272, May 1988.

[84] A. Miler and M. Dewe, "The Application of Multi-Rate Digital Signal Processing Techniques to the Measurement of Power System Harmonic Levels," *IEEE Transactions on Power Delivery*, vol. 8, no. 2, pp. 531–539, Feb. 1993.

[85] J. Castelló, J. M. Espí, and R. García-Gil, "A New Generalized Robust Predictive Current Control for Grid-Connected Inverters Compensates Anti-Aliasing Filters Delay," *IEEE Transactions on Industrial Electronics*, vol. 63, no. 7, pp. 4485–4494, Jul. 2016.

[86] R. Martins, J. Franca, and F. Maloberti, "An Optimum CMOS Switched-Capacitor Antialiasing Decimating Filter," *IEEE Journal of Solid-State Circuits*, vol. 28, no. 9, pp. 962–970, Sep. 1993.

[87] A. Shahsavari, M. Farajollahi, E. Stewart, C. Roberts, and H. Mohsenian-Rad, "A Data-Driven Analysis of Lightning-Initiated Contingencies at a Distribution Grid with a PV Farm Using Micro-PMU Data," in *Proceedings of the IEEE North American Power Symposium*, Sep. 2017.

[88] Center for Environmental Research and Technology, "Sustainable Integrated Grid Initiative." www.cert.ucr.edu/laboratoryservices/sustainable-integrated-grid-initiative.

[89] H. Mohsenian-Rad and E. Cortez, "Smart Grid for Smart City Activities in the California City of Fiverside," in *Proceedings of the International Conference on Smart Grid for Smart Cities*, Toronto, ON, Oct. 2015.

[90] Z. Taylor, H. Akhavan-Hejazi, E. Cortez, L. Alvarez, S. Ula, M. Barth, and H. Mohsenian-Rad, "Customer-Side SCADA-Assisted Large Battery Operation Optimization for Distribution Feeder Peak Load Shaving," *IEEE Transactions on Smart Grid*, vol. 10, no. 1, pp. 992–1004, Jan. 2019.

[91] D. W. Scott, *Multivariate Density Estimation: Theory, Practice, and Visualization, Second Edition Histograms: Theory and Practice*, 2nd ed. Hoboken, NJ: John Wiley & Sons Inc., 2015.

[92] A. Shahsavari, M. Farajollahi, E. Stewart, C. Roberts, F. Megala, L. Alvarez, E. Cortez, and H. Mohsenian-Rad, "Autopsy on Active Distribution Networks: A Data-Driven Fault Analysis Using Micro-PMU Data," in *Proceedings of the IEEE North American Power Symposium*, Sep. 2017.

[93] P. Khaledian, A. Aligholian, and H. Mohsenian-Rad, "Event-Based Analysis of Solar Power Distribution Feeder Using Micro-PMU Measurements," in *Proceedings of the IEEE PES Conference on Innovative Smart Grid Technologies Conference (ISGT)*, Washington, DC, Feb. 2021.

[94] B. Pal and B. Chaudhuri, *Robust Control in Power Systems*. New York, NY: Springer, 2005.

[95] J. F. Hauer, C. J. Demeure, and L. L. Scharf, "Initial Results in Prony Analysis of Power System Response Signals," *IEEE Transactions on Power Systems*, vol. 5, no. 1, pp. 80–89, Feb. 1990.

[96] B. Pierre, R. Elliott, D. Schoenwald, J. Neely, R. Byrne, D. Trudnowski, and J. Colwell, "Supervisory System for a Wide Area Damping Controller Using PDCI Modulation and Real-Time PMU Feedback," in *Proceedings of the IEEE Power and Energy Society General Meeting*, Bostob, MA, Jul. 2016.

[97] D. Trudnowski, "Properties of the Dominant Inter-Area Modes in the WECC Interconnect," Jan. 2012. www.wecc.org/Reliability/WECCmodesPaper130113Trudnowski.pdf.

[98] B. J. Pierre, F. Wilches-Bernal, D. A. Schoenwald, R. T. Elliott, J. C. Neely, R. H. Byrne, and D. Trudnowski, "Open-Loop Testing Results for the Pacific DC Intertie Wide Area Damping Controller," in *Proceedings of the IEEE Manchester PowerTech*, Manchester, UK, Jun. 2017.

[99] E. Jury, *Theory and Application of the Z-transform Method*. Huntington, NY: Krieger Publishing, 1964.

[100] A. D. Poularikas, "The Z-Transform," in *The Transforms and Applications Handbook*, A. D. Poularikas, Ed. Boca Raton, FL: CRC Press, 2000.

[101] J. F. Hauer, C. J. Demeure, and L. L. Scharf, "Initial Results in Prony Analysis of Power System Response Signals," *IEEE Transactions on Power Systems*, vol. 5, pp. 80–89, Feb. 1.

[102] MathWorks, "Help Center–lsqlin." www.mathworks.com/help/optim/ug/lsqlin.html.

[103] S. Boyd and L. Vandenberghe, *Convex Optimization*. Cambridge, UK: Cambridge University Press, 2004.

[104] MathWorks, "Help Center–roots" www.mathworks.com/help/matlab/ref/roots.html.

[105] J. Ma, P. J. Matuszyk, R. K. Mallan, C. Torres-Verdín, and B. C. Voss, "Joint Processing of Forward and Backward Extended Prony and Weighted Spectral Semblance Methods

for Robust Extraction of Velocity Dispersion Data," in *Proceedings of the SPWLA 51st Annual Logging Symposium*, Perth, Australia, Jun. 2010.

[106] D. I. Trudnowski, "Order Reduction of Large-Scale Linear Oscillatory Models," *IEEE Transactions Power Systems*, vol. 9, no. 1, pp. 451–458, Feb. 1994.

[107] J. Xiao, X. Xie, Y. Han, and J. Wu, "Dynamic Tracking of Low-Frequency Oscillations with Improved Prony Method in Wide-Area Measurement System," in *Proceedings of the IEEE Power and Energy Society General Meeting*, Denver, CO, Jun. 2004.

[108] J. C. H. Peng and N. K. C. Nair, "Adaptive Sampling Scheme for Monitoring Oscillations Using Prony Analysis," *IET Generation, Transmission and Distribution*, vol. 3, no. 12, pp. 1052–1060, Dec. 2009.

[109] R. Kumaresan and D. W. Tufts, "Estimating the Parameters of Exponentially Damped Sinusoids and Pole-Zero Modeling in Noise," *IEEE Transactions on Acoustic, Speech, and Signal Processing*, vol. 30, no. 6, pp. 833–840, Dec. 1982.

[110] D. Ruiz-Vega, A. R. Messina, and M. Pavella, "Online Assessment and Control of Transient Oscillations Damping," *IEEE Transactions on Power Systems*, vol. 19, no. 2, pp. 1038–1047, May 2004.

[111] F. F. Costa, A. L. de Almeida, F. A. Wegelin, and E. G. da Costa, "Recursive Prony's Method for Improving the Monitoring of Electrical Machines," in *Proceedings of the Instrumentation and Measurement Technology Conference*, Ottawa, Canada, May 2005.

[112] H. Guoqiang, H. Renmu, Y. Huachun, W. Peng, and M. Rui, "Iterative Prony Method Based Power System Low Frequency Oscillation Mode Analysis and PSS Design," in *Proceedings of the IEEE/PES Transmission and Distribution Conference and Exposition: Asia and Pacific*, Dalian, China, Aug. 2005.

[113] A. Allen, S. Santoso, and E. Muljadi, "Algorithm for Screening Phasor Measurement Unit Data for Power System Events and Categories and Common Characteristics for Events Seen in Phasor Measurement Unit Relative Phase-Angle Differences and Frequency Signals," National Renewable Energy Laboratory, Aug. 2013. www.nrel.gov/docs/fy13osti/58611.pdf

[114] MathWorks, "Help Center–fft." www.mathworks.com/help/matlab/ref/fft.html.

[115] P. J. Huber and E. M. Ronchetti, *Robust Statistics*. Hoboken, NJ: Wiley, 2009.

[116] E. Perez and J. Barros, "Voltage Event Detection and Characterization Methods: A Comparative Study," in *Proceedings of the IEEE PES Transmission and Distribution Conference and Exposition: Latin America*, Caracas, Venezuela, Aug. 2006.

[117] D.-I. Kim, T. Y. Chun, S. H. Yoon, G. Lee, and Y. J. Shin, "Wavelet-Based Event Detection Method Using PMU Data," *IEEE Transactions on Smart Grid*, vol. 8, no. 3, pp. 1154–1162, May 2017.

[118] A. Shahsavari, M. Farajollahi, E. Stewart, E. Cortez, and H. Mohsenian-Rad, "Situational Awareness in Distribution Grid Using Micro-PMU Data: A Machine Learning Approach," *IEEE Transactions on Smart Grid*, vol. 10, no. 6, pp. 6167–6177, Nov. 2019.

[119] N. L. Tasfi, W. A. Higashino, K. Grolinger, and M. A. M. Capretz, "Deep Neural Networks with Confidence Sampling for Electrical Anomaly Detection," in *Proceedings of IEEE International Conference on Internet of Things*, Exeter, UK, Jun. 2017.

[120] D. B. Araya, K. Grolinger, H. F. El-Yamany, M. A. Capretz, and G. Bitsuamlak, "An Ensemble Learning Framework for Anomaly Detection in Building Energy Consumption," *Energy and Buildings*, vol. 144, pp. 191–206, Jun. 2017.

[121] M. Kahl, T. Kriechbaumer, D. Jorde, A. U. Haq, and H. A. Jacobsen, "Appliance Event Detection—A Multivariate, Supervised Classification Approach," in *Proceedings of the ACM International Conference on Future Energy Systems*, Phoenix, AZ, Jun. 2019.

[122] D. Nguyen, R. Barella, S. A. Wallace, X. Zhao, and X. Liang, "Smart Grid Line Event Classification Using Supervised Learning over PMU Data Streams," in *Proceedings of the International Green and Sustainable Computing Conference*, Las Vegas, NV, Dec. 2015.

[123] A. Aligholian, M. Farajollahi, and H. Mohsenian-Rad, "Unsupervised Learning for Online Abnormality Detection in Smart Meter Data," in *Proceedings of the IEEE PES General Meeting*, Atlanta, GA, Aug. 2019.

[124] A. Aligholian, A. Shahsavari, E. Cortez, E. Stewart, and H. Mohsenian-Rad, "Event Detection in Micro-PMU Data: A Generative Adversarial Network Scoring Method," in *Proceedings of the IEEE PES General Meeting*, Montreal, QC, Canada, Aug. 2020.

[125] J. Luo, T. Hong, and M. Yue, "Real-Time Anomaly Detection for Very Short-Term Load Forecasting," *Journal of Modern Power Systems and Clean Energy*, vol. 6, no. 2, pp. 235–243, Mar. 2018.

[126] A. Saad and N. Sisworahardjo, "Data Analytics-Based Anomaly Detection in Smart Distribution Network," in *Proceedings of the International Conference on High Voltage Engineering and Power Systems*, Sanur, Indonesia, Oct. 2017.

[127] A. Shahsavari, M. Farajollahi, E. Stewart, E. Cortez, and H. Mohsenian-Rad, "A Machine Learning Approach to Event Analysis in Distribution Feeders Using Distribution Synchrophasors," in *Proceedings of the IEEE International Conference on Smart Grid Synchronized Measurements and Analytics (SGSMA)*, College Station, TX, May 2019.

[128] A. von Jouanne and B. Banerjee, "Assessment of Voltage Unbalance," *IEEE Transactions on Power Delivery*, vol. 16, pp. 782–790, Oct. 2001.

[129] T. A. Short, *Electric Power Distribution Handbook*. Boca Raton, FL: CRC Press, 2004.

[130] H. Pezeshki and P. J. Wolfs, "Consumer Phase Identification in a Three Phase Unbalanced LV Distribution Network," in *Proceedings of the IEEE Innovative Smart Grid Technologies*, Berlin, Germany, May 2012.

[131] B. K. Seal and M. F. McGranaghan, "Automatic Identification of Service Phase for Electric Utility Customers," in *Proceedings of the IEEE Power and Energy Society General Meeting*, Detroit, MI, Jul. 2011.

[132] North American Electric Reliability Corporation, "Standard BAL-003-1—Frequency Response and Frequency Bias Setting Reliability Standard." www.federalregister.gov/documents/2014/01/23/2014-01218/frequency-response-and-frequency-bias-setting-reliability-standard.

[133] J. Dong, J. Zuo, L. Wang, K. S. Kook, I. Y. Chung, Y. Liu, S. Affare, B. Rogers, and M. Ingram, "Analysis Power System Disturbances Based on Wide-Area Frequency Measurements," in *Proceedings of the IEEE Power and Energy Society General Meeting*, Jul. 2007.

[134] North American Electric Reliability Corporation, "Balancing and Frequency Control," Princeton, NJ, Jan. 2011.

[135] D. P. Chassin, Z. Huang, M. K. Donnelly, C. Hassler, E. Ramirez, and C. Ray, "Estimation of WECC System Inertia Using Observed Frequency Transients," *IEEE Transactions on Power Systems*, vol. 20, pp. 1190–1192, 2005.

[136] S. Sharma, S. H. Huang, and N. Sarma, "System Inertial Frequency Response Estimation and Impact of Renewable Resources," in *Proceedings of the IEEE Power and Energy Society General Meeting*, Jul. 2011.

[137] Institute of Electrical and Electronics Engineers, "Standard Requirements for Instrument Transformers, C57.13-2016." https://ieeexplore.ieee.org/document/7501435.

[138] R. Moghe, "Smart Sensors for Utility Assets," doctoral thesis, Georgia Institute of Technology, Atlanta, GA, 2012.

[139] H. Mohsenian-Rad, E. Stewart, and E. Cortez, "Distribution Synchrophasors: Pairing Big Data with Analytics to Create Actionable Information," *IEEE Power and Energy Society Magazine*, vol. 16, no. 3, pp. 26–34, May 2018.

[140] A. G. Phadke and J. S. Thorp, *Synchronized Phasor Measurements and Their Applications*. New York, NY: Springer, 2008.

[141] Institute of Electrical and Electronics Engineers, "Standard for Synchrophasor Measurements for Power Systems, C37.118.1-2011." https://ieeexplore.ieee.org/document/6111222.

[142] D. L. Mills, "Internet Time Synchronization: The Network Time Protocol," *IEEE Transactions on Communications*, vol. 39, no. 10, pp. 1482–1493, Oct. 1991.

[143] A. Derviškadić, R. Razzaghi, Q. Walger, and M. Paolone, "The White Rabbit Time Synchronization Protocol for Synchrophasor Networks," *IEEE Transactions on Smart Grid*, vol. 11, no. 1, pp. 726–738, Jan. 2020.

[144] Schweitzer Engineering Laboratories, "It's About Time: Synchrophasor Measurements Require a Precise, Absolute Time Reference." https://selinc.com/solutions/synchrophasors/report/115267/.

[145] A. G. Phadke, "Synchronized Phasor Measurements-a Historical Overview," in *Proceedings of the IEEE PES Transmission and Distribution Conference and Exhibition*, Oct. 2002.

[146] S. Nuthalapati, *Power System Grid Operation Using Synchrophasor Technology*. New York, NY: Springer, 2019.

[147] A. von Meier, E. Stewart, A. McEachern, M. Andersen, and L. Mehrmanesh, "Precision Micro-Synchrophasors for Distribution Systems: A Summary of Applications," *IEEE Transactions on Smart Grid*, vol. 8, no. 6, pp. 2926–2936, Nov. 2017.

[148] Institute of Electrical and Electronics Engineers, "Guide for Phasor Data Concentrator Requirements for Power System Protection, Control, and Monitoring, C37.244-2013." https://ieeexplore.ieee.org/document/6514039.

[149] P. Kundur, *Power System Stability and Control*. New York, NY: McGraw-Hill, 1994.

[150] B. J. Pierre, F. Wilches-Bernal, D. A. Schoenwald, R. T. Elliott, D. J. Trudnowski, R. H. Byrne, and J. C. Neely, "Design of the Pacific DC Intertie Wide Area Damping Controller," *IEEE Transactions on Power Systems*, vol. 34, no. 5, pp. 3594–3604, Sep. 2019.

[151] M. Farajollahi, A. Shahsavari, and H. Mohsenian-Rad, "Location Identification of Distribution Network Events Using Synchrophasor Data," in *Proceedings of the IEEE North American Power Symposium*, Morgantown, WV, Sep. 2017.

[152] A. Shahsavari, M. Farajollahi, E. Stewart, A. Von-Meier, L. Alvarez, E. Cortez, and H. Mohsenian-Rad, "Data-Driven Analysis of Capacitor Bank Operation at a Distribution Feeder Using Micro-PMU Data," in *Proceedings of the IEEE Power and Energy Society Conference on Innovative Smart Grid Technologies*, Apr. 2017.

[153] M. Farajollahi, A. Shahsavari, and H. Mohsenian-Rad, "Tracking State Estimation in Distribution Networks Using Distribution-Level Synchrophasor Data," in *Proceedings of the IEEE Power and Energy Society General Meeting*, Aug. 2018.

[154] A. Akrami, M. S. Asif, and H. Mohsenian-Rad, "Sparse Distribution System State Estimation: An Approximate Solution against Low Observability," in *Proceedings of the IEEE Power and Energy Society Conference on Innovative Smart Grid Technologies*, Feb. 2020.

[155] M. Farajollahi, A. Shahsavari, E. Stewart, and H. Mohsenian-Rad, "Locating the Source of Events in Power Distribution Systems Using Micro-PMU Data," *IEEE Transactions on Power Systems*, vol. 33, no. 6, pp. 6343–6354, Nov. 2018.

[156] K. S. Kumar, *Electric Circuits and Networks*. Chennai, India: Pearson, 2009.

[157] V. K. Gaur and B. R. Bhalja, "Synchrophasor Based Fault Distance Estimation Method for Tapped Transmission Line," in *Proceedings of the IEEE International Conference on Smart Grid Synchronized Measurements and Analytics (SGSMA)*, College Station, TX, May 2019.

[158] Q. Jiang, X. Li, B. Wang, and H. Wang, "PMU-Based Fault Location Using Voltage Measurements in Large Transmission Networks," *IEEE Transactions on Power Delivery*, vol. 27, no. 3, 1644.

[159] Q. Jiang, B. Wang, and X. Li, "An Efficient PMU-Based Fault-Location Technique for Multiterminal Transmission Lines," *IEEE Transactions on Power Delivery*, vol. 29, no. 4, pp. 1675–1682, Aug. 2014.

[160] J. Ren, S. S. Venkata, and E. Sortomme, "An Accurate Synchrophasor Based Fault Location Method for Emerging Distribution Systems," *IEEE Transactions on Power Delivery*, vol. 29, no. 1, pp. 297–298, Feb. 2014.

[161] K. Mei, S. M. Rovnyak, and C. Ong, "Clustering-Based Dynamic Event Location Using Wide-Area Phasor Measurements," *IEEE Transactions on Power Systems*, vol. 23, no. 2, pp. 673–679, May 2008.

[162] M. Pignati, L. Zanni, P. Romano, R. Cherkaoui, and M. Paolone, "Fault Detection and Faulted Line Identification in Active Distribution Networks Using Synchrophasors-Based Real-Time State Estimation," *IEEE Transactions on Power Delivery*, vol. 32, no. 1, pp. 381–392, Feb. 2017.

[163] G. Feng and A. Abur, "Fault Location Using Wide-Area Measurements and Sparse Estimation," *IEEE Transactions on Power Systems*, vol. 31, no. 4, pp. 2938–2945, Jul. 2016.

[164] W. Li, D. Deka, M. Chertkov, and M. Wang, "Real-Time Faulted Line Localization and PMU Placement in Power Systems Through Convolutional Neural Networks," *IEEE Transactions on Power Systems*, vol. 34, no. 6, pp. 4640–4651, Nov. 2019.

[165] J. Blanco, J. F. Petit, and G. Ordóñez, "Algorithm for Relative Location of Voltage Sags and Capacitor Switching Transients Based on Voltage Measurement Only," in *Proceedings of the IEEE International Conference on Harmonics and Quality of Power (ICHQP)*, Bucharest, Romania, May 2014.

[166] C. D. Le, M. H. J. Bollen, and I. Y. H. Gu, "Analysis of Power Disturbances from Monitoring Multiple Levels and Locations in a Power System," in *Proceedings of the IEEE International Conference on Harmonics and Quality of Power (ICHQP)*, Bergamo, Italy, Sep. 2010.

[167] Powerside Inc., "Synchrophasor Micro-PMU." https://powerside.com/products/monitoring/micropmu/

[168] J. L. Blackburn and T. J. Domin, *Protective Relaying Principles and Applications*. Boca Raton, FL: CRC Press, 2007.

[169] P. M. Anderson, *Analysis of Faulted Power Systems*. Ames, IA: Iowa State University Press, 1973.

[170] E. O. Schweitzer and S. E. Zocholl, "Introduction to Symmetrical Components," in *Proceedings of the Annual Georgia Tech Protective Relaying Conference*, Atlanta, GA, Apr. 2004.

[171] T. Smith and C. Wester, "Fundamentals of Modern Protective Relaying," 2011. https://ewh.ieee.org/r3/atlanta/ias/2011-2012_Presentations/IEEE%20Seminar%20-%20Fundamentals%20of%20Modern%20Protective%20Relaying%20-%20Part1.pdf.

[172] N. Cristianini and J. Shawe-Taylor, *An Introduction to Support Vector Machines and Other Kernel-Based Learning Methods*. Cambridge, UK: Cambridge University Press, 2000.

[173] C. Hsu and C. Lin, "A Comparison of Methods for Multi-Class Support Vector Machines," *IEEE Transactions on Neural Networks*, vol. 13, no. 2, pp. 415–425, Mar. 2002.

[174] S. Suthaharan, *Machine Learning Models and Algorithms for Big Data Classification*. New York, NY: Springer, 2016.

[175] G. Madzarov, D. Gjorgjevikj, and I. Chorbev, "A Multi-Class SVM Classifier Utilizing Binary Decision Tree," *Informatica*, vol. 33, no. 2, pp. 233–241, 2009.

[176] A. Shahsavari, M. Farajollahi, and H. Mohsenian-Rad, "Individual Load Model Parameter Estimation in Distribution Systems using Load Switching Events," *IEEE Transactions on Power Systems*, vol. 34, no. 6, pp. 4652–4664, Nov. 2019.

[177] MathWorks "Help Center–quadprog." https://www.mathworks.com/help/optim/ug/quadprog.html.

[178] CVX Research, "CVX: Matlab Software for Disciplined Convex Programming." http://cvxr.com/cvx/.

[179] B. E. Boser, I. M. Guyon, and V. N. Vapnik, "A Training Algorithm for Optimal Margin Classifiers," in *Proceedings of the ACM Annual Workshop on Computational Learning Theory*, Pittsburgh, PA, Jul. 1992.

[180] Z. Fu, A. Robles-Kelly, and J. Zhou, "Mixing Linear SVMs for Nonlinear Classification," *IEEE Transactions on Neural Networks*, vol. 21, no. 12, pp. 1963–1975, Dec. 2010.

[181] F. Perez-Cruz, A. Navia-Vazquez, P. L. Alarcon-Diana, and A. Artes-Rodriguez, "Support Vector Classifier with Hyperbolic Tangent Penalty Function," in *Proceedings of the IEEE International Conference on Acoustics, Speech, and Signal Processing*, Istanbul, Turkey, Jun. 2000.

[182] B. V. Dasarathy, *Nearest Neighbor (NN) Norms: NN Pattern Classification Rechniques*. Los Alamitos, CA: IEEE Computer Society Press, 1991.

[183] L. Breiman, J. H. Friedman, R. A. Olshen, and C. J. Stone, *Classification and Regression Trees*. Monterey, CA: Wadsworth and Brooks, 1984.

[184] S. Shalev-Shwartz and S. Ben-David, *Understanding Machine Learning*. Cambridge, UK: Cambridge University Press, 2014.

[185] N. H. Abbasy and H. M. Ismail, "A Unified Approach for the Optimal PMU Location for Power System State Estimation," *IEEE Transactions on Power Systems*, vol. 24, no. 2, pp. 806–813, May 2009.

[186] S. Chakrabarti and E. Kyriakides, "Optimal Placement of Phasor Measurement Units for Power System Observability," *IEEE Transactions on Power Systems*, vol. 23, no. 3, pp. 1433–1440, Aug. 2008.

[187] N. M. Manousakis, G. N. Korres, and P. S. Georgilakis, "Taxonomy of PMU Placement Methodologies," *IEEE Transactions on Power Systems*, vol. 27, no. 2, pp. 1070–1077, May 2012.

[188] B. Xu and A. Abur, "Observability Analysis and Measurement Placement for Systems with PMUs," in *Proceedings of the IEEE PES Power Systems Conference and Exposition*, Oct. 2004.

[189] B. Gou, "Generalized Integer Linear Programming Formulation for Optimal PMU Placement," *IEEE Transactions on Power Systems*, vol. 23, no. 3, pp. 1099–1104, Aug. 2008.

[190] Y. Xiao, J. C. Maun, H. B. Mahmoud, T. Detroz, and S. Do, "Harmonic Impedance Measurement Using Voltage and Current Increments from Disturbing Loads," in *Proceedings of the IEEE International Conference on Harmonics and Quality of Power*, Oct. 2000.

[191] M. Farajollahi, A. Shahsavari, and H. Mohsenian-Rad, "Topology Identification in Distribution Systems Using Line Current Sensors: An MILP Approach," *IEEE Transactions on Smart Grid*, vol. 11, no. 2, pp. 1159–1170, Mar. 2020.

[192] Z. Tian, W. Wu, and B. Zhang, "A Mixed Integer Quadratic Programming Model for Topology Identification in Distribution Network," *IEEE Transactions on Smart Grid*, vol. 31, no. 1, pp. 823–824, Jan. 2016.

[193] S. Affijulla and P. Tripathy, "Development of Phasor Estimation Algorithm for P-Class PMU Suitable in Protection Applications," *IEEE Transactions on Smart Grid*, vol. 9, no. 2, pp. 1250–1260, Mar. 2018.

[194] S. Ajulla and P. Tripathy, "Development of Dictionary-Based Phasor Estimator Suitable for P-Class Phasor Measurement Unit," *IEEE Transactions on Instrumentation and Measurement*, vol. 67, no. 11, pp. 2603–2615, Nov. 2018.

[195] A. J. Roscoe, I. F. Abdulhadi, and G. M. Burt, "Filters for M Class Phasor Measurement Units," in *Proceedings of the IEEE International Workshop on Applied Measurements for Power Systems*, Aachen, Germany, Sep. 2012.

[196] A. J. Roscoe, I. F. Abdulhadi, and G. M. Burt, "Filter Design Masks for C37.118.1a-Compliant Frequency-Tracking and Fixed-Filter M-Class Phasor Measurement Units," *IEEE Transactions on Instrumentation and Measurement*, vol. 64, no. 8, pp. 2096–2107, Aug. 2015.

[197] M. Kamal, M. Farajollahi, and H. Mohsenian-Rad, "Analysis of Cyber Attacks Against Micro-PMUs: The Case of Event Source Location Identification," in *Proceedings of the IEEE Power and Energy Society Conference on Innovative Smart Grid Technologies*, Feb. 2020.

[198] MathWorks, "Help Center–smoothdata." www.mathworks.com/help/matlab/ref/smoothdata.html.

[199] C. Mullins, "Adjusting Waveform Capture Sampling Rates," White Paper, Power Monitors Inc., Jan. 2016.

[200] D. Carnovale, "Power Quality Monitoring: Waveform Analysis," https://ewh.ieee.org/r3/nashville/events/2008/2008.05.06.pdf,May2008.

[201] J. H. R. Enslin and P. J. M. Heskes, "Harmonic Interaction between a Large Number of Distributed Power Inverters and the Distribution Network," *IEEE Transactions on Power Electronics*, vol. 19, no. 6, pp. 1586–1593, Nov. 2004.

[202] European Standards, "Standard EN 50160: Voltage Characteristics in Public Distribution Systems," http://copperalliance.org.uk/uploads/2018/03/542-standard-en-50160-voltage-characteristics-in.pdf.

[203] C. Payne, "Understanding Crest Factor," White Paper, Power Monitors Inc., Mar. 2016.

[204] R. Arghandeh, A. Onen, J. Jung, D. Cheng, R. P. Broadwater, and V. Centeno, "Phasor-Based Assessment for Harmonic Sources in Distribution Networks," *Electric Power Systems Research*, vol. 116, pp. 94–105, Jul. 2014.

[205] Institute of Electrical and Electronics Engineers, "519-2014–IEEE Recommended Practice and Requirements for Harmonic Control in Electric Power Systems," 2014. https://ieeexplore.ieee.org/document/6826459

[206] International Electrotechnical Commission, "Electromagnetic Compatibility (EMC)–Part 2-2: Environment–Compatibility Levels for Low-Frequency Conducted Disturbances and Signalling in Public Low-Voltage Power Supply Systems," 2002. https://webstore.iec.ch/publication/63116.

[207] International Electrotechnical Commission, "Electromagnetic Measurement Techniques–General Guide on Harmonics and Interharmonics Measurements and Instrumentation, for Power Supply Systems and Equipment Connected Thereto," 2002. https://webstore.iec.ch/publication/4226.

[208] C. Mullins, "Measuring PLC AMR Signals with the Revolution," White Paper, Power Monitors Inc., May 2013.

[209] C. Andrus, "Voltage Notching in IEEE Std 519-2014," White Paper, Power Monitors Inc., Jan. 2015.

[210] C. Andrus, "Power Factor Correction Capacitors and Resonances," White Paper, Power Monitors Inc., Nov. 2014.

[211] C. Mullins, "Transient Capture versus Waveform Capture," White Paper, Power Monitors Inc., Jul. 2015.

[212] B. Li, R. Torquato, W. Freitas, D. D. Sabin, C. Li, M. J. Mousavi, W. Xu, G. MacLeod, T. E. Grebe, J. Yong, T. Laughner, A. Murphy, and T. A. Cooke, "Electric Signatures of Power Equipment Failures," IEEE PES Working Group on Power Quality Data Analytics, 2018.

[213] B. Li, "Abnormality Detection Methods for Utility Equipment Condition Monitoring," master's thesis, University of Alberta, Edmonton, Alberta, Canada, 2016.

[214] S. Kulkarni, S. Santoso, and T. A. Short, "Incipient Fault Location Algorithm for Underground Cables," *IEEE Transactions on Smart Grid*, vol. 5, no. 3, pp. 1165–1174, May 2014.

[215] T. E. Grebe, "Effective Collection and Management of Power Quality Data for Analysis and Detection of Incipient Distribution System Components Faults and Identification of Their Locations," CEATI Report Number T124700-5159, Sep. 2013.

[216] L. A. Kojovic and C. W. Williams, "Sub-cycle Detection of Incipient Cable Splice Faults to Prevent Cable Damage," in *Proceedings of the IEEE Power Engineering Society Summer Meeting*, Jul. 2000.

[217] S. Kulkarni, A. J. Allen, S. Chopra, S. Santoso, and T. A. Short, "Waveform Characteristics of Underground Cable Failures," in *Proceedings of the IEEE Power Engineering Society General Meeting*, Jul. 2010.

[218] U.S. Department of Energy and Electric Power Research Institute, "DOE/EPRI National Database Repository of Power System Events," https://pqmon.epri.com/see_all.html.

[219] H. Blume, "PG&E Reaches $13.5-Billion Settlement with Victims of Devastating California Wildfires," *Los Angeles Times*, Dec. 2019.

[220] L. A. Irwin, "Real Experience Using Power Quality Data to Improve Power Distribution Reliability," in *Proceedings of the IEEE International Conference on Harmonics and Quality of Power*, Sep. 2010.

[221] S. Santoso and D. D. Sabin, "Power Quality Data Analytics: Tracking, Interpreting, and Predicting Performance," in *Proceedings of the IEEE Power and Energy Society General Meeting*, Jul. 2012.

[222] B. Kasztenny, I. Voloh, A. Depew, and J. Wolete, "Re-Strike and Breaker Failure Conditions for Circuit Breakers Connecting Capacitor Banks," in *Proceedings of the IEEE Annual Conference for Protective Relay Engineers*, Apr. 2008.

[223] C. L. Benner, B. D. Russell, and A. Sundaram, "Feeder Interruptions Caused by Recurring Faults on Distribution Feeders: Faults You Don't Know," in *Proceedings of the IEEE Annual Conference for Protective Relay Engineers*, Apr. 2008.

[224] C. L. Benner and B. D. Russell, "Distribution Incipient Faults and Abnormal Events: Case Studies from Recorded Field Data," in *Proceedings of the IEEE Annual Conference for Protective Relay Engineers*, Apr. 2004.

[225] Electric Power Research Institute, "DPQ Event: Back-to-Back Capacitor Switching," Report Number 1017221, Palo Alto, CA, 2003. www.epri.com/research/products/000000000001017221.

[226] Electric Power Research Institute, "DPQ Event: Arcing Switch Contacts during Capacitor Energization," EPRI Technical Report Number 1017218, Palo Alto, CA, 2003. www.epri.com/research/products/000000000001017225.

[227] M. A. Eltawil and Z. Zhao, "Grid-Connected Photovoltaic Power Systems: Technical and Potential Problems—A Review," *Renewable and Sustainable Energy Reviews*, vol. 14, no. 1, pp. 112–129, Jan. 2010.

[228] C. Hochgraf and R. H. Lasseter, "Statcom Controls for Operation with Unbalanced Voltages," *IEEE Transactions on Power Delivery*, vol. 13, no. 2, pp. 538–544, Apr. 1998.

[229] M. Hagiwara, K. Wada, H. Fujita, and H. Akagi, "Dynamic Behavior of a 21-Level BTB-Based Power Flow Controller under Single Line-to-Ground Fault Conditions," *IEEE Transactions on Industry Applications*, vol. 43, no. 5, pp. 1379–1387, Sep. 2007.

[230] A. Hoke, A. Nelson, S. Chakraborty, J. Chebahtah, T. Wang, and M. McCarty, "Inverter Ground Fault Overvoltage Testing," National Renewable Energy Laboratory, Technical Report NREL/TP-5D00-64173, Aug. 2015.

[231] M. Ropp, A. Hoke, S. Chakraborty, D. Schutz, C. Mouw, A. Nelson, M. McCarty, T. Wang, and A. Sorenson, "Ground Fault Overvoltage with Inverter-Interfaced Distributed Energy Resources," *IEEE Transactions on Power Delivery*, vol. 32, no. 2, pp. 890–899, Apr. 2017.

[232] Joint NERC and WECC Staff Report, "900 mw Fault Induced Solar Photovoltaic Resource Interruption Disturbance Report, Southern California Event: October 9, 2017," Feb. 2018. www.nerc.com/pa/rrm/ea/October%209%202017%20Canyon%202%20Fire%20Disturbance%20Report/900%20MW%20Solar%20Photovoltaic%20Resource%20Interruption%20Disturbance%20Repo

[233] North American Electric Reliability Corporation, "1,200 mw Fault Induced Solar Photovoltaic Resource Interruption Disturbance Report," 2017. www.nerc.com/pa/rrm/ea/1200_MW_Fault_Induced_Solar_Photovoltaic_Resource_/1200_MW_Fault_Induced_Solar_Photovoltaic_Resource_Interruption_Final.pdf.

[234] North American Electric Reliability Corporation, "Reliability Guideline: BPS-Connected Inverter-Based Resource Performance," Sep. 2018. www.nerc.com/comm/PC_Reliability_Guidelines_DL/Inverter-Based_Resource_Performance_Guideline.pdf.

[235] V. B. Nunez, S. Kulkarni, S. Santoso, and M. F. Joaquim, "Feature Analysis and Classification Methodology for Overhead Distribution Fault Events," in *Proceedings of the IEEE Power Engineering Society General Meeting*, Jul. 2010.

[236] D. Sabin, G. MacLeod, and M. Wojdan, "Distribution Fault Location and Grid Analytics at Hydro Ottawa," in *Proceedings of the IEEE International Conference on Harmonics and Quality of Power*, May 2018.

[237] IEEE, "IEEE Recommended Practice for Monitoring Electric Power Quality," IEEE Standard 1159–2009.

[238] T. Keppler, N. Watson, and J. Arrillaga, "Computation of the Short-Term Flicker Severity Index," *IEEE Transactions on Power Delivery*, vol. 15, no. 4, pp. 1110–1115, Oct. 2000.

[239] S. Rahman, M. Moghaddami, A. I. Sarwat, T. Olowu, and M. Jafaritalarposhti, "Flicker Estimation Associated with PV Integrated Distribution Network," in *Proceedings of the IEEE Southeastcon*, Apr. 2018.

[240] J. Mora-Flòrez, J. Melóndez, and G. Carrillo-Caicedo, "Comparison of Impedance Based Fault Location Methods for Power Distribution Systems," *Electric Power Systems Research*, vol. 78, no. 4, pp. 657–666, Apr. 2008.

[241] S. S. Kulkarni, "Fault Location and Characterization in AC and DC Power Systems," PhD thesis, University of Texas at Austin, Dec. 2012.

[242] J. Goodfellow, "Investigating Tree-Caused Faults," *Distribution World*, Nov. 2005.

[243] S. Kulkarni, D. Lee, A. J. Allen, S. Santoso, and T. A. Short, "Waveform Characterization of Animal Contact, Tree Contact, and Lightning Induced Faults," in *Proceedings of the IEEE Power Engineering Society General Meeting*, Jul. 2010.

[244] Y. Cai and M. Y. Chow, "Small World Stratification for Distribution Fault Diagnosis," in *Proceedings of the IEEE PES Power Systems Conference and Exposition*, Mar. 2011.

[245] A. Carta, N. Locci, and C. Muscas, "A PMU for the Measurement of Synchronized Harmonic Phasors in Three-Piece Distribution Networks," *IEEE Transactions on Instrumentation and Measurement*, vol. 58, no. 10, pp. 3723–3730, Oct. 2009.

[246] B. Zeng, Z. Teng, Y. Cai, S. Guo, and B. Qing, "Harmonic Phasor Analysis Based on Improved FFT Algorithm," *IEEE Transactions on Smart Grid*, vol. 2, no. 1, pp. 51–59, Mar. 2011.

[247] M. Chakir, I. Kamwa, and H. L. Huy, "Extended C37.118.1 PMU Algorithms for Joint Tracking of Fundamental and Harmonic Phasors in Stressed Power Systems and Microgrids," *IEEE Transactions on Power Delivery*, vol. 29, no. 3, pp. 1465–1480, Jun. 2014.

[248] S. K. Jain, P. Jain, and S. N. Singh, "A Fast Harmonic Phasor Measurement Method for Smart Grid Applications," *IEEE Transactions on Smart Grid*, vol. 8, no. 1, pp. 493–502, Jan. 2017.

[249] L. Chen, W. Zhao, F. Wang, and S. Huang, "Harmonic Phasor Estimator for P Class Phasor Measurement Units," *IEEE Transactions on Instrumentation and Measurement*, vol. 58, no. 10, pp. 1–10, May 2019.

[250] M. Elad, *Sparse and Redundant Representations: From Theory to Applications in Signal and Image Processing*. New York, NY: Springer, 2010.

[251] H. Liao, "Power System Harmonic State Estimation and Observability Analysis via Sparsity Maximization," *IEEE Transactions on Power Systems*, vol. 22, no. 1, pp. 15–23, Feb. 2007.

[252] L. Chen, M. Farajollahi, M. Ghamkhari, W. Zhao, S. Huang, and H. Mohsenian-Rad, "Switch Status Identification in Distribution Networks Using Harmonic Synchrophasor Measurements," *IEEE Transactions on Smart Grid*, vol. 12, no. 3, pp. 2413–2424, May 2021.

[253] C. L. Benner and B. D. Russell, "Practical High-Impedance Fault Detection on Distribution Feeders," *IEEE Transactions on Industry Applications*, vol. 33, no. 3, pp. 635–640, May 1997.

[254] M. Farajollahi, A. Shahsavari, and H. Mohsenian-Rad, "Location Identification of High Impedance Faults Using Synchronized Harmonic Phasors," in *Proceedings of the IEEE Power and Energy Society Conference on Innovative Smart Grid Technologies*, Apr. 2017.

[255] A. F. Bastos, S. Santoso, W. Freitas, and W. Xu, "Synchrowaveform Measurement Units and Applications," in *Proceedings of the IEEE Power and Energy Society General Meeting*, Aug. 2019.

[256] M. Izadi and H. Mohsenian-Rad, "Event Location Identification in Distribution Networks Using Waveform Measurement Units," in *Proceedings of the IEEE PES Innovative Smart Grid Technologies Conference*, The Hague, Netherlands, Oct. 2020.

[257] G. Ziegler, *Numerical Differential Protection: Principles and Applications*. Erlangen, Germany: Publicis Publishing, 2012.

[258] M. Izadi and H. Mohsenian-Rad, "Synchronous Waveform Measurements to Locate Transient Events and Incipient Faults in Power Distribution Networks," *IEEE Transactions on Smart Grid*, May 2021.

[259] Power Monitors Inc., "Revolution Power Quality Recorder," https://powermonitors.com/product/revolution-power-quality-recorder/.

[260] S. Misak, J. Fulnecek, T. Vantuch, and L. Prokop, "Towards the Character and Challenges of Partial Discharge Pattern Data Measured on Medium Voltage Overhead Lines," in *Proceedings of the IEEE International Scientific Conference on Electric Power Engineering*, Kouty nad Desnou, Czech Republic, May 2019.

[261] S. Misak, J. Fulnecek, T. Vantuch, T. Burianek, and T. Jezowicz, "A Complex Classification Approach of Partial Discharges from Covered Conductors in Real Environment," *IEEE Transactions on Dielectrics and Electrical Insulation*, vol. 24, no. 2, pp. 1097–1104, Apr. 2017.

[262] L. A. Barclay, *Propagation of Radio Waves*. London: Institution of Engineering and Technology, 2003.

[263] H. A. Illias, M. A. Tunio, A. H. A. Baker, H. Mokhlis, and G. Chen, "Partial Dicharge Phenomena within an Artificial Void in Cable Insulation Geometry: Experimental Validation and Simulation," *IEEE Transactions on Dielectrics and Electrical Insulation*, vol. 23, no. 1, pp. 451–459, Feb. 2016.

[264] F. Alvarez, J. Ortega, F. Garnacho, and M. A. Sanchez-Uran, "A Clustering Technique for Partial Discharge and Noise Sources Identification in Power Cables by Means of Waveform Parameters," *IEEE Transactions on Dielectrics and Electrical Insulation*, vol. 23, no. 1, pp. 469–481, Feb. 2016.

[265] C. Andrus, "Understanding Variable Frequency Drives," White Paper, Power Monitors Inc., Sep. 2015.

[266] T. Hong, D. Deswal, and F. De-Leon, "An Online Data-Driven Technique for the Detection of Transformer Winding Deformations," *IEEE Transactions on Power Delivery*, vol. 33, pp. 600–609, Apr. 2018.

[267] M. Izadi and H. Mohsenian-Rad, "A Synchronized Lissajous-Based Method to Achieve Situational Awareness Using Synchronized Waveform Measurements," in *Proceedings of the IEEE Power and Energy Society General Meeting*, Aug. 2021.

[268] Pecan Street Project, www.pecanstreet.org/dataport/Austin,TX.

[269] Electric Reliability Council of Texas, "Wind Power Production," www.ercot.com/gridinfo/generation/.

[270] S. Svensson, "Power Measurement Techniques for Non-sinusoidal Conditions," PhD thesis, Chalmers University of Technology, Göteborg, Sweden, 1999.

[271] S. Grijalva and M. U. Tariq, "Prosumer-Based Smart Grid Architecture Enables a Flat, Sustainable Electricity Industry," in *Proceedings of the IEEE PES Conference on Innovative Smart Grid Technologies*, Anaheim, CA, Jan. 2011.

[272] I. Lampropoulos, G. M. A. Vanalme, and W. L. Kling, "A Methodology for Modeling the Behavior of Electricity Prosumers within the Smart Grid," in *Proceedings of the IEEE PES Conference on Innovative Smart Grid Technologies—Europe*, Gothenberg, Sweden, Oct. 2010.

[273] M. Albachrony, D. Ha, Q. Tran, A. Brun, and M. Petit, "Coordinated Prosumer Transaction Based on Load Shifting and Optimization," in *Proceedings of the IEEE PES Conference on Innovative Smart Grid Technologies—Europe*, Bucharest, Romania, Sep. 2019.

[274] Southern California Edison, "Net Energy Metering," www.sce.com/residential/generating-your-own-power/net-energy-metering.

[275] California Public Utilities Commission, "CPUC Approves Feed-in Tariffs to Support Development of Onsite Renewable Generation." https://docs.cpuc.ca.gov/PUBLISHED/NEWS_RELEASE/78824.htm

[276] E. Hinds, C. Matsuishi, and B. Schoradt, "Feed-In Tariffs Emerge as Key Driver for Solar Development," *Solar Power International Show Preview Guide*, 2012.

[277] K. S. K. Weranga, S. Kumarawadu, and D. P. Chandima, *Smart Metering Design and Applications*. New York, NY: Springer, 2014.

[278] T&D World, "Smart Meter Deployment Projected to Reach 107 Million as of Year-End 2020." www.tdworld.com/grid-innovations/smart-grid/article/21120206/smart-meter-deployment-projected-to-reach-107-million-as-of-yearend-2020.

[279] M. Albadi and E. El-Saadany, "A Summary of Demand Response in Electricity Markets," *Electric Power Systems Research*, vol. 78, no. 11, pp. 1989–1996, Nov. 2008.

[280] A monthly bill issued by Riverside Public Utilities in 2017.

[281] H. Mohsenian-Rad and A. Leon-Garcia, "Optimal Residential Load Control with Price Prediction in Real-Time Electricity Pricing Environments," *IEEE Transactions on Smart Grid*, vol. 1, no. 2, pp. 120–133, Sep. 2010.

[282] X. Chen, T. Wei, and S. Hu, "Uncertainty-Aware Household Appliance Scheduling Considering Dynamic Electricity Pricing in Smart Home," *IEEE Transactions on Smart Grid*, vol. 4, no. 2, pp. 932–941, Jun. 2013.

[283] Z. Xu, R. Diao, S. Lu, J. Lian, and Y. Zhang, "Modeling of Electric Water Heaters for Demand Response: A Baseline PDE Model," *IEEE Transactions on Smart Grid*, vol. 5, no. 5, pp. 2203–2210, Sep. 2014.

[284] A. Gholian, H. Mohsenian-Rad, Y. Hua, and J. Qin, "Optimal Industrial Load Control in Smart Grid: A Case Study for Oil Refineries," in *Proceedings of the IEEE PES General Meeting*, Vancouver, Canada, Jul. 2013.

[285] M. H. Shoreh, P. Siano, M. Shafie-khah, V. Loia, and J. P. S. Catalão, "A Survey of Industrial Applications of Demand Response," *Electric Power Systems Research*, vol. 141, pp. 31–49, Dec. 2016.

[286] Z. Liu, I. Liu, S. Low, and A. Wierman, "Pricing Data Center Demand Response," in *Proceedings of the ACM Sigmetrics*, Austin, TX, Jun. 2014.

[287] G. Marks, E. Wilcox, D. Olsen, and S. Goli, "Opportunities for Demand Response in California Agricultural Irrigation: A Scoping Study," *Technical Report, Lawrence Berkeley National Laboratory*, 2013. www.osti.gov/servlets/purl/1172125.

[288] S. Meyn, P. Barooah, A. Busic, and J. Ehren, "Ancillary Service to the Grid from Deferrable Loads: The Case for Intelligent Pool Pumps in Florida," in *Proceedings of the Conference on Decision and Control*, Florence, Italy, Dec. 2013.

[289] A. Gholian, H. Mohsenian-Rad, and Y. Hua, "Optimal Industrial Load Control in Smart Grid," *IEEE Transactions on Smart Grid*, vol. 7, no. 5, pp. 2305–2316, Sep. 2016.

[290] N. Lessem, A. Faruqui, S. Sergici, and D. Mountain, "The Impact of Time-of-Use Rates in Ontario," *Public Utilities Fortnightly*, pp. 56–87, Feb. 2017.

[291] EnerNOC Inc., "The Demand Response Baseline," White Paper, 2009. https://www.naesb.org/pdf4/dsmee_group3_100809w3.pdf.

[292] XENERGY Inc., "Protocol Development for Demand Response Calculation – Draft Findings and Recommendations," Aug. 2002. www.calmac.org/publications/2002-08-02_XENERGY_REPORT.pdf.

[293] J. Oyedokun, S. Bu, Z. Han, and X. Liu, "Customer Baseline Load Estimation for Incentive-Based Demand Response Using Long Short-Term Memory Recurrent Neural Network," in *Proceedings of the IEEE PES Innovative Smart Grid Technologies Europe*, Bucharest, Romania, Sep. 2019.

[294] T. K. Wijaya, M. Vasirani, and K. Aberer, "When Bias Matters: An Economic Assessment of Demand Response Baselines for Residential Customers," *IEEE Transactions on Smart Grid*, vol. 5, no. 4, pp. 1755–1763, Jul. 2014.

[295] Y. Weng and R. Rajagopal, "Probabilistic Baseline Estimation via Gaussian Process," in *Proceedings of the IEEE Power and Energy Society General Meeting*, Denver, CO, Jul. 2015.

[296] Y. Zhang, W. Chen, R. Xu, and J. Black, "A Cluster-Based Method for Calculating Baselines for Residential Loads," *IEEE Transactions on Smart Grid*, vol. 7, no. 5, pp. 2368–2377, Sep. 2016.

[297] A. Albert and R. Rajagopal, "Smart Meter Driven Segmentation: What Your Consumption Says About You," *IEEE Transactions on Power Systems*, vol. 28, no. 4, pp. 4019–4030, Nov. 2013.

[298] J. Kwac, J. Flora, and R. Rajagopal, "Household Energy Consumption Segmentation Using Hourly Data," *IEEE Transactions on Smart Grid*, vol. 5, no. 1, pp. 420–430, Jan. 2014.

[299] M. Chaouch, "Clustering-Based Improvement of Nonparametric Functional Time Series Forecasting: Application to Intra-Day Household-Level Load Curves," *IEEE Transactions on Smart Grid*, vol. 5, no. 1, pp. 411–419, Jan. 2014.

[300] T. Teeraratkul, D. O'Neill, and S. Lall, "Shape-Based Approach to Household Electric Load Curve Clustering and Prediction," *IEEE Transactions on Smart Grid*, vol. 9, no. 5, pp. 5196–5206, Sep. 2018.

[301] C. Dinesh, S. Makonin, and I. V. Bajić, "Residential Power Forecasting Using Load Identification and Graph Spectral Clustering," *IEEE Transactions on Circuits and Systems II*, vol. 66, no. 11, pp. 1900–1904, Nov. 2019.

[302] K. Hopf, M. Sodenkamp, I. Kozlovkiy, and T. Staake, "Feature Extraction and Filtering for Household Classification Based on Smart Electricity Meter Data," *Computer Science Research and Development*, vol. 31, no. 3, pp. 141–148, Aug. 2016.

[303] D. McClendon, "American Electric Power—Meter Remote Connect/Disconnect," 2011. https://smartgrid.epri.com/UseCases/Meter%20Remote%20Connect%20Disconnect_ph2add.pdf.

[304] B. Mullenmaster, "American Electric Power—Outage Notification," 2011. https://smartgrid.epri.com/UseCases/Outage%20Notification_ph2add.pdf.

[305] Y. Jiang, C.-C. Liu, M. Diedesch, E. Lee, and A. K. Srivastava, "Outage Management of Distribution Systems Incorporating Information from Smart Meters," *IEEE Transactions on Power Systems*, vol. 31, no. 5, pp. 4144–4154, Sep. 2016.

[306] PR Newswire, "World Loses $89.3 Billion to Electricity Theft Annually, $58.7 Billion in Emerging Markets," Dec. 2014. www.prnewswire.com/news-releases/world-loses-893-billion-to-electricity-theft-annually-587-billion-in-emerging-markets-300006515.html.

[307] J. Nagi, K. S. Yap, S. K. Tiong, S. K. Ahmed, and M. Mohamad, "Nontechnical Loss Detection for Metered Customers in Power Utility Using Support Vector Machines," *IEEE Transactions on Power Delivery*, vol. 25, no. 2, pp. 1162–1171, Apr. 2010.

[308] S. Sahoo, D. N. Nikovski, T. Muso, and K. Tsuru, "Electricity Theft Detection Using Smart Meter Data," in *Proceedings of the IEEE PES COnference on Innovative Smart Grid Technologies*, Washington, DC, Feb. 2015.

[309] Y. Wang, Q. Chen, T. Hong, and C. Kang, "Review of Smart Meter Data Analytics: Applications, Methodologies, and Challenges," *IEEE Transactions on Smart Grid*, vol. 10, no. 3, pp. 3125–3148, May 2019.

[310] M. E. H. Dyson, S. D. Borgeson, M. D. Tabone, and D. S. Callaway, "Using Smart Meter Data to Estimate Demand Response Potential with Application to Solar Energy Integration," *Energy Policy*, vol. 73, pp. 607–619, Oct. 2014.

[311] A. Albert and R. Rajagopal, "Finding the Right Consumers for Thermal Demand-Response: An Experimental Evaluation," *IEEE Transactions on Smart Grid*, vol. 9, no. 2, pp. 564–572, Mar. 2018.

[312] K. Dehghanpour, Z. Wang, J. Wang, Y. Yuan, and F. Bu, "A Survey on State Estimation Techniques and Challenges in Smart Distribution Systems," *IEEE Transactions on Smart Grid*, vol. 10, no. 2, pp. 2312–2322, Mar. 2019.

[313] A. Alimardani, F. Therrien, D. Atanackovic, J. Jatskevich, and E. Vaahedi, "Distribution System State Estimation Based on Nonsynchronized Smart Meters," *IEEE Transactions on Smart Grid*, vol. 6, no. 6, pp. 2919–2928, Nov. 2015.

[314] X. Feng, F. Yang, and W. Peterson, "A Practical Multi-Phase Distribution State Estimation Solution Incorporating Smart Meter and Sensor Data," in *Proceedings of the IEEE Power and Energy Society General Meeting*, San Diego, CA, Jul. 2012.

[315] J. Peppanen, M. J. Reno, M. Thakkar, S. Grijalva, and R. G. Harley, "Leveraging AMI Data for Distribution System Model Calibration and Situational Awareness," *IEEE Transactions on Smart Grid*, vol. 6, no. 4, pp. 2050–2059, Jul. 2015.

[316] S. J. Pappu, N. Bhatt, R. Pasumarthy, and A. Rajeswaran, "Identifying Topology of Low Voltage Distribution Networks Based on Smart Meter Data," *IEEE Transactions on Smart Grid*, vol. 9, no. 5, pp. 5113–5122, Sep. 2018.

[317] B. Shah, A. Bose, and A. Srivastava, "Load Modeling and Voltage Optimization Using Smart Meter Infrastructure," in *Proceedings of the IEEE PES Innovative Smart Grid Technologies Conference*, Washington, DC, Feb. 2013.

[318] U.S. Department of Energy, "Advanced Metering Infrastructure and Customer Systems: Results from the Smart Grid Investment Grant Program," Sep. 2016.

[319] W. Luan and W. Li, "Smart Metering and Infrastructure," in *Smart Grids: Clouds, Communications, Open Source, and Automation*, D. Bakken, Ed. Boca Raton, FL: CRC Press, 2014, pp. 399–420.

[320] S. Bavarian, L. Lampe, C. Siew, S. Lancashire, and K. Adeleye, "Leveraging the Smart Metering Infrastructure in Distribution Automation," in *Proceedings of the IEEE International Conference on Smart Grid Communications*, Tainan, Taiwan, Nov. 2012.

[321] Itron Inc, "Meter Data Management: A Key to the Utility of the Future," Dec. 2012, www.smart-energy.com/regional-news/north-america/meter-data-management-a-key-to-the-utility-of-the-future/

[322] M. Zeifman and K. Roth, "Nonintrusive Appliance Load Monitoring: Review and Outlook," *IEEE Transactions on Consumer Electronics*, vol. 57, no. 1, pp. 76–84, Feb. 2011.

[323] G. W. Hart, "Nonintrusive Appliance Load Monitoring," *Proceedings of the IEEE*, vol. 80, no. 12, pp. 1870–1891, Dec. 1992.

[324] D. He, W. Lin, N. Liu, R. G. Harley, and T. G. Habetler, "Incorporating Non-Intrusive Load Monitoring Into Building Level Demand Response," *IEEE Transactions on Smart Grid*, vol. 4, no. 4, 1870.

[325] J. Liang, S. K. K. Ng, G. Kendall, and J. W. M. Cheng, "Load Signature Study–Part II: Disaggregation Framework, Simulation, and Applications," *IEEE Transactions on Power Delivery*, vol. 25, no. 2, pp. 561–569, Apr. 2010.

[326] T. Spanyol, "Smart Meter Data: A Behavioural Marketer's Dream?" *Journal of Direct, Data and Digital Marketing Practice*, vol. 14, pp. 66–69, Aug. 2012.

[327] A. Armel, A. Gupta, G. Shrimali, and A. Albert, "Is Disaggregation the Holy Grail of Energy Efficiency? The Case of Electricity," *Energy Policy*, vol. 52, pp. 213–234, Jan. 2013.

[328] A. Zoha, A. Gluhak, M. A. Imran, and S. Rajasegarar, "Non-Intrusive Load Monitoring Approaches for Disaggregated Energy Sensing: A Survey," *Sensors*, vol. 12, pp. 16838–16866, Dec. 2012.

[329] J. Liang, S. K. K. Ng, G. Kendall, and J. W. M. Cheng, "Load Signature Study–Part I: Basic Concept, Structure, and Methodology," *IEEE Transactions on Power Delivery*, vol. 25, no. 2, pp. 551–560, Apr. 2010.

[330] G. C. Koutitas and L. Tassiulas, "Low Cost Disaggregation of Smart Meter Sensor Data," *IEEE Sensors Journal*, vol. 16, no. 6, pp. 1665–1673, Mar. 2016.

[331] D. Chen and D. Irwin, "SunDance: Black-box Behind-the-Meter Solar Disaggregation," in *Proceedings of the ACM International Conference on Future Energy Systems*, Shatin, Hong Kong, May 2017.

[332] E. Vrettos, E. C. Kara, E. M. Stewart, and C. Roberts, "Estimating PV Power from Aggregate Power Measurements Within the Distribution Grid," *Journal of Renewable and Sustainable Energy*, vol. 11, no. 2, Apr. 2019.

[333] M. Tabone, S. Kiliccote, and E. C. Kara, "Disaggregating Solar Generation Behind Individual Meters in Real Time," in *Proceedings of the ACM Conference on Systems for Built Environments*, Shenzhen, China, Nov. 2018.

[334] M. Ghonima, B. Urquhart, C. Chow, J. Shields, A. Cazorla, and J. Kleissl, "A Method for Cloud Detection and Opacity Classification Based on Ground Based Sky Imagery," *Atmospheric Measurement Techniques*, vol. 5, no. 11, 2881.

[335] C. M. Cheung, W. Zhong, C. Xiong, A. Srivastava, R. Kannan, and V. K. Prasanna, "Behind-the-Meter Solar Generation Disaggregation Using Consumer Mixture Models," in *Proceedings of the IEEE International Conference on Communications, Control, and Computing Technologies for Smart Grids*, Aalborg, Denmark, Oct. 2018.

[336] F. Kabir, N. Yu, W. Yao, R. Yang, and Y. Zhang, "Estimation of Behind-the-Meter Solar Generation by Integrating Physical with Statistical Models," in *Proceedings of the IEEE International Conference on Communications, Control, and Computing Technologies for Smart Grids*, Beijing, China, Oct. 2019.

[337] J. Z. Kolter and M. J. Johnson, "REDD: A Public Data Set for Energy Disaggregation Research," in *Proceedings of the ACM SustKDD Workshop on Data Mining Applications in Sustainability*, San Diego, CA, Aug. 2011.

[338] X. M. Milla, "Building Energy Management Systems Technology for HVAC, Lighting, and Outlets with Smart Sensor Networks," master's thesis, Department of Electrical Engineering, California State University, Long Beach, CA, 2018.

[339] C. K. Metallidou, K. E. Psannis, and E. A. Egyptiadou, "Energy Efficiency in Smart Buildings: IoT Approaches," *IEEE Access*, vol. 8, pp. 63679–63699, Mar. 2020.

[340] D. H. Tran, M. H. Nazari, A. Sadeghi-Mobarakeh, and H. Mohsenian-Rad, "Smart Building Design: A Framework for Optimal Placement of Smart Sensors and Actuators," in *Proceedings of the IEEE PES Conference on Innovative Smart Grid Technologies*, Washington, DC, Jan. 2019.

[341] D. Kosterev, A. Meklin, J. Undrill, B. Lesieutre, W. Price, D. Chassin, R. Bravo, and S. Yang, "Load Modeling in Power System Studies: WECC Progress Update," in *Proceedings of the IEEE Power and Energy Society General Meeting*, Pittsburgh, PA, Jul. 2008.

[342] H. Renmu, M. Jin, and D. J. Hill, "Composite Load Modeling via Measurement Approach," *IEEE Transactions on Power Systems*, vol. 21, no. 2, pp. 663–672, May 2006.

[343] A. Shahsavari, M. Farajollahi, and H. Mohsenian-Rad, "Individual Load Model Parameter Estimation in Distribution Systems Using Load Switching Events," *IEEE Transactions on Power Systems*, vol. 34, no. 6, pp. 4652–4664, Nov. 2019.

[344] X. Zhang, S. Grijalva, and M. J. Reno, "A Time-Variant Load Model Based on Smart Meter Data Mining," in *Proceedings of the IEEE PES General Meeting*, National Harbor, MD, Jul. 2014.

[345] L. Zhu, X. Li, H. Ouyang, Y. Wang, W. Liu, and K. Shao, "Research on Component-Based Approach Load Modeling Based on Energy Management System and Load Control Strength," in *Proceedings of the IEEE PES Innovative Smart Grid Technologies, Asia*, Tianjin, China, May 2012.

[346] A. Gaikwad, P. Markham, and P. Pourbeik, "Implementation of the WECC Composite Load Model for Utilities Using the Component-Based Modeling Approach," in *Proceedings of the IEEE PES Transmission and Distribution Conference and Exposition*, Dallas, TX, May 2016.

[347] A. Arif, Z. Wang, J. Wang, B. Mather, H. Bashualdo, and D. Zhao, "Load Modeling — A Review," *IEEE Transactions on Smart Grid*, vol. 9, no. 6, pp. 5986–5999, Nov. 2018.

[348] A. Bokhari, A. Alkan, R. Dogan, M. Diaz-Aguiló, F. de León, D. Czarkowski, Z. Zabar, L. Birenbaum, A. Noel, and R. E. Uosef, "Experimental Determination of the ZIP Coefficients for Modern Residential, Commercial, and Industrial Loads," *IEEE Transactions on Power Delivery*, vol. 29, no. 3, pp. 1372–1381, Jun. 2014.

[349] IEEE Task Force on Load Representation for Dynamic Performance, "Load Representation for Dynamic Performance Analysis," *IEEE Transactions on Power Systems*, vol. 8, no. 2, pp. 472–482, May 1993.

[350] V. Vignesh, S. Chakrabarti, and S. C. Srivastava, "Power System Load Modelling under Large and Small Disturbances Using Phasor Management Units Data," *IET Generation, Transmission & Distribution*, vol. 9, no. 12, pp. 1316–1323, Apr. 2015.

[351] IEEE Task Force on Load Representation for Dynamic Performance, "Standard Load Models for Power Flow and Dynamic Performance Simulation," *IEEE Transactions on Power Systems*, vol. 10, no. 3, pp. 1302–1313, Aug. 1995.

[352] D. J. Hill, "Nonlinear Dynamic Load Models with Recovery for Voltage Stability Studies," *IEEE Transactions on Power Systems*, vol. 8, no. 1, pp. 166–176, Feb. 1993.

[353] J. Milanović, K. Yamashita, S. M. Villanueva, S. Djokic, and L. Korunović J. V. Milanović, K. Yamashita, S. M. Villanueva, S. Å¡. Djokic, and L. M. Korunović, "International Industry Practice on Power System Load Modeling," *IEEE Transactions on Power Systems*, vol. 28, no. 3, pp. 3038–3046, Aug. 2013.

[354] Electric Power Research Institute, "Measurement-Based Load Modeling," Report Number 1014402, Palo Alto, CA, 2006.

[355] G. W. Chang, C.I. Chen, Y. J. Liu, "A Neural-Network-Based Method of Modeling Electric Arc Furnace Load for Power Engineering Study," *IEEE Transactions on Power Systems*, vol. 25, no. 1, pp. 138–146, Feb. 2010.

[356] A. Abur and A. G. Exposito, *Power System State Estimation: Theory and Implementation.* New York, NY: Marcel Dekker, 2004.

[357] Mathworks, "Help Center–lsqnonlin," www.mathworks.com/help/optim/ug/lsqnonlin.html.

[358] F. F. Wu, "Power System State Estimation: A Survey," *International Journal of Electrical Power and Energy Systems*, vol. 12, no. 2, pp. 80–87, Apr. 1990.

[359] M. Ahmad, *Power System State Estimation*, Boston, MA: Artech House, 2013.

[360] A. Monticelli, *State Estimation in Electric Power Systems: A Generalized Approach.* Boston, MA: Kluwer Academic Publishers, 1999.

[361] M. L. Crow, *Computational Methods for Electric Power Systems*, 3rd ed. Boca Raton, FL: CRC Press, 2016.

[362] B. Stott, J. Jardim, and O. Alsac, "DC Power Flow Revisited," *IEEE Transactions on Power Systems*, vol. 24, no. 3, pp. 1290–1300, Aug. 2009.

[363] J. Lavaei and S. H. Low, "Zero Duality Gap in Optimal Power Flow Problem," *IEEE Transactions on Power Systems*, vol. 27, no. 1, pp. 92–107, Feb. 2012.

[364] R. A. Jabr, "Radial Distribution Load Flow Using Conic Programming," *IEEE Transactions on Power Systems*, vol. 21, no. 3, pp. 1458–1459, Aug. 2006.

[365] D. Molzahn, J. Holzer, B. Lesieutre, and C. DeMarco, "Implementation of a Large-Scale Optimal Power Flow Solver Based on Semidefinite Programming," *IEEE Transactions on Power Systems*, vol. 28, no. 4, pp. 3987–3998, Nov. 2013.

[366] S. Low, "Convex Relaxation of Optimal Power Flow–Part I: Formulations and Equivalence," *IEEE Transactions on Control of Network Systems*, vol. 1, no. 1, pp. 15–27, Mar. 2014.

[367] S. Low, "Convex Relaxation of Optimal Power Flow–Part II: Exactness," *IEEE Transactions on Control of Network Systems*, vol. 1, no. 2, pp. 177–189, Jun. 2014.

[368] Hao Zhu and Georgios B. Giannakis, "Power System Nonlinear State Estimation Using Distributed Semidefinite Programming," *IEEE Journal of Selected Topics in Signal Processing*, vol. 8, no. 6, pp. 1039–1050, Dec. 2014.

[369] Y. Weng, Q. Li, R. Negi, and M. Ilić, "Distributed Algorithm for SDP State Estimation," in *Proceedings of the IEEE PES Innovative Smart Grid Technologies Conference*, Washington, DC, Feb. 2013.

[370] R. Yang and Y. Zhang, "Three-Phase AC Optimal Power Flow Based Distribution Locational Marginal Price," in *Proceedings of the IEEE North American Power Symposium*, Charlotte, NC, Oct. 2015.

[371] Y. Yao, X. Liu, D. Zhao, and Z. Li, "Distribution System State Estimation: A Semidefinite Programming Approach," *IEEE Transactions on Smart Grid*, vol. 10, no. 4, pp. 4369–4378, Jul. 2019.

[372] J. Zhang, G. Welch, G. Bishop, and Z. Huang, "A Two-Stage Kalman Filter Approach for Robust and Real-Time Power System State Estimation," *IEEE Transactions on Sustainable Energy*, vol. 5, no. 2, pp. 629–636, Apr. 2014.

[373] G. Valverde and V. Terzija, "Unscented Kalman Filter for Power System Dynamic State Estimation," *IET Generation, Transmission & Distribution* , vol. 5, no. 1, pp. 29–37, Jan. 2011.

[374] Newtons4th Ltd, "3 Phase 2 Wattmeter Power Measurements Explained—Application Note 14," Mar. 2012, www.newtons4th.com/wp-content/uploads/2010/03/APP014-3-Phase-2-Wattmeter-Explained.pdf.

[375] V. Arya, D. Seetharam, S. Kalyanaraman, K. Dontas, C. Pavlovski, S. Hoy, and J. R. Kalagnanam, "Phase Identification in Smart Grids," in *Proceedings of IEEE International Conference on Smart Grid Communications*, Oct. 2011.

[376] CVX Research, "Mixed-Integer Support in CVX 2.0." http://cvxr.com/news/2012/08/midcp/.

[377] International Business Machines, "IBM CPLEX Optimizer: High-Performance Mathematical Programming Solver for Linear Programming, Mixed-Integer Programming and Quadratic Programming." www.ibm.com/analytics/cplex-optimizer.

[378] K. Zipp, "Three Become One: The Rise of Three-Phase Solar String Inverters," Sep. 2014, www.solarpowerworldonline.com/2014/09/three-become-one-rise-three-phase-solar-string-inverters/.

[379] Jiaqi Chen and Ye Guo and Wenchuan Wu, "Optimal Dispatch Scheme for DSO and Prosumers by Implementing Three-Phase Distribution Locational Marginal Prices," *IET Generation, Transmission & Distribution*, vol. 14, pp. 2138–2146, Jun. 2020.

[380] R. Yang and Y. Zhang, "Three-Phase AC Optimal Power flow Based Distribution Locational Marginal Price," in *Proceedings of the IEEE PES Conference on Innovative Smart Grid Technologies Conference (ISGT)*, Washington, DC, Apr. 2017.

[381] M. N. Faqiry and L. Wang and H. Wu, "HEMS-Enabled Transactive Flexibility in Three-Phase Unbalanced Distribution Systems," *Journal of Modern Power Systems and Clean Energy*, vol. 7, no. 6, pp. 1434–1449, Jul. 2019.

[382] M. E. Baran and A. W. Kelley, "State Estimation for Real-Time Monitoring of Distribution Systems," *IEEE Transactions on Power Systems*, vol. 9, no. 3, pp. 1601–1609, Aug. 1994.

[383] Y. Yao, X. Liu, D. Zhao, and Z. Li, "Distribution System State Estimation: A Semidefinite Programming Approach," *IEEE Transactions on Smart Grid*, vol. 10, no. 4, pp. 4369–4378, Jul. 2019.

[384] Y. Deng, and Y. He, and B. Zhang, "A Branch-estimation-Based State Estimation Method for Radial Distribution Systems," *IEEE Transactions on Power Systems*, vol. 17, no. 4, pp. 1057–1062, Oct. 2002.

[385] G. Wang, G. B. Giannakis, J. Chen, and J. Sun, "Distribution System State Estimation: An Overview of Recent Developments," *Frontiers of Information Technology & Electronic Engineering*, vol. 20, no. 1, pp. 4–17, 2019.

[386] L. S. Czarnecki and P. M. Haley, "Unbalanced Power in Four-Wire Systems and Its Reactive Compensation," *IEEE Transactions on Power Systems*, vol. 30, no. 1, pp. 53–63, Feb. 2015.

[387] AIEE Committee, "Apparent Power in Three-Phase Systems," *Transactions of AIEE*, vol. 39, pp. 1450–1455, 1920.

[388] F. Buchholz, "The Apparent Three-Phase Power in the Event of an Uneven Load on the Three Branches," *Light Power*, pp. 9–11, Jan. 1922.

[389] American National Standards Institute, "ANSI C12.20-2015: American National Standard for Electricity Meters—0.1, 0.2, and 0.5 Accuracy Classes," Feb. 2017.

[390] T. Mooney and D. Martin, "Increasing Metering Accuracy by Optimizing the Analog to Digital Converter Characteristics," in *Proceedings of the IEEE Power Systems Conference: Advanced Metering, Protection, Control, Communication, and Distributed Resources*, Clemson, SC, Mar. 2006.

[391] S. Bell, "A Beginner's Guide to Uncertainty of Measurement," *IEEE Transactions on Smart Grid*, vol. 11, no. 2, pp. 1–33, Aug. 1999.

[392] SATEC Power Solutions Ltd, "Accuracy Class: A Small "S" That Makes a Big Difference," Jul. 2018, www.satec-global.com/sites/default/files/Application-note_Accuracy-Class_July2018.pdf

[393] Electric Power Research Institute, "Accuracy of Digital Electricity Meters," Report Number 1020908, Palo Alto, CA, 2010.

[394] P. WÄŹgierek and M. Konarski, "The Temperature Effect on Measurement Accuracy of the Smart Electricity Meter," *Electrical Review*, vol. 92, no. 8, pp. 148–150, 2016.

[395] EnerNOC Inc., "The Demand Response Baseline," 2011. https://library.cee1.org/sites/default/files/library/10774/CEE_EvalDRBaseline_2011.pdf.

[396] S. Bhela, V. Kekatos, and S. Veeramachaneni, "Enhancing Observability in Distribution Grids Using Smart Meter Data," *IEEE Transactions on Smart Grid*, vol. 9, no. 6, pp. 5953–5961, Nov. 2018.

[397] S. Bhela, V. Kekatos, L. Zhang, and S. Veeramachaneni, "Enhancing Observability in Power Distribution Grids," *IEEE International Conference on Acoustics, Speech and Signal Processing (ICASSP)*, Mar. 2017.

[398] K. J. Caird, "Meter Phase Identification," U.S. Patent Number 0164473. https://patents.google.com/patent/US20100164473A1/en.

[399] D. B. Arnold, M. Negrete-Pincetic, M. D. Sankur, D. M. Auslander, and D. S. Callaway, "Model-Free Optimal Control of VAR Resources in Distribution Systems: An Extremum

Seeking Approach," *IEEE Transactions on Power Systems*, vol. 31, no. 5, pp. 3583–3593, Sep. 2016.

[400] M. L. Shelton, P. F. Winkelman, W. A. Mittelstadt, and W. J. Bellerby, "Bonneville Power Administration 1400-MW Braking Resistor," *IEEE Transactions on Power Apparatus and Systems*, vol. 94, no. 2, pp. 602–611, Mar. 1975.

[401] J. F. Hauer, D. J. Trudnowski, and J. G. DeSteese, "A Perspective on WAMS Analysis Tools for Tracking of Oscillatory Dynamics," in *Proceedings of the IEEE Power Engineering Society General Meeting*, Tampa, FL, Jun. 2007.

[402] International Council on Large Electric Systems, "Impact of Interactions among Power System Controls," *CIGRE Technical Brochure*, May 2000.

[403] J. F. Hauer, W. A. Mittelstadt, K. E. Martin, J. Burns, H. Lee, J. Pierre, and D. J. Trudnowski, "Use of the WECC WAMS in Wide-Area Probing Tests for Validation of System Performance and Modeling," *IEEE Transactions on Power Systems*, vol. 24, no. 1, pp. 250–257, Feb. 2009.

[404] L. Ljung, *System Identification Theory for the User*. Upper Saddle River, NJ: Prentice Hall, 1999.

[405] L. Lampe, A. Tonello, and D. Shaver, "Power Line Communications for Automation Networks and Smart Grid," *IEEE Communications Magazine*, vol. 49, no. 12, pp. 26–27, Dec. 2011.

[406] M. Yigit, V. C. Gungor, G. Tuna, M. Rangoussi, and E. Fadel, "Power Line Communication Technologies for Smart Grid Applications: A Review of Advances and Challenges," *Computer Networks*, vol. 70, no. 9, pp. 366–383, Sep. 2014.

[407] R. Rao, S. Akella, and G. Guley, "Power Power Line Carrier (PLC) Signal Analysis of Smart Meters for Outlier Detection," in *Proceedings of the IEEE International Conference on in Smart Grid Communications*, Oct. 2011.

[408] A. Sendin, I. Peña, and P. Angueira, "Strategies for Power Line Communications Smart Metering Network Deployment," *Energies*, vol. 7, pp. 2377–2420, 2014.

[409] L. T. Berger, A. Schwager, and J. J. Escudero-Garzás, "Power Line Communications for Smart Grid Applications," *Journal of Electrical and Computer Engineering*, pp. 1–16, Mar. 2013.

[410] S. Galli, A. Scaglione, and Z. Wang, "For the Grid and through the Grid: The Role of Power Line Communications in the Smart Grid," in *Proceedings of the IEEE*, vol. 99, no. 6, pp. 998–1027, Jun. 2011.

[411] L. G. S. Costa, A. C. M. Queiroz, B. Adebisi, V. L. R. Costa, and M. V. Ribeiro, "Coupling for Power Line Communication: A Survey," *Journal of Communication and Information System*, vol. 32, no. 1, pp. 8–22, 2017.

[412] H. Ferreira, L. Lampe, J. Newbury, and T. Swart, "Industrial and International Standards on PLC Base Networking Technologies," in *Power Line Communications: Theory and Applications for Narrowband and Broadband Communications over Power Lines*, S. Galli, M. Koch, H. Latchman, S. Lee, and V. Oksman, Eds. New York: Wiley, 2010, pp. 377–426.

[413] D. Fink and R. J. Jeung, "Feasible Connectivity Solutions of PLC for Rural and Remote Areas," *Proceedings of the IEEE International Symposium on Power Line Communications and Its Applications*, Apr. 2008.

[414] W. Liu, M. Sigle, and K. Dostert, "Channel Characterization and System Verification for Narrowband Power Line Communication in Smart Grid Applications," *IEEE Communications Magazine*, vol. 49, no. 12, pp. 28–35, Dec. 2011.

[415] Y. Huo, G. Prasad, L. Atanackovic, L. Lampe, and V. C. M. Leung, "Cable Diagnostics with Power Line Modems for Smart Grid Monitoring," *IEEE Access*, vol. 7, pp. 60206–60220, 2019.

[416] L. Forstel and L. Lampe, "Grid Diagnostics: Monitoring Cable Aging Using Power Line Transmission," in *Proceedings of the IEEE International Symposium on Power Line Communications and its Applications*, Apr. 2017.

[417] F. Yang, W. Ding, and J. Song, "Non-intrusive Power Line Quality Monitoring Based on Power Line Communications," in *Proceedings of the IEEE International Symposium on Power Line Communications and Its Applications*, Mar. 2013.

[418] T. Erseghe, S. Tomasin, S. Member, and A. Vigato, "Topology Estimation for Smart Micro Grids via Powerline Communications," *IEEE Transactions on Signal Processing*, vol. 61, no. 13, pp. 3368–3377, Jul. 2013.

[419] M. O. Ahmed and L. Lampe, "Power Line Communications for Low-Voltage Power Grid Tomography," *IEEE Transactions on Communications*, vol. 61, no. 12, pp. 5163–5175, Dec. 2013.

[420] L. Lampe and M. O. Ahmed, "Power Grid Topology Inference Using Power Line Communications," in *Proceedings of the IEEE International Conference on Smart Grid Communications*, Oct. 2013.

[421] Y. Huo, G. Prasad, L. Atanackovic, L. Lampe, and V. C. Leung, "Grid Surveillance and Diagnostics using Power Line Communications," in *Proceedings of the IEEE International Symposium on Power Line Communications and its Applications*, Apr. 2018.

[422] R. Rao, S. Akella, and G. Guley, "Power Line Carrier (PLC) Signal Analysis of Smart Meters for Outlier Detection," in *Proceedings of the IEEE International Conference on Smart Grid Communications*, Oct. 2011.

[423] H. J. Byun and S. G. Shon, "Cable and Phase Identification Based on Power Line Communication," *International Journal of Control and Automation*, vol. 8, no. 9, pp. 63–74, 2015.

[424] J. J. Woo, H. J. Byun, and S. G. Shon, "Cable Identification as a New Application of Power Line Communication Technology," *Advanced Science and Technology Letters*, vol. 95, pp. 17–21, 2015.

[425] F. Passerini and A. Tonello, "Smart Grid Monitoring Using Power Line Modems: Anomaly Detection and Localization," *IEEE Transactions on Smart Grid*, vol. 10, no. 6, pp. 6178–6186, Nov. 2019.

[426] F. Passerini and A. Tonello, "Power Line Network Topology Identification Using Admittance Measurements and Total Least Squares Estimation," in *Proceedings of the IEEE International Conference on Communications*, Paris, France, May 2017.

[427] A. M. Lehmann, K. Raab, F. Gruber, E. Fischery, R. Muller, and J. B. Huber, "A Diagnostic Method for Power Line Networks by Channel Estimation of PLC Devices," *Proceedings of the IEEE International Conference on Smart Grid Communications*, Nov. 2016.

[428] M. Wang, A. J. Vandermaar, and K. D. Srivastava, "Review of Condition Assessment of Power Transformers in Service," *IEEE Electrical Insulation Magazine*, vol. 18, pp. 12–25, Nov. 2002.

[429] A. Jahromi, R. Piercy, S. Cress, J. Service, and W. Fan, "An Approach to Power Transformer Asset Management Using Health Index," *IEEE Electrical Insulation Magazine*, vol. 25, no. 2, pp. 20–34, Mar. 2009.

[430] C. Sun, P. R. Ohodnicki, and E. M. Stewart, "Chemical Sensing Strategies for Real-Time Monitoring of Transformer Oil: A Review," *IEEE Sensors Journal*, vol. 17, no. 18, pp. 5786–5806, Aug. 2017.

[431] N. Moodley and C. T. Gaunt, "Low Energy Degradation Triangle for Power Transformer Health Assessment," *IEEE Transactions on Dielectrics and Electrical Insulation*, vol. 24, no. 1, pp. 639–646, Feb. 2017.

[432] General Electric, "Multilin DGCC Capacitor Bank Controller," 2019. www.gegridsolutions.com/multilin/catalog/dgcc.htm.

[433] Allen-Bradley, "Bulletin 1413 Capacitor Bank Controller," 2019. https://rexel-cdn.com/products/1413-cap-me.pdf?i=8DB02F34-16CA-42C3-8131-30BB08CD0A68.

[434] Tollgrade Communications, "Capacitor Bank Monitoring: Application Guide," 2014. www.enghousenetworks.com/tollgrade/smart-grid-resources/application-guides/.

[435] J. Liu, "Dynamic Line Rating in Power Systems," Master's Thesis, Department of Electrical and Computer Engineering, University of California at Riverside, Riverside, CA, Mar. 2017.

[436] CIGRE Study Committee B2, "Technical Brochure 324: Sag-Tension Calculation Methods for Overhead Lines," Jun. 2007.

[437] D. Douglass, W. Chisholm, and G. Davidson, "Real-Time Overhead Transmission-Line Monitoring for Dynamic Rating," *IEEE Transactions on Power Delivery*, vol. 31, pp. 921–927, Jun. 2016.

[438] F. P. G. Márquez, A. M. Tobias, J. M. P. Pérez, and M. Papaelias, "Condition Monitoring of Wind Turbines: Techniques and Methods," *Renewable Energy*, vol. 46, pp. 169–178, 2012.

[439] ONYX InSight Inc., "CASE STUDY 1: Main Bearing Fault Detection and Grease Flushing," https://onyxinsight.com/monitoring-software/.

[440] S. Mekhilef, R. Saidur, and M. Kamalisarvestani, "Effect of Dust, Humidity and Air Velocity on Efficiency of Photovoltaic Cells," *Renewable and Sustainable Energy Reviews*, vol. 16, no. 5, pp. 2920–2925, Jun. 2012.

[441] K. Tsamaase, T. Ramasesane, I. Zibani, E. Matlotse, and K. Motshidisi, "Automated Dust Detection and Cleaning System of PV Module," *IOSR Journal of Electrical and Electronics Engineering*, vol. 12, no. 6, pp. 93–98, Nov. 2017.

[442] B. Guo, W. Javed, B. W. Figgis, and T. Mirza, "Effect of Dust and Weather Conditions on Photovoltaic Performance in Doha, Qatar," in *Proceedings of the IEEE Workshop on Smart Grid and Renewable Energy*, Mar. 2015.

[443] Z. Taylor, H. Akhavan-Hejazi, and H. Mohsenian-Rad, "Optimal Operation of Grid-Tied Energy Storage Systems Considering Detailed Device-Level Battery Models," *IEEE Transactions on Industrial Informatics*, pp. 1–12, Aug. 2019.

[444] U.S. Department of Energy, "An Assessment of Energy Technologies and Research Opportunities—Chapter 5: Increasing Efficiency of Building Systems and Technogies," Sep. 2015. www.energy.gov/sites/default/files/2017/03/f34/qtr-2015-chapter5.pdf.

[445] Z. Xu, X. Guan, Q.-S. Jia, J. Wu, D. Wang, and S. Chen, "Performance Analysis and Comparison on Energy Storage Devices for Smart Building Energy Management," *IEEE Transactions on Smart Grid*, vol. 3, no. 4, pp. 2136–2147, Dec. 2012.

[446] C. Chen, J. Wang, Y. Heo, and S. Kishore, "MPC-Based Appliance Scheduling for Residential Building Energy Management Controller," *IEEE Transactions on Smart Grid*, vol. 4, no. 3, pp. 1401–1410, Sep. 2013.

[447] L. A. Hurtado, P. H. Nguyen, and W. L. Kling, "Agent-Based Control for Building Energy Management in the Smart Grid Framework," in *Proceedings of the IEEE PES Innovative Smart Grid Technologies*, Istanbul, Turkey, Oct. 2014.

[448] C. Dilouie, "All about occupancy and vacancy sensors," Aug. 2017. https://lightingcontrolsassociation.org/2017/08/21/all-about-occupancy-and-vacancy-sensors/.

[449] E. Samani, P. Khaledian, A. Aligholian, E. Papalexakis, S. Cun, M. H. Nazari, and H. Mohsenian-Rad, "Anomaly Detection in IoT-Based PIR Occupancy Sensors to Improve Building Energy Efficiency," in *Proceedings of the IEEE Innovative Smart Grid Technologies*, Washington, DC, Feb. 2020.

[450] CSULB Research Foundation, "Internet of Things and Ubiquitous Sensing in University Building Energy Management: Design Optimization and Technology Demonstration." https://ceqanet.opr.ca.gov/2017038493.

[451] SensMax Inc., "Real-Time Wireless Bidirectional People Counter," https://sensmax.eu/devices/.

[452] B. Howard, S. Acha, N. Shah, and J. Polak, "Implicit Sensing of Building Occupancy Count with Information and Communication Technology Data Sets," *Building and Environment*, vol. 157, pp. 297–308, Jun. 2019.

[453] Enlighted Inc., "Smart Sensors." www.enlightedinc.com/system-and-solutions/iot-system/smart-sensors/.

[454] Leviton Inc., "Daylighting Control Products." www.leviton.com/en/products/commercial/lighting-controls/daylighting-controls.

[455] A. Akrami, P. Khaledian, H. Akhavan-Hejazi, A. Aligholian, E. Samani, and H. Mohsenian-Rad, "Impact on EMS on Academic Building Operation Report: CEC Project Report—EPIC-16-033," 2019.

[456] E. Veldman and R. A. Verzijlbergh, "Distribution Grid Impacts of Smart Electric Vehicle Charging from Different Perspectives," *IEEE Transactions on Smart Grid*, vol. 6, no. 1, pp. 333–342, Jan. 2015.

[457] S. Semsar, T. Soong, and P. W. Lehn, "On-Board Single-Phase Integrated Electric Vehicle Charger with V2G Functionality," *IEEE Transactions on Power Electronics*, vol. 35, no. 11, pp. 12 072–12 084, Nov. 2020.

[458] F. S. Tidjani, A. Hamadi, A. Chandra, B. Saghir, B. Mounir, and M. Garoum, "Energy Management of Micro Grid Based Electrical Vehicle to the Building (V2B)," in *Proceedings of the IEEE International Renewable and Sustainable Energy Conference*, Agadir, Morocco, Nov. 2019.

[459] F. Xie, M. Huang, W. Zhang, and J. Li, "Research on Electric Vehicle Charging Station Load Forecasting," in *Proceedings of the IEEE International Conference on Advanced Power System Automation and Protection*, Beijing, China, Oct. 2011.

[460] M. Majidpour, C. Qiu, P. Chu, R. Gadh, and H. R. Pota, "Modified Pattern Sequence-Based Forecasting for Electric Vehicle Charging Stations," in *Proceedings of the IEEE International Conference on Smart Grid Communications*, Venice, Italy, Nov. 2014.

[461] Pacific Gas and Electric, "Time of Use Pricing," www.portlandgeneral.com/residential/power-choices/time-of-use/time-of-use-pricing,2020.

[462] A. Primadianto and C.-N. Lu, "A Review on Distribution System State Estimation," *IEEE Transactions on Power Systems*, vol. 32, no. 5, pp. 3875–3883, Sep. 2017.

[463] R. Hoffman, "Practical State Estimation for Electric," in *Proceedings of the IEEE PES Power Systems Conference and Exposition*, Atlanta, GA, Nov. 2006.

[464] California Independent System Operator. www.caiso.com.

[465] California Independent System Operator, "California ISO Open Access Same-Time Information System." http://oasis.caiso.com/.

[466] H. Mohsenian-Rad, "Optimal Bidding, Scheduling, and Deployment of Battery Systems in California Day-Ahead Energy Market," *IEEE Transactions on Power Systems*, vol. 31, no. 1, pp. 442–453, Jan. 2016.

[467] H. S. Jang, K. Y. Bae, H.-S. Park, and D. K. Sung, "Solar Power Prediction Based on Satellite Images and Support Vector Machine," *IEEE Transactions on Sustainable Energy*, vol. 7, no. 3, pp. 1255–1263, 2016 2016.

[468] Google Maps. https://www.google.com/maps/@66.1495066,-150.2543188,2643m/data=!3m1!1e3.

[469] NASA Earth Observatory, "Camp Fire Rages in California." https://earthobservatory.nasa.gov/images/144225/camp-fire-rages-in-california.

[470] C.Wai-Chow, B. Urquhart, M. Lave, A. Dominguez, J. Kleissl, J. Shields, and B. Washom, "Intra-Hour Forecasting with a Total Sky Imager at the UC San Diego Solar Energy Testbed," *Solar Energy*, vol. 85, no. 11, pp. 2881–2893, Nov. 2011.

[471] A. Molaei, H. D. Taghirad, and J. Dargahi, "Extracting of Sagging Profile of Overhead Power Transmission Line Via Image Processing," in *Proceedings of the IEEE Canadian Conference on Electrical & Computer Engineering*, Quebec City, QC, May 2018.

[472] J. Xie, A. Stefanov, and C.-C. Liu, "Physical and Cyber Security in a Smart Grid Environment," *Wiley Interdisciplinary Reviews: Energy and Environment*, vol. 5, no. 5, pp. 519–542, Mar. 2016.

[473] P. Dong and Q. Chen, *LiDAR Remote Sensing and Applications*. Boca Raton, FL: CRC Press, 2018.

[474] M. Frank, Z. Pan, B. Raber, and C. Lenart, "Vegetation Management of Utility Corridors Using High-Resolution Hyperspectral Imaging and LiDAR," in *Proceedings of the IEEE Workshop on Hyperspectral Image and Signal Processing: Evolution in Remote Sensing*, Reykjavik, Iceland, Jun. 2010.

[475] L. Matikainen, M. Lehtomäki, E. Ahokas, J. Hyyppä, M. Karjalainen, A. Jaakkola, A. Kukko, and T. Heinonen, "Remote Sensing Methods for Power Line Corridor Surveys," *ISPRS Journal of Photogrammetry and Remote Sensing*, vol. 119, pp. 10–31, Sep. 2016.

[476] R. Ishino and F. Tsutsumi, "Detection System of Damaged Cables Using Video Obtained from an Aerial Inspection of Transmission Lines," in *Proceedings of the IEEE PES General Meeting*, Denver, CO, Jun. 2004.

[477] R. McLaughlin, "Extracting Transmission Lines From Airborne LIDAR Data," *IEEE Geoscience and Remote Sensing Letters*, vol. 3, no. 2, pp. 222–226, Apr. 2006.

[478] H. Ha, S. Han, and J. Lee, "Fault Detection on Transmission Lines Using a Microphone Array and an Infrared Thermal Imaging Camera," *IEEE Transactions on Instrumentation and Measurement*, vol. 61, no. 1, pp. 267–275, Jan. 2012.

[479] B. Jiang, A. P. Sample, R. M. Wistort, and A. V. Mamishev, "Autonomous Robotic Monitoring of Underground Cable Systems," in *Proceedings of the IEEE International Conference on Advanced Robotics*, Seattle, WA, Jul. 2005.

[480] W. Hongguang, J. Yong, L. Aihua, F. Lijin, and L. Lie, "Research of Power Transmission Line Maintenance Robots in SIACAS," in *Proceedings of the IEEE International Conference on Applied Robotics for the Power Industry*, Montreal, QC, Oct. 2010.

[481] J. Rocha and J. Sequeira, "The Development of a Robotic System for Maintenance and Inspection of Power Lines," in *Proceedings of the International Symposium on Robotics*, Paris, France, Mar. 2004.

[482] O. Menendez, F. A. A. Cheein, M. Perez, and S. Kouro, "Robotics in Power Systems: Enabling a More Reliable and Safe Grid," *IEEE Industrial Electronics Magazine*, vol. 11, no. 2, pp. 22–34, Jun. 2017.

[483] B. Jiang and A. Mamishev, "Robotic Monitoring of Power Systems," *IEEE Transactions on Power Delivery*, vol. 19, no. 3, pp. 912–918, Jul. 2004.

[484] J. Katrasnik, F. Pernus, and B. Likar, "A Survey of Mobile Robots for Distribution Power Line Inspection," *IEEE Transactions on Power Delivery*, vol. 25, no. 1, pp. 485–493, Jan. 2010.

[485] K. L. Cummins, E. P. Krider, and M. D. Malone, "The USNational Lightning Detection NetworkTM and Applications of Cloud-to-Ground Lightning Data by Electric Power Utilities," *IEEE Transactions on Electromagnetic Compatibility*, vol. 40, no. 4, pp. 465–480, Nov. 1998.

[486] C. Wang and S. Wang, "The Automatic Routing System of Urban Mid-Voltage Distribution Network Based on Spatial GIS," *Proceedings of the International Conference on Power System Technology*, Nov. 2004.

[487] T. Hong, P. Pinson, S. Fan, H. Zareipour, A. Troccoli, and R. J. Hyndman, "Probabilistic Energy Forecasting: Global Energy Forecasting Competition 2014 and Beyond," *International Journal of Forecasting*, vol. 32, no. 3, pp. 914–938, Jul. 2016.

[488] J. Aparicio, J. Rosca, M. Mediger, A. Essl, K. Arzig, and C. Develder, "Exploiting Road Traffic Data for Very Short Term Load Forecasting in Smart Grids," in *Proceedings of the IEEE PES Conference on Innovative Smart Grid Technologies*, Washington, DC, Feb. 2014.

[489] A. Nejat and H. Mohsenian-Rad, "Electric Load Forecasting: A Multi-Agent Systems Approach," in *Proceedings of the IEEE International Conference on Intelligent Transportation Systems*, Anchorage, AK, Sep. 2012.

[490] Q. Guo, S. Xin, H. Sun, Z. Li, and B. Zhang, "Rapid-Charging Navigation of Electric Vehicles Based on Real-Time Power Systems and Traffic Data," *IEEE Transactions on Smart Grid*, vol. 5, no. 4, pp. 1969–1979, Jul. 2014.

[491] M. Aiello and G. A. Pagani, "The Smart Grid's Data Generating Potentials," in *Proceedings of the Federated Conference on Computer Science and Information Systems*, Warsaw, Poland, Sep. 2014.

[492] Y. Huang, M. Warnier, F. Brazier, and D. Miorandi, "Social Networking for Smart Grid Users," in *Proceedings of the IEEE International Conference on Networking, Sensing and Control*, Taipei, Taiwan, Apr. 2015.

[493] Alternative Energy Institute, West Texas A & M University, www.windenergy.org/.

[494] H. Akhavan-Hejazi, H. Mohsenian-Rad, and A. Nejat, "Developing a Test Data Set for Electric Vehicle Applications in Smart Grid Research," *Proceedings of the IEEE Vehicular Technology Conference*, Sep. 2014.

Index

Printed in the United States
By Bookmasters

Printed in the United States
by Baker & Taylor Publisher Services